AF771934

NUCLEAR MAGNETIC RESONANCE

FOR ORGANIC CHEMISTS

NUCLEAR MAGNETIC RESONANCE

FOR ORGANIC CHEMISTS

EDITED BY

D. W. MATHIESON

School of Pharmacy, University of London, England

1967

Published in Association with the Royal Institute of Chemistry By Academic Press, London and New York

ACADEMIC PRESS INC. (LONDON) LTD
Berkeley Square House
Berkeley Square
London, W.1

U.S. Edition published by
ACADEMIC PRESS INC.
111 Fifth Avenue
New York, New York 10003

Copyright © 1967 by ACADEMIC PRESS INC. (LONDON) LTD

All Rights Reserved

No part of this book may be reproduced in any form by photostat, microfilm, or any other means, without written permission from the publishers

Library of Congress Catalog Card Number: 66-27813

Printed in Great Britain by
Spottiswoode, Ballantyne and Company Limited, London and Colchester

LIST OF CONTRIBUTORS

R. J. ABRAHAM, *Department of Organic Chemistry, The Robert Robinson Laboratories, Oxford Street, Liverpool 7, England* (p. 129)

C. N. BANWELL, *Chemistry Laboratory, University of Sussex, near Brighton, Sussex, England* (p. 85)

E. O. BISHOP, *Chemistry Laboratory, University of Sussex, near Brighton, Sussex, England* (pp. 61, 103)

J. A. ELVIDGE, *Battersea College of Science and Technology, Battersea Park Road, London, S.W.11, England* (pp. 13, 29, 43, 179)

J. FEENEY, *Varian Associates Limited, Russell House, Molesey Road, Walton on Thames, Surrey, England* (p. 151)

N. SHEPPARD, *School of Molecular Science, University of Norwich, Wilberforce Road, Norwich, England* (p. 1)

PREFACE

The chapters which comprise this book were delivered as a series of lectures at a summer school organized by the Royal Institute of Chemistry in 1964. The course, which was designated a NATO Advanced Study Institute, was run specifically for Organic Chemists and was primarily concerned with the interpretation of NMR spectra in terms of molecular structure. In addition to the lectures, a series of seminars was arranged and the spectra used as examples, together with the necessary correlation tables, are printed as an appendix: answers to the spectra are also provided.

We have to acknowledge with thanks permission to reproduce material as follows: The National Research Council of Canada for Figure 7.2, the American Institute of Physics for Figures 7.3, 7.4, 7.5, 7.6, 7.7 and 7.8 and also Table 9.6; the American Chemical Society for Figure 5.7 and Tables 9.1 and 9.4; the Council of the Royal Society for Figure 9.3. Thanks are also due to the many authors cited in the references to the above mentioned material: they also consented to the use of original material from their papers.

DAVID W. MATHIESON

London
August 1966

CONTENTS

CHAPTER 1

INTRODUCTION TO THE PRINCIPLES OF NUCLEAR MAGNETIC RESONANCE SPECTROSCOPY

N. SHEPPARD

The Magnetic-resonance Experiment

The first measurements in nuclear magnetic resonance spectroscopy were made in 1945; in that year two groups of physicists, one under Professor Bloch at Stanford University and the other under Professor Purcell at Harvard University, first obtained nuclear magnetic resonance signals. The majority of nuclei, though not all, have a magnetic moment and when such a nucleus is placed in a magnetic field it takes up one of a number of quantized orientations with respect to that field. Each nuclear orientation corresponds to an energy level, the lowest corresponding to the orientation in which the nuclear magnetic moment is most closely aligned with the field. Nuclear magnetic resonance spectroscopy consists in inducing transitions between such energy levels (energy separation ΔE) by means of fluctuating magnetic fields of the correct frequency, ν, so as to obey the Bohr relationship $\Delta E = \mathbf{h}\nu$, where $\mathbf{h}$ is Planck's constant.

Although its volume is small, a magnetic nucleus gives rise to a finite spatial distribution of positive charge. If, in addition, the nucleus has spin angular momentum then the spinning of the nucleus will cause the positive charge distribution to rotate and give rise to the equivalent of a current flowing in a circular path (see Fig. 1.1). This in turn produces a magnetic field which is parallel to the axis of spin, and the nucleus is said to possess a magnetic moment. The quantum-mechanical expression for the angular momentum is given by $I(I+1)\mathbf{h}/2\pi$, where I is the quantum number for the spinning motion. Different nuclei have different half-integral values of the spin quantum number I. Some nuclei have $I=0$; these have no spin angular momentum and therefore no magnetic moment. Other nuclei have $I=\frac{1}{2}$, $I=1$, $I=3/2$, etc. A number of very important nuclei are included in the class which has $I=0$, and for which nuclear magnetic resonance experiments cannot be made. All of these have their mass numbers divisible by four; they include the normal abundant isotopes of carbon, ^{12}C; oxygen, ^{16}O; magnesium, ^{24}Mg; silicon, ^{28}Si; and sulphur, ^{32}S. Examples of nuclei which have $I = \frac{1}{2}$ are the normal nuclei of hydrogen ^{1}H, fluorine, ^{19}F and phosphorus, ^{31}P; other less abundant nuclei, such as those of ^{13}C and ^{15}N, also have $I = \frac{1}{2}$.

The nuclear magnetic spectra of nuclei with $I = \frac{1}{2}$ are usually the most highly resolved and the most readily interpreted. However, a number of other nuclei which have $I > \frac{1}{2}$ do give useful n.m.r. spectra. These include the ^{14}N nucleus and the deuterium nucleus, ^{2}D, both of which have a spin quantum number of unity, and the ^{11}B nucleus, which has a spin of 3/2. In organic chemistry by far the most important nucleus is that of hydrogen because this gives rise to well-resolved spectra, and the signals are strong. After hydrogen, phosphorus and fluorine find useful applications. At first sight, it may seem disappointing to the organic chemist

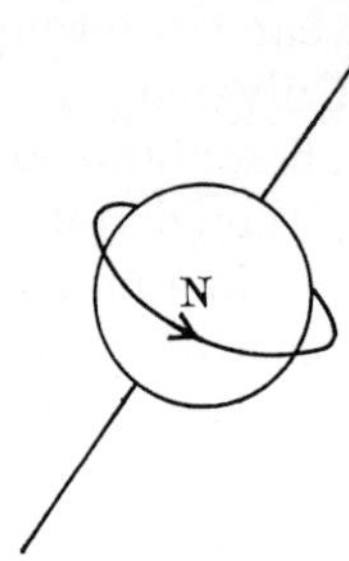

FIG. 1.1. The magnetic moment μ resulting from the rotation of the positive charge distribution of a nucleus N.

that ^{12}C and ^{16}O are non-magnetic nuclei. In fact, as we shall see, if such nuclei had been magnetic the n.m.r. spectra of hydrogen nuclei, of which such tremendous use is now made, would have been too complicated to interpret in the majority of cases.

We shall first of all discuss nuclear magnetic resonance in terms of the concepts of classical physics. By analogy with the behaviour of an electrical dipole in an electric field, it might be imagined that when a magnetic nucleus is placed in an applied magnetic field it will line itself up, or tend to line itself up, with that field. In fact, because of the spin angular momentum that the nucleus possesses, a better analogy for the behaviour of the nucleus is that of a spinning gyroscopic top in an applied field which is tending to cause it to reorient—for example, the ordinary model gyroscope in the earth's gravitational field. Under such conditions, the spinning top does not change its mean orientation with respect to the field, but its axis of spin precesses around the direction of the applied field. A magnetic nucleus in a magnetic field behaves in a similar way (see Fig.

1.2), and it is the frequency of this precessional motion that we measure in magnetic resonance experiments. The frequency of precession ν_0 is directly proportional to the magnetic moment μ of the nucleus, and also to the applied magnetic field H_0; it is inversely proportional to the spin angular momentum. Reorientation of the precessing nucleus with respect to H_0 can be brought about by causing the nucleus to interact with a second fluctuating magnetic field H_1 which is applied at right angles to H_0 and which is of the same frequency and phase as the precessional motion of the nucleus. Under these conditions, the precessing nucleus now changes its angle of precession relative to H_0; if it is caused to tilt away from alignment with H_0, energy is absorbed. The essential feature of the n.m.r. experiment is a measurement of the change in impedance of the coil that provides the fluctuating magnetic field H_1. This change in impedance is dependent on the energy absorbed.

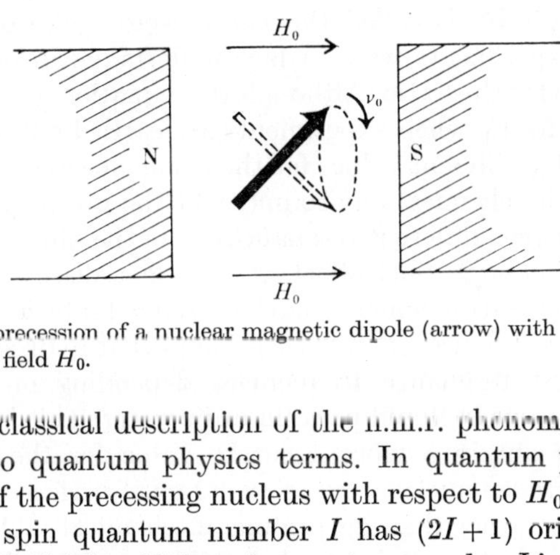

FIG. 1.2. The precession of a nuclear magnetic dipole (arrow) with frequency ν_0 in an applied magnetic field H_0.

The above classical description of the n.m.r. phenomenon has to be translated into quantum physics terms. In quantum physics not all orientations of the precessing nucleus with respect to H_0 are allowed. A nucleus with spin quantum number I has $(2I+1)$ orientations with respect to the field. Thus if the spin quantum number I is equal to $\frac{1}{2}$ then only two precessional orientations are possible. One of them has a component magnetic moment parallel to H_0, the other antiparallel to H_0. Energy is absorbed as the precessing nucleus is caused to change from the "parallel" to the "anti-parallel" orientation. If $\mathbf{h}\nu_0$ is the quantum of energy involved, the frequency ν_0 of precession is given by the equation

$$\nu_0 = (\mu/I\mathbf{h})H_0$$

This contains the same factors as the classical equation except that the component of the spin angular momentum has been replaced by its quantum expression $(I\mathbf{h}/2\pi)$.

The n.m.r. experiment can be carried out in practice by varying the frequency of H_1 for a fixed magnetic field H_0, this last being provided by a powerful electromagnet or permanent magnet. Alternatively, it is usually and more conveniently carried out by keeping the frequency H_1 fixed and varying H_0.

The Chemical Shift

An important practical point is that high resolution n.m.r. spectra are obtained only when the molecule concerned can rotate or reorientate rapidly in comparison with the nuclear precession frequency. Thus, high resolution spectra are limited to experiments in fluid media, i.e. in liquids and in solutions, or (when signal strengths permit) in the gas phase.

So far we have described only an experiment in physics, for the resonance frequency given by the previous equation is determined merely by H_0, the applied magnetic field, and by the ratio (μ/I) which, when multiplied by $(2\pi/\mathbf{h})$, is called the *magnetogyric ratio* of the nucleus. Early n.m.r. experiments were in fact undertaken in order to obtain accurate values for this ratio. Although this equation gives a reasonably accurate value for ν_0, when experiments are carried out with very high precision, slightly different values for the resonance frequency are found depending on the chemical environment of the given type of nucleus. Separations between different resonances from the same type of nucleus are called *chemical shifts*. Such effects were, in fact, discovered for protons when experimental arrangements had been refined sufficiently to measure ν_0 to six significant figures. It was then found that hydrogen nuclei give slightly different resonance frequencies depending on whether, for example, they occur in liquid paraffin or in water. This led to the realization that, to obtain a more accurate value for the magnetogyric ratio of the hydrogen nucleus, experiments must perforce be carried out on bare protons rather than on protons in chemical molecules. In one sense, therefore, the term "chemical shift" is a physicist's expression of exasperation! At this point, however, it seemed possible that there could be chemical applications in n.m.r. The n.m.r. lines are exceedingly sharp, and even when the frequencies observed for a fixed field differ only by one part in 10^6, it is still possible to measure these differences with precision. Indeed by the use of a variety of techniques, modern n.m.r. spectrometers can be made to give the very high resolution of one part in 10^8.

The resonance frequency of a given type of nucleus depends on its chemical location, because the surrounding electrons which constitute the chemical environment shield the nucleus from the applied magnetic field H_0. This is because the applied magnetic field sets up current circulations in the electron orbitals that surround the nucleus (Fig. 1.3 (a)). These are of a nature such as to produce at the nucleus a slight field δH which is in opposition to the incident field H_0. In different chemical environments, this effect occurs to differing extents. The previous equation has then to be replaced by one in which the magnetic field is given by $(H_0 - \delta H)$. Alternatively, this can be written in terms of the field-independent shielding parameter σ, as follows:

$$\nu = (\mu/I\mathbf{h})(H_0 - \delta H) = (\mu H_0/I\mathbf{h})(1-\sigma)$$

In the remainder of this chapter, explicit consideration will be given only to the chemical shift pattern found for hydrogen nuclei. It should be

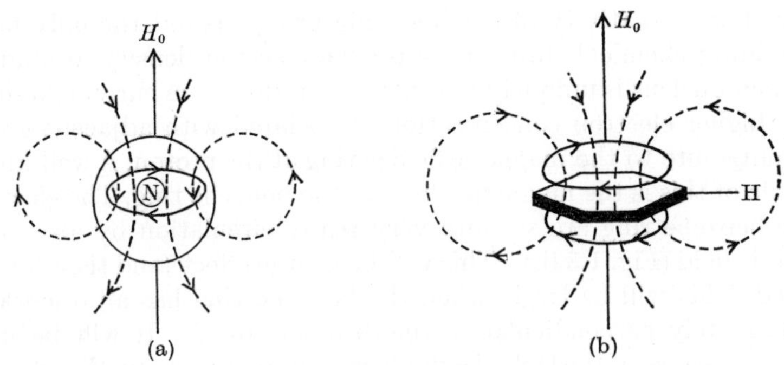

FIG. 1.3. The current circulation, and resulting magnetic field, caused by the effect of a magnetic field, H_0, on (a) the electron-distribution of an atom, (b) the π-electron orbitals of a benzene ring.

made clear at this point that because of the uniquely low electron density around hydrogen nuclei, such chemical shifts are smaller than for any other nucleus. With hydrogen nuclei the absolute magnitude of σ is of the order of 3×10^{-5}, with values *differing* by several parts in 10^7 or 10^6 between differing environments; for other nuclei σ may be as high as 10^{-4} or 10^{-3}.

An obvious factor that would be expected to cause changes in the shielding is the electron density around the nucleus, although this is only one of several factors that determine the observed pattern of chemical shifts. The study of the chemical shifts of hydrogen nuclei in methyl groups attached to atoms of different electronegativity reveals a pattern of chemical shifts that closely reflects the electronegativity of the attached atom; thus, it is clear in this case that variation in electron density is important. For example, the degree of shielding of the hydrogen nuclei of methyl groups increases from OCH_3 to NCH_3 to CCH_3 to $SiCH_3$ groups. In this sequence oxygen is the most electronegative substituent and its high electron-withdrawing power would be expected to reduce the electron density around the hydrogen atom of the CH_3 group more than would N, C or Si; the silicon atom is the least electronegative. Indeed, the electropositive nature of silicon with respect to carbon has the effect that the resonance from $SiCH_3$ groups occurs at higher fields than most resonances of other organic groupings. As a result, silicon tetramethyl $Si(CH_3)_4$ is a convenient and widely used internal reference standard for measuring chemical shifts. Although the precision used in n.m.r. work is such that it would be very difficult to measure the absolute value of the field at each nucleus, it is quite adequate to measure the position of a peak relative to that of a reference compound such as tetramethyl silane.

The electronegativity of neighbouring groups is not the only factor determining chemical shifts. Because the electron density around the hydrogen nucleus is uniquely low, fields resulting from currents induced in the higher electron concentrations associated with adjacent groups also contribute to the magnetic field acting at the proton. A well-known example of this is the magnetic effect of the benzene ring. The electrons of the benzene ring are set into very ready circulation by an applied magnetic field (Fig. 1.3 (b)). This will be a large effect (and therefore the induced fields will be large) when the benzene ring has an orientation approximately perpendicular to the direction of H_0. It will be much reduced, because of the lack of cylindrical symmetry about H_0, when the benzene is orientated with its plane parallel to H_0. Thus, the magnitude of the induced magnetic field in the molecule varies with orientation, and this effect is called magnetic anisotropy. When anisotropy is present such induced fields have a net effect at nuclei that are not in the centre of the electron distribution concerned. When anisotropy is not present this effect of the neighbouring group cancels out. The anisotropy associated with the aromatic ring results in the protons attached to such rings having a chemical shift of 1 to 2 parts per million to low field of those of sp^2 ethylenic CH bonds. As the whole range of chemical shifts between protons in organic molecules rarely exceeds 10 or 15 parts per million, the aromatic effect mentioned above is an important one. It is clear, therefore, that an adequate understanding of chemical shifts can be achieved only if the magnetic as well as the electrical properties of molecules are taken into account. Only the electrical properties are usually familiar to the chemist in the form of inductive and mesomeric effects.

Typical values of chemical shift of different types of group in organic chemistry are given in Tables 1–7, p.179. For the present, it is sufficient to note that sp^3, sp^2, and sp hybridized CH bonds can readily be distinguished from each other on the basis of observed chemical shifts.

Intensities of Resonances

Another very important property of n.m.r. spectra is the fact that, for any given type of nucleus, the *strength of a signal* depends only on the number of nuclei involved and is independent of the nature of the chemical environment. This leads to many important quantitative applications of n.m.r., and it is of interest to cite an early example of a structural problem solved on these grounds alone. The structure of Feist's acid was considered on the basis of organic chemical reactions to have one of the following two alternative formulae:

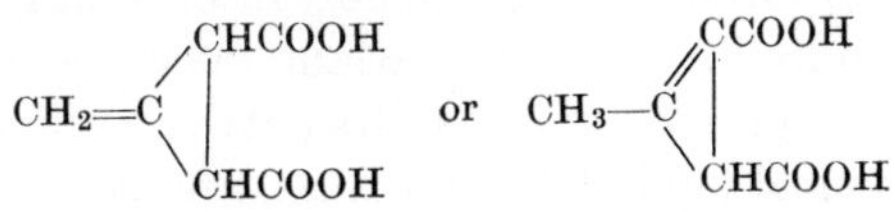

(A) (B)

As can be seen, these differ in that for formula (A) there are two ethylenic CH bonds and two cyclopropane CH bonds, whereas in formula (B), there is only one cyclopropane CH bond and three CH bonds of the sp^3 hybridized methyl group. N.M.R. experiments were carried out with Feist's acid in D_2O solution (so as to eliminate resonances from the hydrogen nuclei of the COOH groups): the intensity distribution showed at once that the first formula was correct; two resonances were found which were of equal intensity. Had the second formula been correct, the two resonances would have had an intensity ratio of 1 to 3.

Spin–Spin Splitting

The third parameter that contributes to the great subtlety of n.m.r. as a tool in chemical structure determination is the fine structure often associated with individual resonances. This occurs because of mutual interaction between different magnetic nuclei within the same molecule. With a solid sample, where the molecule is fixed in its orientation with respect to the applied magnetic field, the effects of direct magnetic interactions between such nuclei affect the spectrum. Because the field at one nucleus caused by a neighbour can be very considerable, and also because any given nucleus has a large variety of magnetic neighbours at different distances and orientations with respect to H_0, this phenomenon can lead to very broad resonances. In the liquid state rapid molecular motion leads to such direct magnetic interaction being averaged to zero. However, there remains a much smaller, but still measurable, effect due to indirect interaction between magnetic nuclei via the intervening electrons which does not average to zero in the liquid state or in solution. The full name of this phenomenon is "indirect electron-coupled nuclear spin–spin splitting". For brevity, the resolved structure in the resonances is usually termed spin–spin splitting.

Let us imagine how this type of electron-coupled interaction takes place for the very simple case of two magnetic nuclei joined by an electron distribution in an electron-pair single bond. This would happen, for example, with a P—H bond, where the directly bonded nuclei are both magnetic, or with a C—H bond when the non-magnetic ^{12}C nucleus is replaced by a magnetic ^{13}C nucleus. In such cases, the magnitude of the coupling is considerable because of the short intervening path and is of the order of several hundred to several thousand cycles per second (c/s). This is often larger than typical chemical shifts between different nuclei of the same type (at a typical field of 14 kilogauss, and a hydrogen n.m.r. resonance frequency of 60×10^6 c/s, a chemical shift of 1 part in 10^6 corresponds to a frequency difference of 60 c/s). Figure 1.4 may be used to indicate the nature of this internuclear coupling. In this the nuclear magnet moments are represented by short arrows and the much greater magnetic moments of the electrons by long arrows.

The right-hand hydrogen nucleus is shown in Fig. 1.4 with a parallel orientation of its magnetic moment with respect to H_0. The electron

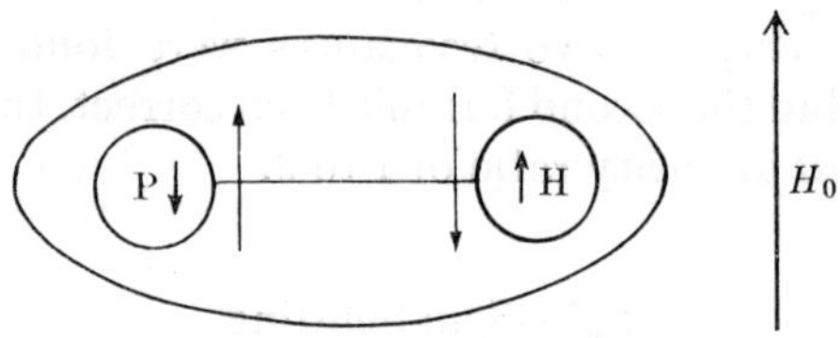

FIG. 1.4. The interaction of the nuclear magnetic dipoles of phosphorus and hydrogen nuclei (short arrows) via the magnetic dipoles of the electrons (long arrows) in an electron-pair bond.

closest to it then has its magnetic moment polarized in the opposite direction by mutual interaction with the nucleus, because (as with two bar magnets) an anti-parallel orientation of adjacent magnetic dipoles leads to the lowest energy. In this particularly simple case of an electron-pair bond the two intervening electrons have, of course, anti-parallel spins. The resulting effect is that the electron which is at the same instant adjacent to the left hand phosphorus nucleus has a resulting polarization parallel with respect to H_0, and it exerts an appropriate magnetic field on the P nucleus. Because of the very small energy difference between parallel and anti-parallel orientations of the hydrogen nucleus with respect to H_0, these two orientations are virtually equally probable, and so the P nucleus occurs in one local field or another depending on the orientation of the nearby hydrogen nucleus.

The result of this is that the phosphorus nucleus gives a doublet resonance, the separation between the component lines being equal to the difference of the two local fields. When this is expressed in frequency units (cycles/second), the separation is termed the inter-nuclear coupling constant and is usually given the symbol J. Because the interaction is a mutual one, and because in this example we have chosen an $I = \frac{1}{2}$ nucleus in each case, the hydrogen resonance is also split into a doublet with the same energy difference, i.e. the same frequency separation, and the complete spectrum is shown in Fig. 1.5. If the phosphorus nucleus were

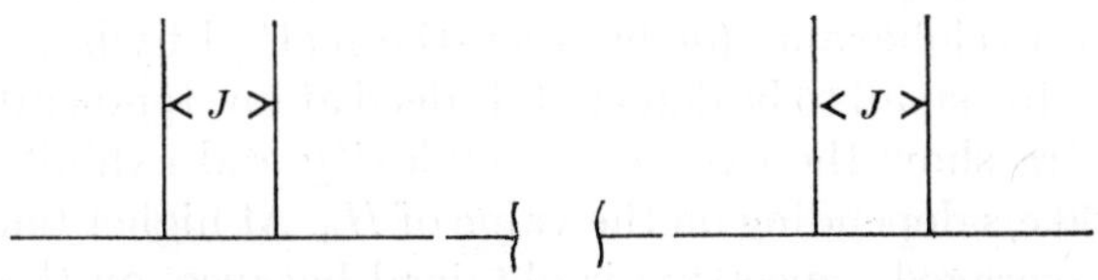

FIG. 1.5. The doublets caused by mutual spin-spin coupling of two $I = \frac{1}{2}$ nuclei which have well-separated resonances.

replaced by ^{14}N ($I = 1$), the hydrogen resonance would then be split into three lines of equal intensity corresponding to the three possible orientations of the nitrogen nucleus with respect to H_0 for the component values of I of 1, 0, −1. Once again, however, the separations between adjacent lines in the hydrogen triplet and the nitrogen doublet would be equal in frequency units. The same effect will occur even if the nuclei are of the same species as, for example, two hydrogen nuclei in different environments, provided that there is a considerable chemical shift between them, i.e. provided that they have well-separated resonances. An example of this latter type could be the two hydrogens in 1,1-dibromo-2,2-dichloroethane, $Cl_2CHCHBr_2$. This aspect of n.m.r. spectra with other more advanced aspects of spin–spin splitting is more fully discussed in Chapters 4, 5 and 6.

These will include examples with more than two sets of coupled nuclei; the case of chemically equivalent but magnetically non-equivalent nuclei will also be discussed. Finally, it should be mentioned that the magnitudes of coupling constants are often highly diagnostic of stereochemical situations with respect to pairs of C—H bonds.

The Time-scale of n.m.r. Spectra; Simplified Spectra

In a considerable number of cases the n.m.r. spectrum of a substance is simpler than might be expected if the molecule is thought of in terms of its instantaneous configuration. For example, it might be expected that there would be a chemical shift between the axial and equatorial hydrogen nuclei in cyclohexane. If so, this should result in a complex spectrum because of spin–spin splitting between these two types of nuclei. In fact, at room temperature, the n.m.r. spectrum of cyclohexane consists of a single sharp line. Again, it might be thought that in the staggered configuration of ethyl chloride the hydrogen of the methyl group in the plane of the heavy atoms would have a different chemical shift from the two mutually equivalent methyl hydrogens on either side of that plane. However, the n.m.r. spectrum shows that all these three methyl hydrogens are equivalent.

It is not that the chemical shifts between the axial and equatorial hydrogens in cyclohexane, (or between the methyl hydrogens in ethyl chloride), are too small to be detected. Indeed at low temperatures cyclohexane spectra show the expected complexity and exhibit a chemical shift of 10–20 c/s depending on the value of H_0. At higher temperatures, a simpler "averaged" spectrum is obtained because, on the time-scale of this particular spectroscopic experiment, the same nucleus goes to and fro many times between the different locations, and thus an average chemical shift is measured. In the cyclohexane case, the chair-form of the ring inverts sufficiently rapidly for an average $\frac{1}{2}(\sigma_{\text{equatorial}} + \sigma_{\text{axial}})$ chemical shift to be obtained. In the case of the methyl group in ethyl chloride, the internal rotation about the C—C bond is likewise very rapid.

In each of these examples the time-scale of the experiment is set by the chemical shift between the separate resonances for a fixed configuration of the molecule. If we assume that in the cyclohexane case the chemical shift between axial and equatorial protons is 15 c/s, then when ring inversion occurs much more slowly than 15 times per second (as it does at sufficiently low temperatures), the complex spectrum will be obtained. If ring inversion occurs much more rapidly than this speed, the simplified "average" spectrum will be obtained. Similarly, the internal rotation of the CH_3 group in ethyl chloride occurs at room temperature much more rapidly than the relevant chemical shift difference expressed in c/s, and an average spectrum is obtained. In general, internal rotational motions have to be exceptionally slow (corresponding usually to barriers to rotation of more than 10 kcal/mole) for averaged n.m.r. spectra *not* to be obtained. On the other hand rigidly fixed cyclohexane rings, such as occur in steroids, exhibit the full complexity expected in their proton spectra.

Another class of rate processes that commonly give rise to averaged n.m.r. spectra are proton-transfer processes. Thus a mixture of carefully purified water and ethyl alcohol exhibits separate OH resonances from the H_2O and C_2H_5OH molecules. However if a trace of acid is added to catalyse proton-transfer processes, a single combined resonance is obtained from the OH groups of water and alcohol. For this reason it is commonly found that when a substance is dissolved in a solvent with active hydrogens (such as water), active hydrogens in the solute (OH or NH groups) may combine with the solvent to give a single peak. Rapid proton-transfer processes of this kind can also lead to collapse of spin–spin fine structure. Ethyl alcohol provides a good example of this latter phenomenon. When it is carefully purified, the OH resonance is a triplet due to coupling with the adjacent CH_2 group. On the other hand a trace of acid, quite insufficient to cause a separate resonance, will so catalyse proton-transfer between different alcohol molecules that the OH resonance is reduced to a single sharp peak. In this case, the time-scale is set by the OH—CH_2 coupling constant of about 5 c/s. If the proton transfer is much more rapid than J, expressed in c/s, the OH proton does not stay coupled to a CH_2 group long enough for fine structure to appear. At the same time as the fine-structure disappears from the OH resonance, the CH_2 resonance is converted to a quartet, as expected for the observed coupling with the CH_3 group alone.

Typical separations between lines in infrared or ultraviolet spectra might be 1 to 10 cm^{-1}, i.e. 3×10^{10} to 3×10^{11} c/s. The internal rotation motions and proton transfer processes discussed above are much too slow to "average out" these spectra, and the "instantaneous" molecular configuration is relevant when interpreting them. But the tremendous resolution routinely obtained in n.m.r. spectroscopy (1 c/s $\equiv 3 \times 10^{-11}$ cm^{-1}.) means that we must consider the molecular motions and processes from the point of view of an altogether much longer time-scale.

Conclusion

Whenever a molecule contains a reasonable number of magnetic nuclei, such as ^{1}H, ^{19}F, ^{31}P, ^{11}B etc., the n.m.r. spectrum, when considered in the appropriate "time-scale", can give very important information about the structure of the molecule. Much of the power of the method is due to the fact that three different parameters—the position of resonances (or the chemical shift), their intensities and their fine-structure (spin–spin splitting)—all contribute directly to the structure-determination process.

In the context of organic chemistry, the spectral simplification caused by the non-magnetic nature of ^{12}C and ^{16}O nuclei enables very great use to be made of many of the features of the strong n.m.r. spectra from the hydrogen nuclei. Because hydrogen atoms are located in many places in typical organic molecules, the overall pattern of information obtained can be a rich one. Even with fairly complex organic natural products, at least some parts of the spectra are sufficiently free of overlapping so that considerable useful information can be obtained. Indeed, when it can be obtained, the n.m.r. spectrum may reasonably be claimed to be the most penetrating of all spectroscopic techniques from the point of view of organic structure determination. The experimental requirements are not very high—usually about 1 to 10 mg of sample are needed, and a suitable solvent has to be found from a fairly wide range of possibilities including CCl_4, $CHCl_3$ (or $CDCl_3$) and D_2O. The most prominent resonances in organic n.m.r. spectra are those from CH_3 groups (because of the combined intensity provided by the 3 protons) and in general it is the spectra of CH groups that can be systematically interpreted. Hydroxyl resonances, and to a lesser extent NH resonances, can be difficult to locate because of exchange processes mentioned above and for other reasons, including the effects of hydrogen-bonding on resonance positions and line broadening due to quadrupole processes involving nitrogen atoms. For such groupings, the infrared spectrum is often more informative, and it goes without saying that *infrared and ultraviolet spectra should always be used in addition to n.m.r.* so as to determine in addition the presence of double or triple bonds or conjugated systems. However, because n.m.r. spectra depend on the *magnetic* properties of molecules, they provide much information additional to that available from the use of these longer-established spectroscopic methods and others such as mass spectrometry.

CHAPTER 2

THE CHEMICAL SHIFT[1-5]

Part I

J. A. ELVIDGE

In the preceding section it was shown that many atomic nuclei possess spin angular momentum, characterized by a spin quantum number I, and that as a result they have a magnetic moment μ, pictured in classical terms as the result of the spinning of the charge. Such nuclei interact with an externally applied magnetic field, and we have gathered that these interactions may be detected and displayed in the form of a spectrum. The different positions of the spectral lines, termed chemical shifts, can be correlated with functional groupings and so n.m.r. spectra can be used for the deduction of molecular structure. Tables of chemical shifts (Tables 1–7, p.179) are an essential aid.

We have noted that the interaction of a magnetic nucleus with an external magnetic field H is quantized*, as is general with atomic phenomena, and only $(2I+1)$ energy states (or spin states) are allowed. These correspond in number to the number of allowed values for the magnetic quantum number m, which can have the values $I, I-1, \ldots -I+1, -I$. Note that these values proceed in integral steps from $+I$ to $-I$. The energies are equally spaced, the separation in energy ΔE, between any two adjacent levels being $\mu H/I$ (Fig. 2.1). Transitions between *adjacent* levels are stimulated by electromagnetic radiation of the correct frequency ν, given by $\Delta E = \mathbf{h}\nu$ where $\mathbf{h}$ is Planck's constant. We thus obtain as the condition for resonance interaction that (i) $\mathbf{h}\nu = \mu H/I$ ergs where H is in gauss (strictly speaking in oersteds, although in air this is equivalent to gauss), μ is in multiples of βN, the nuclear magneton $e\mathbf{h}/4\pi Mc$, and I is in units of $\mathbf{h}/2\pi$, or (ii) $\nu = \gamma H$ c/s, where $\gamma = \gamma/2\pi$ and γ is the nuclear proportionality constant known as the magnetogyric ratio in radians gauss^{-1} sec^{-1}. At room temperature in a magnetic field, lower spin states are slightly more populated than upper,† so although stimulated

* The symbol H denotes the field experienced *at* the nucleus. H_0 denotes the magnitude of the field applied by the spectrometer: because of local screening (see p. 15) H may differ from H_0.

† In a magnetic field there is a Boltzmann distribution with temperature such that at room temperature there is a small excess of protons, amounting to 1 in 10^5, in the lower spin state.

transitions are equally probable up and down, there are on balance more upward than downward transitions. Hence there is a net absorption of energy and this can be detected and shown on a recorder as a spectral absorption line.

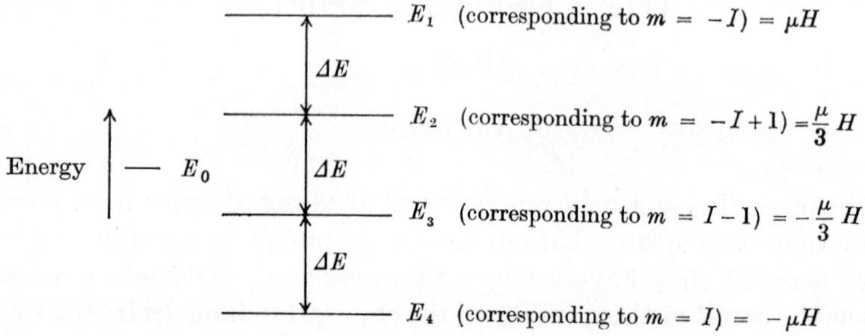

FIG. 2.1. Energies [given by $-(m\mu/I)H$] of a nucleus with $I=\frac{3}{2}$ in a magnetic field H, relative to the energy in zero field (E_0). There are $(2I+1)$ states (i.e. 4 in this case, corresponding to the allowed values of m) equally separated in energy by $\Delta E = \mu H/I$. In classical terms, the uppermost level corresponds to full alignment of the nuclear moment *against* the applied field and the lowest level to full alignment *with* the field.

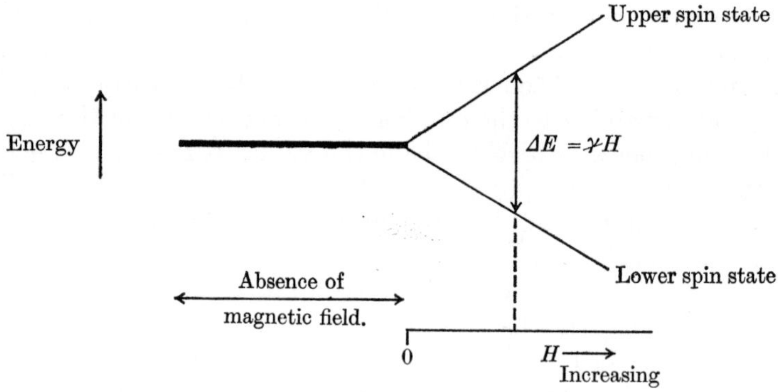

FIG. 2.2. Interaction of a proton ($I=\frac{1}{2}$) with a magnetic field: the energy separation between the two allowed spin states is proportional to the magnetic field strength, H. In the absence of a field the two spin states of the proton are equal in energy: the energy (or the system) is then said to be doubly degenerate.

For the proton, with $I=\frac{1}{2}$, there are two energy states only in a magnetic field (Fig. 2.2) and so only one transition can be effected and detected. With a field strength of 14,100 gauss, this occurs at 60 Mc/s in the r.f. region of the spectrum; at 23,500 gauss it is at 100 Mc/s.

Nuclear properties are fixed so that, at a constant operating frequency ν, every proton in our sample would be expected to come to resonance at one and the same field strength H given by the above relationships. It might be expected that all compounds of hydrogen would give only a single proton magnetic resonance line and that other kinds of magnetic nuclei, similarly, would give their own characteristic lines. In practice, the spectrum from a compound is much more complicated and consequently more informative. Thus, with an organic compound it is found that separate lines are obtained from each of the various chemically distinct hydrogen nuclei in the molecule.

Protons (and indeed other magnetic nuclei) which constitute different functional groups come to resonance at different, characteristic, values of the applied field. The reason is that the field they experience is a modified one. It is the applied field modified by secondary magnetic fields arising from the motions, in the applied field, of valency electrons, either in the bonds or about adjacent atomic groupings within the molecule. The nuclei are thereby screened magnetically, the magnitude (and sign) depending on the local electron density and on neighbouring groups. For the proton in organic compounds, the extent of the local screening is very small since only two electrons circulate about a proton which is covalently bound to another atom. This weak screening enables a proton to experience other small modifications in the field, arising from electron circulations about atoms or groups in more distant parts of the molecule.

Each magnetic nucleus in a molecule thus experiences a field H which differs from the applied field H_0 by a small fraction. The fraction is known as the screening constant for that nucleus; its value depends on the molecular environment in which the nucleus is situated. Thus, $H_{\text{at nucleus}} = H_{0\,\text{applied}} \times (1-\sigma)$. The degree of screening determines the amount by which the applied field H_0 has to be increased (at constant operating frequency) to reach the one correct field H at the nucleus to satisfy the resonance condition $\nu = \gamma H$. For several nuclei, each with different screening constants σ, σ', etc., we have $H = H_0(1-\sigma)$, $H = H_0'(1-\sigma')$, etc. Hence,

$$\frac{H_0}{H_0'} = \frac{(1-\sigma')}{(1-\sigma)}$$

and so if $\sigma' > \sigma$, then $H_0' > H_0$. This means that the more highly screened the nucleus, the more the applied field must be increased for resonance to occur. With the operating frequency constant, the field is therefore swept (between limits) and at various values of the applied field, absorption of energy occurs, thus giving a series of resonance lines. For example, *p*-xylene (I) gives a two-line spectrum,[6a] one line from the 4 chemically equivalent protons of the ring and another from the 6 protons of the methyl groups (Fig. 2.3). It is to be noted that the integrated intensities of these lines are in the same ratio as the numbers of protons responsible so that, provided that the total number of protons is known, the spectrum effectively gives a hydrogen analysis. With correct spectrometer operation, this proportionality is always obtained, because there is only

(I)

one transition for protons and so the transition probability is always the same. Herein lies an important difference from infrared or ultraviolet spectrometry.

Our spectra then have a low-field end (on the left, by convention) where absorption lines from less shielded or "deshielded" nuclei appear, and a

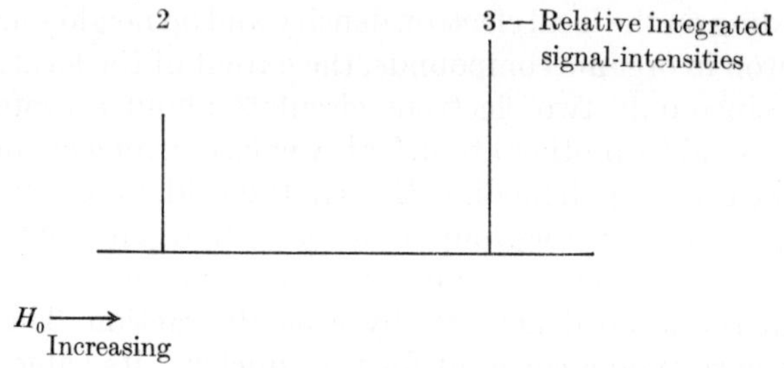

FIG. 2.3. Proton resonance lines from *p*-xylene (I) in $CDCl_3$.

high field side (on the right) where the signals from the more highly shielded nuclei appear. The higher the electron density about a nucleus, the more highly shielded it becomes. Thus, in the spectrum of *p*-xylene (Fig. 2.3), the methyl resonance line is at higher field than that from the aromatic ring protons.

These differences in the positions of resonance lines are termed *chemical shifts*, because they depend upon structural features within the molecule. Shifts between lines from different kinds of nuclei can be extremely large. Normally, we are concerned with only one kind of nucleus at a time, e.g. with the protons in an organic molecule. In this case, for the reason already given, chemical shifts are rather small and range over only about 10 ppm. Thus, at 60 Mc/s the resonance lines for the majority of hydrogen-containing groupings lie within a range of 600 c/s. Fortunately, radio frequency techniques enable *differences* in line positions to be measured with great accuracy, to 0·1 c/s. Chemical shifts, then, which are energy differences—differences in applied field strength—are measured in frequency units rather than in units of magnetic field strength, although it is the magnetic field that is altered during the spectral measurements, while the operating frequency is kept constant.

It is important that spectral lines should be as sharp as possible, for accurate measurement of their positions. Experimentally, non-viscous liquids and solutions give sharp spectra with line widths down to 0·3 c/s or so. Solids give very broad lines, e.g. tens of *kilo*cycles wide. This is because the molecules are orientated and any one nucleus is subjected to a variety of magnetic fields due to the other orientated nuclei, the spins of which, and hence the associated magnetic fields, are changing, through spin–spin coupling. Resonance therefore occurs over a broad range of applied field strengths. In a mobile liquid or solution, however, the random tumbling of the molecules averages these local fields to zero. Each nucleus then experiences, in general, only its own screened value of the applied field, and resonance occurs at a single characteristic value of the latter. Dissipation of the energy of an upper spin state of a nucleus to the rotational and vibrational modes of the molecules requires special conditions and is slow. So the lifetime of an upper spin state of a nucleus in a liquid sample is relatively long by other spectral standards, and the natural line width, proportional to the reciprocal of the upper-state lifetime, is correspondingly small. Hence the sharpness of the lines. A factor which would contribute to broadening would be inhomogeneity in the applied magnetic field. By instrumentation, the field is made as homogeneous as possible. Effectively, to increase the field homogeneity to perfection, the liquid or solution sample (contained in a sample tube) is spun at 200–300 rev/min. The time of passage of nuclei in the sample between points of slightly different field strength is then made so short that, by the uncertainty principle, these field inhomogeneities cannot be detected. A small energy difference, ΔE (i.e. difference in field strength) can be measured or can otherwise become apparent only if the time interval available, Δt, is such that $\Delta E . \Delta t \gtrsim \mathbf{h}/2\pi$.

For comparative purposes, it is convenient to measure chemical shifts from some arbitrary, virtually invariant line position, provided by a standard reference substance. This may be used as an *external* or as an *internal* reference. In the former case, the substance is added in a sealed capillary to the sample. Such a reference experiences a field which is independent of the sample, so that the position of the reference signal is constant. Because the bulk magnetic susceptibility of different samples varies however, the difference between the fields experienced by sample and reference will not be constant but will depend on the sample. Hence, the protons in a particular grouping in the same chemical environment in a series of different compounds, e.g. in the α–CH_2 in an homologous series of acids, will not give a resonance line at a constant position—they will not have a single characteristic chemical shift. This is highly inconvenient. However it is possible to obtain reliable shielding values by making measurements against an external standard at a series of dilutions with each sample and extrapolating the chemical shifts to infinite dilution. By using an *internal* reference this end-result can be obtained directly, and so is much to be preferred. As an internal reference the standard substance is added in small amounts ($<1\%$) to the dilute solution (5–10%) of the solute under examination. Being together in a common medium, both reference and solute molecules experience an identical field. The relative separation of the resonance lines will therefore be constant, even though absolute positions will vary from sample to sample owing to their different susceptibilities. For this to hold true, an internal reference must be chemically inert, so that it does not associate even weakly either with itself or with other molecules. As a reference substance for proton resonance spectra, tetramethylsilane (TMS), Me_4Si, first proposed by Tiers[7], is highly suitable. It is a spherically symmetrical, inert molecule; it gives a single, strong sharp resonance line from its 12 equivalent protons and so can be used at 0·1–0·5% concentration. The line occupies a unique position. Silicon is less electronegative than carbon, so the electron density around the methyl groups in silicon tetramethyl is higher than that round the methyl groups in, for example, tetramethylmethane. Hence the position of the line from TMS is at higher field than the proton resonance lines in the majority of organic compounds. The TMS line is at once recognizable at the high-field end of the spectrum, and is unlikely to be obscured.

The ideal solvent is an inert mobile liquid with a molecular structure which is magnetically isotropic, or nearly so. The solvent, which should preferably not contain hydrogen, must be capable of giving a 5–10% (w/w) solution of the compound under examination. Carbon tetrachloride is an ideal, but not a sufficiently powerful, general solvent. Carbon disulphide, chloroform and deuterochloroform are better general solvents and give closely similar results. Of these, deuterochloroform (which as normally prepared contains a small amount of $CHCl_3$) is perhaps most widely useful. Solvents less suitable from the n.m.r. viewpoint, but which must sometimes be employed for solubility reasons are benzene, pyridine, dimethylsulphoxide, water or their deuterated analogues. These are all liquids with polar or magnetically anisotropic molecules, and it is important that the chemical shift measurements should be accompanied by a statement of the concentrations and temperature, so that any other future measurements may be made under comparable conditions.

As an internal reference for the proton magnetic resonance spectroscopy of aqueous (or D_2O) solutions, dioxan (τ 6·25) and acetonitrile (τ 7·94) are more suitable than t-butanol[8]. The latter tends to associate with some compounds, so that its relative line position is not invariant. The use of the water line as a general reference point is strongly to be deprecated. The position of this line is temperature dependent but, perhaps worse, water is strongly associating, so that the relative line position cannot be expected to remain invariant from sample to sample. Observations on aqueous or D_2O solutions with acetonitrile as internal standard indicate that the position of the water line may vary over a region as large as 1 ppm. A widely useful internal standard for aqueous solutions is sodium 2,2-dimethyl-2-silapentane 5-sulphonate (DSS), $Me_3SiCH_2CH_2$-CH_2SO_3Na[9, 10] At the low concentrations used only the line from the Me_3 group shows. This comes at the TMS position (δ 0·0 or τ 10·0),[11a] so that the chemical shifts are obtained directly on the TMS scales. In aqueous solutions more strongly acid than 5N, the DSS line is at a position 0·03 ppm to low field of the TMS line. For strongly acid solutions then, we have to add 0·03 ppm to the apparent chemical shifts (measured in ppm from the internal reference as zero) to obtain correct δ values; alternatively, we must subtract 0·03 ppm from the apparent chemical shifts (in ppm from the internal reference line as 10·0) to obtain true τ values.

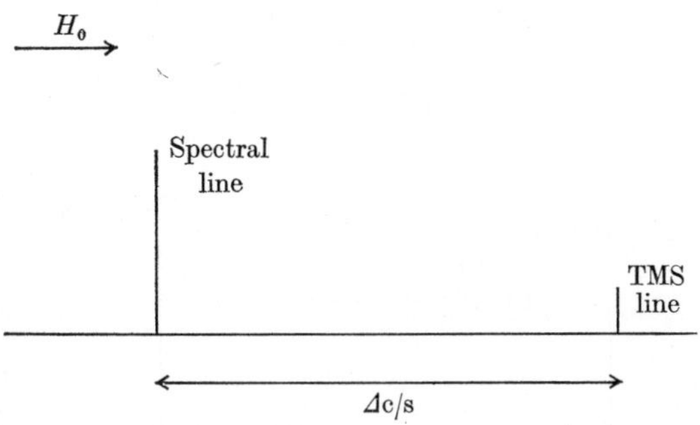

FIG. 2.4. The position of a spectral line, measured in c/s from the internal standard.

The study of magnetic nuclei other than the proton requires different internal standards, and these are discussed in Chapter 8.

Although chemical shifts are directly obtained from spectra in frequency units, this is not the best way of recording them. The reason is that the magnetic screening of a nucleus (arising from the induced circulation of electrons) is proportional to the strength of the applied field. Hence, the magnitude of a chemical shift depends upon the field

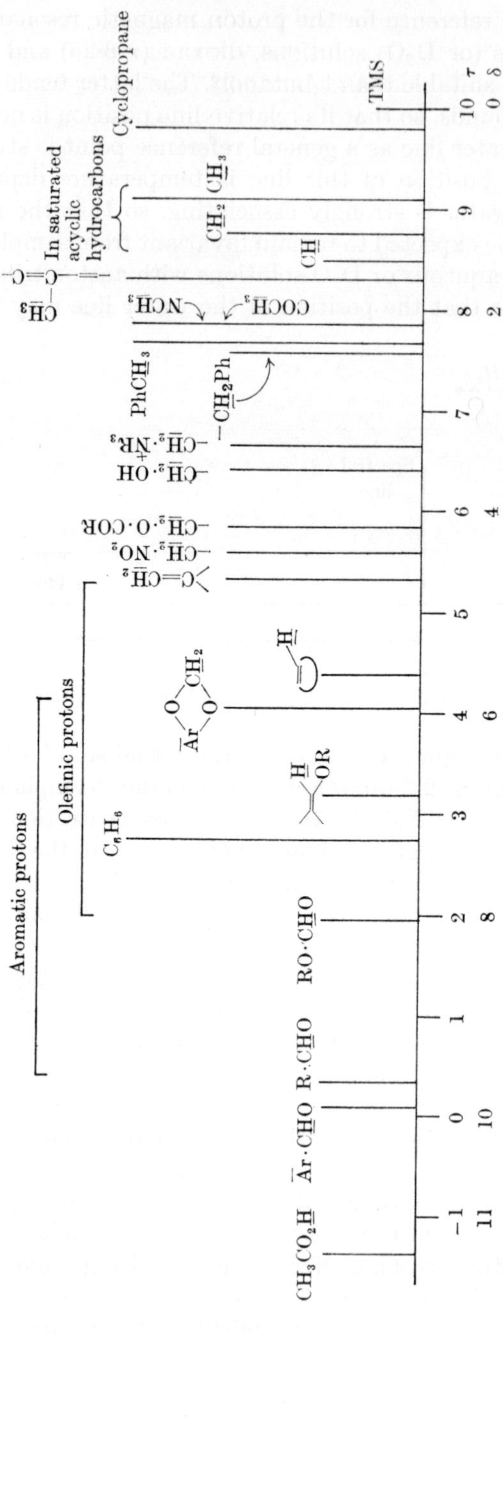

FIG. 2.5. Some common functions in organic compounds and their proton chemical shifts on the τ and δ scales.

strength and therefore on the operating frequency. Thus, two lines 400 c/s apart at 40 Mc/s are 600 c/s apart at 60 Mc/s. It is customary, therefore, to express chemical shifts in a field-independent manner.

The chemical shift, Δ c/s from TMS, measured on the spectral chart (Fig. 2.4), is divided by the spectrometer operating frequency (in c/s) and multiplied by 10^6 to give a shift δ as a convenient number in ppm. On

$$\delta\ (\text{ppm}) = \frac{(\Delta\ (\text{c/s})\ \text{from TMS}) \times 10^6}{\text{operating frequency in c/s}}$$

this system, the TMS line has the arbitrary zero position. We note that as δ values increase, the less shielded becomes the proton.

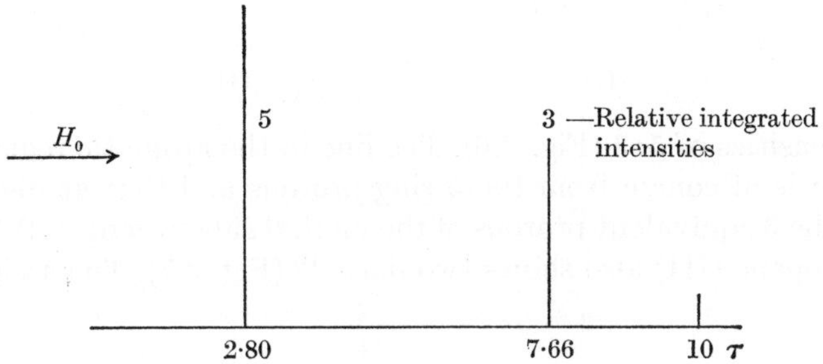

FIG. 2.6. Proton magnetic resonance spectrum of toluene (II) in carbon tetrachloride.

Another system in common use and one which is recommended by the Chemical Society, London, is the τ (tau) scale originated by Tiers[7]. In this, the TMS line is assigned the arbitrary position of 10 ppm. Other lines then have τ values given by $(10-\delta)$ ppm where δ has the significance already given above. On the τ scale, shifts decrease in magnitude with decreasing shielding, and may go to negative values.

Tables of chemical shifts, expressed as τ or δ values are a prerequisite for the interpretation of nuclear magnetic resonance spectra. Tables of proton chemical shifts, given as τ values, will be found on pages 179–186. Figure 2.5 gives a general idea of the proton chemical shifts for some common functional groupings.

Some regions overlap, for example, those where aromatic and olefinic proton lines appear, but in many cases chemical shifts are very characteristic of the chemical groupings and/or their environment. Thus methyl groups in acyclic aliphatic compounds, and removed from other functional groups always have chemical shifts near to τ 9·1. Methyl groups attached to aromatic or pseudo aromatic rings appear at τ 7–8. Also in this latter region we find N.Me and CO.Me lines. However, distinction between such possibilities is usually easy, especially with the help of chemical and other physical data. Information from n.m.r. is complementary to that obtainable in other ways and in general should not be used in isolation.

It is helpful at this point to consider a few simple proton resonance spectra. (i) That given earlier for *p*-xylene (Fig. 2.3) shows two lines as we should now expect, one at τ 7·7 from the protons of the methyl groups and one at τ 2·95 from those of the benzene ring.[6a] The integrated intensities of these lines are in the ratio 6:4 respectively. (ii) The spectrum (at 60 Mc/s) of toluene (II) in deuterochloroform also shows two lines[6b] but

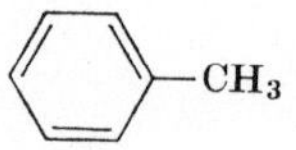

(II)

$ClCH_2—CCl_2—CH_3$

(III)

with intensities of 5:3 (Fig. 2.6). The line in the aromatic region of the spectrum is of course from the 5 ring protons and that at higher field is from the 3 equivalent protons of the methyl substituent. (iii) 1,2,2-trichloropropane (III) also shows two lines[11b] (Fig. 2.7). The assignments

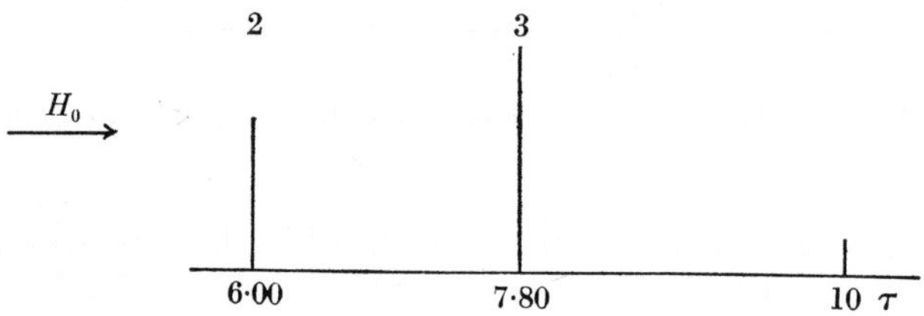

FIG. 2.7. Proton magnetic resonance spectrum of 1,2,2-trichloropropane (III) in deuterochloroform.

should now be obvious, both from the intensities and the chemical shifts. (iv) In the molecule of methyl β-amino-crotonate (IV) there are 4 chemically distinct kinds of hydrogen nuclei, denoted as shown by (A), (B), (C),

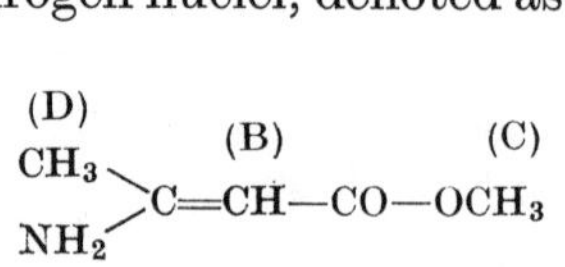

(IV)

(D) and these groups, having different chemical shifts, give rise to a four line spectrum[11c] (Fig. 2.8). Again, the assignments of the signals as shown are readily made from the information already given, i.e. from the chemical shifts and line intensities. The broad nature of the signal (A) from the NH_2 group deserves some comment. Band shapes vary enormously so that heights of lines cannot be used as an approximation for intensity: only the integrated intensities (area) will be proportional to the numbers of protons responsible for the signals. Those groups which can give broadened signals, and signals whose position may vary over a wide region of the spectrum, are collected together in Table 5 (p. 184).

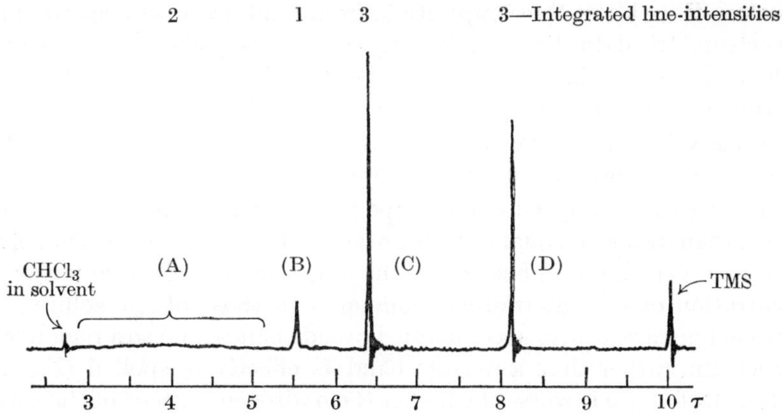

FIG. 2.8. The proton magnetic resonance spectrum of the compound (IV), showing the assignment of the 4 lines to the 4 different groups of protons (A), (B), (C), (D).

There, special characteristics are listed which help in correlating these signals with the possible functional groups, hydroxyl, amino, amido, thiol, and acid functions (i.e. OH in the situations CO_2H, SO_3H, OPOH, etc.). Hydroxyl protons give signals which may occur throughout the whole proton spectral range, from 9·5 to 6 τ for alcohols, through lower values for phenols, to negative values for carboxylic acids (and other acids), down to -6 τ or even lower still for strongly hydrogen-bonded enols (e.g. $-9{\cdot}20$ τ for usnic acid).

Protons which exchange rapidly give sharp signals (as do non-exchanging protons); their environments are well averaged. Protons which exchange slowly are effectively in more than one environment and may give rise to separate signals or to a broadened line, depending on the residence times and rate of exchange.[3] Hence protons attached to oxygen (–OH) and nitrogen ($>$N—H) may give either sharp or broad lines, depending on the chemical structure (Table 5, p. 184). In general, traces of acid or alkali catalyse these proton-exchange processes and so cause line sharpening. Deuteration on the other hand, effected by shaking the sample (usually in $CDCl_3$) with a trace of D_2O, results in the signal disappearing because of the conversion of –OH to –OD, etc.

The splitting of –OH and >N—H signals occasionally observed as a result of spin–spin coupling is mentioned in Chapter 4.

It will now be appreciated that, with the aid of tables of chemical shifts, we can deduce from a spectrum the probable kinds of protons present i.e. the nature of the functional groups in the molecule; from the integrated line intensities we deduce the relative numbers of the different kinds of proton. If we know the composition, from combustion or accurate mass-spectrometric data then, particularly with the help of other chemical and physical information, we may be able to derive the correct structural formula, or at least to eliminate some alternatives with certainty.

Finally in this section, some further remarks on solvents and their effects upon proton magnetic resonance spectra seem appropriate.

If the solvent is a hydrogen compound (e.g. $CHCl_3$, dioxan, methanol, etc.), then its own resonance lines are added to the spectrum. The solvent lines are very intense, because of the much greater effective molar concentration of solvent protons, compared to those of the solute. Very strong lines are always accompanied by side bands, and so a region of the spectrum, rather than a narrow band, is effectively spoiled (Fig. 2.9). The interference is worse the higher the hydrogen content of the solvent molecule. To circumvent this nuisance deuterated solvents are employed although these are somewhat costly. Sometimes it may be necessary to examine a compound in two different solvents which have their resonance lines in different parts of the spectrum, so that the whole of the spectral range can be covered.

The side bands are of two kinds (Fig. 2.9). The spinning side bands result from an amplitude modulation of the main resonance signal and appear in pairs centred about the main signal and spaced from it by the sample rotation frequency or integral mutliples of this. Spinning side bands are easily recognized because of the symmetry of their arrangement about the main signal and the fact that alteration of the speed of spinning of the sample causes them to move in the spectrum. Other side bands, sometimes observed in proton resonance spectra, are invariant in position with spinning rate. These are the so-called carbon-13 satellites.

^{13}C occurs naturally with 1% of the abundance of ^{12}C. The ^{13}C nucleus has a spin $I=\frac{1}{2}$ and so causes splitting of the signal from an attached proton into two components (Chapter 4), and these appear centred on the main proton signal some 120–200 c/s apart. The combined intensities of these satellite bands will be 1% of that of the main signal, corresponding to the ^{13}C abundance in the particular functional grouping. Only in

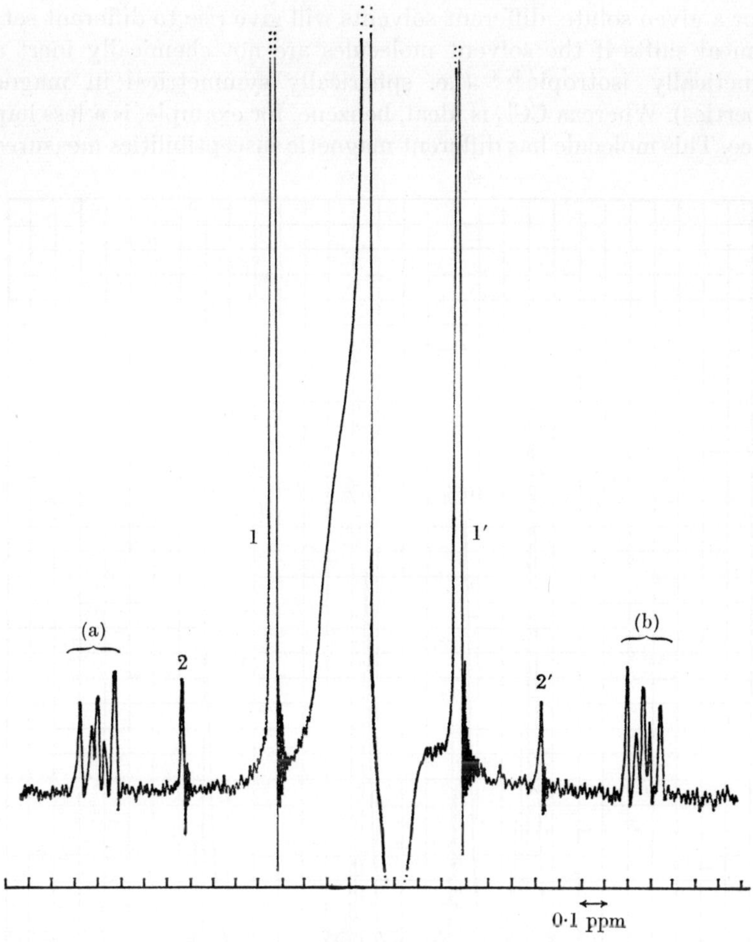

FIG. 2.9. Spinning side bands, 1,1′,2,2′ and ^{13}C—H satellite signals, (a) and (b), surrounding the proton resonance line from liquid dioxan.

special cases do these satellite signals become appreciable in magnitude. This is so with dioxan (V) (Fig. 2.9) where, because of the molecular

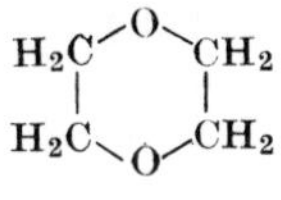

(V)

symmetry, there is a 4% chance of any one molecule containing a $^{13}CH_2$ group.[3]

For a given solute, different solvents will give rise to different sets of chemical shifts if the solvent molecules are not chemically inert and magnetically isotropic[1,3] (i.e. spherically symmetrical in magnetic properties). Whereas CCl_4 is ideal, benzene, for example, is a less happy choice. This molecule has different magnetic susceptibilities measured in

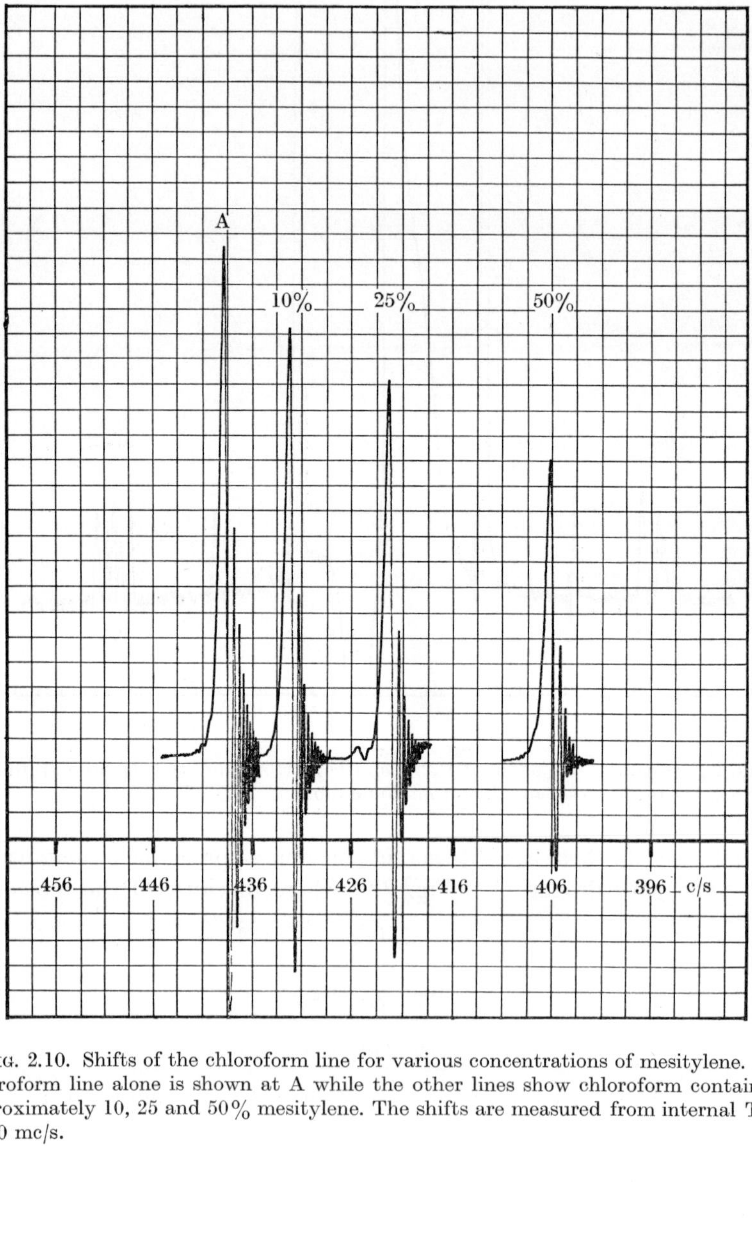

FIG. 2.10. Shifts of the chloroform line for various concentrations of mesitylene. The chloroform line alone is shown at A while the other lines show chloroform containing approximately 10, 25 and 50% mesitylene. The shifts are measured from internal TMS at 60 mc/s.

the plane of the ring and perpendicular to this plane. Solute molecules thus experience different magnetic environments, depending on their average orientations with respect to the solvent molecules; this will vary with concentration. Chemical shifts measured in benzene as solvent are concentration-dependent and differ from those measured in CCl_4, perhaps by 0·1–0·5 ppm or even more. Figure 2.10 shows the apparent shifts for the chloroform line, at various concentrations in mesitylene. Chemical shifts also vary with temperature if there is any association between solvent and solute molecules. This will certainly be so for many solutes in polar solvents such as H_2O, D_2O, MeOH, Me_2SO, pyridine. Different solvents of these kinds will associate specifically with a given solute in different ways, while the internal reference (TMS), being inert, is unaffected. So the differential shielding between reference and solute will vary with the solvent and lead to different chemical shifts in each of such solvents. This should not be overlooked when one is in the habit of using an internal reference. Although solvent-caused variations in chemical shifts can be appreciable, they do not usually cause difficulty because of the latitude which one normally allows in the use of correlation tables. It is important nevertheless always to state precisely the conditions of determination of an n.m.r. spectrum so that any future comparative measurements may be possible and the results may be of use to other workers.

References

1. L. M. Jackman, *Applications of Nuclear Magnetic Resonance Spectroscopy in Organic Chemistry*. Pergamon Press, London (1959).
2. J. D. Roberts, *Nuclear Magnetic Resonance. Applications to Problems in Organic Chemistry*. McGraw-Hill, New York (1959).
3. J. A. Pople, W. G. Schneider and H. J. Bernstein, *High-resolution Nuclear Magnetic Resonance*. McGraw-Hill, New York (1959).
4. J. B. Strothers, "Applications of Nuclear Magnetic Resonance Spectroscopy", Chap. IV of Part I, Vol. XI in the series *Techniques of Organic Chemistry*, ed. by A. Weissberger. Interscience, New York (1963).
5. H. Conroy, "Nuclear Magnetic Resonance in Organic Structural Determination", in *Advances in Organic Chemistry*, ed. by R. A. Raphael, E. C. Taylor and H. Wynberg, Vol. 2, p. 265. Interscience, New York (1960).
6. N. S. Bhacca, L. F. Johnson and J. N. Shoolery, *NMR Spectra Catalog* (a) No. 203, (b) No. 157. Varian Associates, Palo Alto (1962).
7. G. V. D. Tiers, *J. Phys. Chem.* **62**, 1151 (1958).
8. R. A. Y. Jones, A. R. Katritzky, J. N. Murrell and N. Sheppard, *J. Chem. Soc.* 2576 (1962).
9. G. V. D. Tiers and R. I. Coon, *J. Org. Chem.* **26**, 2097 (1961).
10. G. V. D. Tiers and A. Kowalsky, in press.
11. N. S. Bhacca, D. P. Hollis, L. F. Johnson and E. A. Pier, *NMR Spectra Catalog*, Vol. 2 (a) No. 481, (b) No. 383, (c) No. 441. Varian Associates, Palo Alto (1963).

CHAPTER 3

THE CHEMICAL SHIFT

Part 2

J. A. ELVIDGE

Shielding Mechanisms

Matter contains electrons in motion. In an applied magnetic field their motions cause the production of secondary magnetic fields which oppose the applied field. (This is the basis of the diamagnetism of all matter.)[1] For protons it is convenient to recognize three main kinds of diamagnetic screening which may be regarded as resulting from three kinds of induced electron circulation.

1. Local diamagnetic circulations about protons

We may regard each proton bound in a molecule as having two 1*s* orbital electrons, and the induced circulation of these as producing a secondary magnetic field, which at the nucleus, is opposed to the applied field (see Fig. 3.1). The effective field at the proton is then less than the

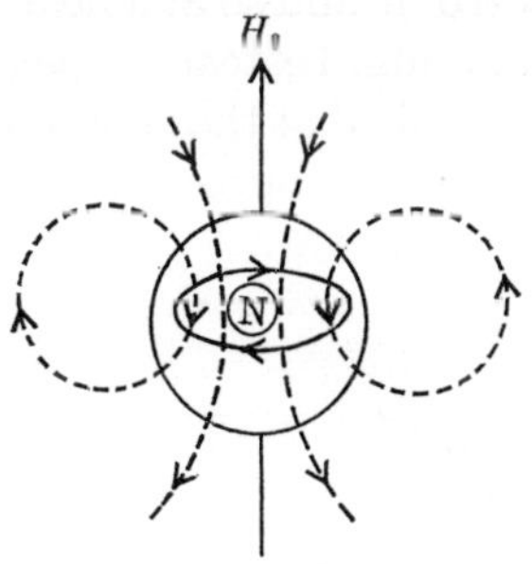

FIG. 3.1. Positive shielding of a proton by the induced diamagnetic circulation of the 1*s* orbital electrons.

applied field. Consequently, local diamagnetic circulations are a source of positive shielding because the applied field has to be increased beyond the normal value to attain resonance. (The resulting spectral line appears at a higher field than it otherwise would.) The only way in which molecular environment can affect this type of shielding is by altering the electron density around the proton concerned. Groups that withdraw electrons inductively decrease the shielding, and vice versa. The following chemical shifts illustrate this point:

$$CH_3—Cl,\ 6{\cdot}95;\ CH_3—H,\ 9{\cdot}1;\ CH_3—Si(CH_3)_3,\ 10{\cdot}0.$$

The proton with its associated pair of 1 *s* electrons has spherical symmetry, and so the induced shielding at the nucleus is independent of orientation with respect to the applied field. At a neighbouring external point, the induced field will depend on the orientation. However, if the relative orientations of the proton and the external point vary randomly, as would occur for molecules free to move in solution, or in the liquid state, then the induced field at the external point will be averaged to zero. Local diamagnetic circulations about protons make no contribution therefore to the shielding of neighbouring protons. The chemical shift of one proton has no effect *per se* on that of another proton across space.

2. Local diamagnetic circulations in anisotropic groupings

Although we may conveniently regard protons and their associated electrons in molecules as being isotropic groups, this is not strictly correct because the electrons are really distributed over bonding orbitals which are not spherically symmetrical. The picture may be corrected by treating the bond framework as an anisotropic assemblage of electrons. The induced secondary field from the electrons in a carbon–carbon bond, for example, will be anisotropic and so at a nearby point will not average to zero with random tumbling. Instead, depending on the orientation with respect to the bond, the field at that point may reinforce or diminish the applied field. Thus, we have a source of shielding at a limited distance

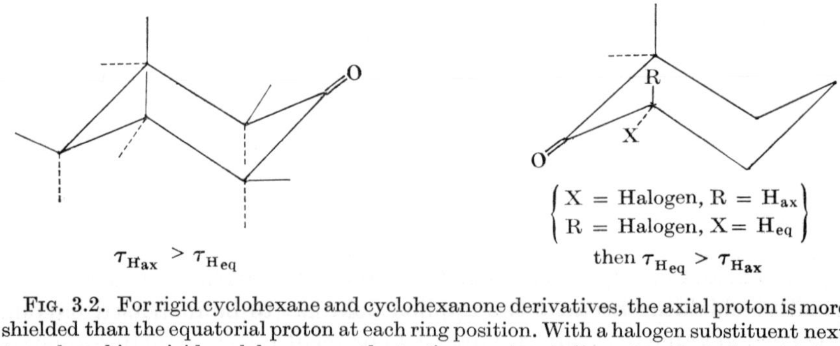

FIG. 3.2. For rigid cyclohexane and cyclohexanone derivatives, the axial proton is more shielded than the equatorial proton at each ring position. With a halogen substituent next to carbonyl in a rigid cyclohexanone, the tertiary proton (adjacent to the halogen) is less shielded in the axial position than in the equatorial position.

which may be negative or positive. In a rigid cyclohexane ring, for example, axially disposed protons are more shielded than the equatorial ones, i.e. $\tau_{H_{ax}} > \tau_{H_{eq}}$ and this effect is a result of the anisotropy of the C—C framework.[2, 3] It must be remembered, of course, that the carbon nuclei themselves (^{12}C) are non-magnetic.

In a rigid cyclohexanone, there is still the same differential shielding between axial and equatorial protons as in a rigid cyclohexane, i.e. $\tau_{H_{ax}} > \tau_{H_{eq}}$ This also holds for halogen mono-substituted rigid cyclohexane rings, where the tertiary proton is at higher field in an axial position than when it is equatorial.

However, when there is a halogen substituent next to the carbonyl group in a rigid cyclohexanone system, we find the anomaly that the tertiary proton is less shielded when it occupies the axial position than when it occupies the equatorial position, i.e. that $\tau_{H_{ax}} < \tau_{H_{eq}}$.[4] With the halogen in an equatorial position, the C=O and C-halogen dipoles are parallel, and by the inductive mechanism, the electron density in the neighbourhood of the axial proton is strongly reduced. These cyclohexane results are summarized in the Fig. 3.2.

Single bonds have axial symmetry and the shielding at a nearby point

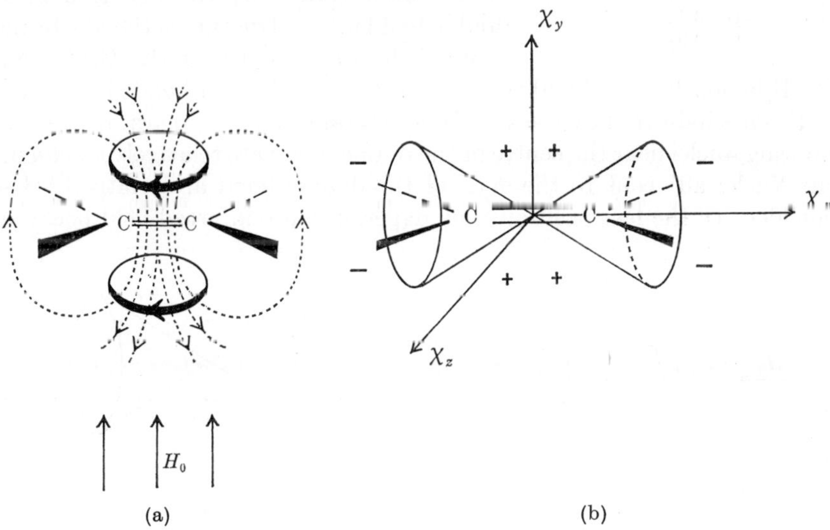

FIG. 3.3. (a) This shows the induced diamagnetic circulation of π-electrons in a double bond for one of the orientations between bond and field. (b) The positive and negative shielding regions are shown as the average effect for all mutual orientations. The pictures apply approximately also for >C=O, >C=S, >C=N. The intensity of the effects in the regions diminished outwards from the multiple bond and radially within each region. The deshielding regions are ellipsoidal cones of aperture 114° and 104° in the XY and XZ planes respectively.

is given by the McConnell equation.[5] The shielding diagram is similar to Fig. 3.4 (b) *but has the signs reversed.*

In contrast to single bonds, double bonds lack axial symmetry. So groups such as >C=C<, >C=O<, >C=N—, present a more complicated magnetic picture. Figure 3.3 (a) shows how the induced circulation of π-electrons in a double bond gives rise to a secondary field, in opposition to that applied, when the latter is at *right angles* to the plane of the double bond. When the applied field is *in* the plane of the double bond, there is still an induced circulation of the π-electrons. A rough picture of the total resultant shielding, after averaging for all mutual orientations of the group and the applied field, is obtained by a continuous rotation of the usually given double shielding cone about the bond axis.

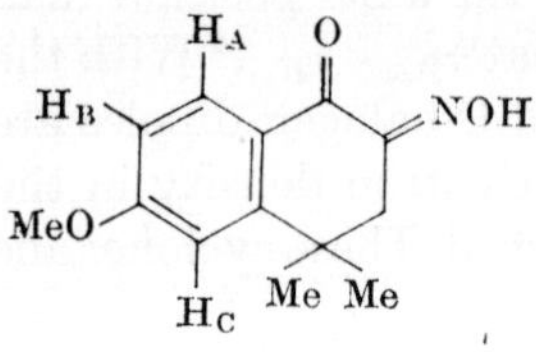

(I)

A 1·83τ
B 3·12
C 3·07

The double cone shielding diagram shown in many current texts appears to be based on an erroneous application of the McConnell equation: the true shielding diagram is as in Fig. 3.3 (b).[6] Although this is the picture for a double bond in *isolation*, it should be remembered that the shielding or deshielding of a proton caused by the insertion of a double bond into a structure is the algebraic sum of the anisotropies of the C—C and C—H bonds displaced and those of the C=C and C=H bonds inserted.[6]

Results indicated by the shielding diagram Fig. 3.3 (b) are: (i) Neighbouring nuclei near the centre of the double bond are relatively shielded. (ii) Nuclei adjacent to the ends of the double bond are relatively deshielded. In the last position they experience a reinforced field because

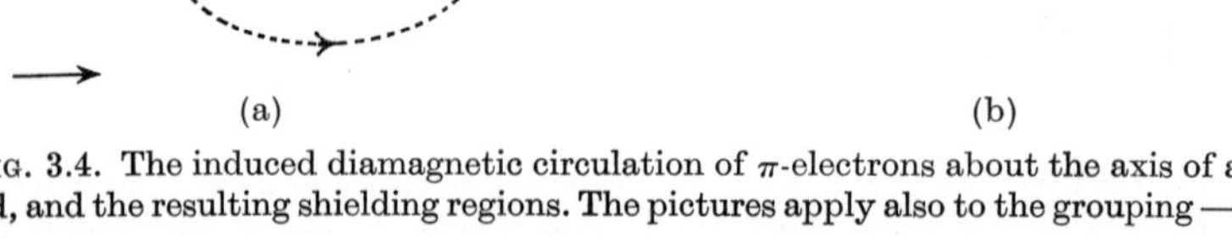

FIG. 3.4. The induced diamagnetic circulation of π-electrons about the axis of a triple bond, and the resulting shielding regions. The pictures apply also to the grouping —C≡N.

the induced field there is parallel to that applied, and so they will come to resonance at a lower applied field than would otherwise be required. This is clearly shown by proton (A) in compound (I).[7]

Triple bonds such as —C≡C— and —C≡N have axial symmetry, and so the induced circulation of π-electrons and the resultant magnetic field shell are as indicated in Fig. 3.4.[8] It follows that nuclei along the axis of the triple bond will be relatively shielded, whereas nuclei in the space surrounding the axis of the bond will be deshielded, for there the induced field is parallel to that applied. Although at first sight the chemical shifts of acetylenic protons (τ 7·6) are anomalous when compared with olefinic protons (*ca.* τ 4–5·4) and with protons in saturated hydrocarbons 8·5–9·1), consideration of Fig. 3.4 provides some explanation. In an acetylene, the attached proton is relatively more shielded than in an olefin. A proton held over the centre of a triple bond, on the other hand, will be deshielded. Comparison of the spectra of compounds (II) and (III)

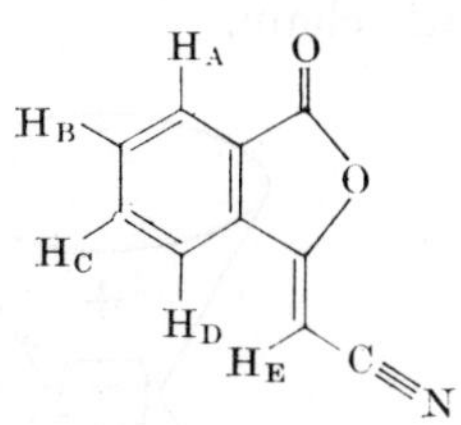

(II)

A–D 2·0–2·2τ
E 4·23

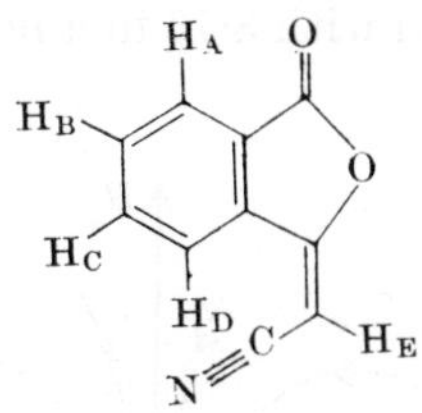

(III)

A–C *ca.* 2·0τ
D 1·6
E 4·28

shows that in the *cis*-nitrile (III) the proton (D) of the benzene ring is relatively deshielded,[9] as expected on the basis of Fig. 3.4.

3. Interatomic diamagnetic currents induced in aromatic rings

Aromatic and pseudo aromatic rings, with their cyclically delocalized π-electrons, allow the induced circulation of these electrons over all the ring atoms. The considerable ring currents thus induced produce correspondingly large magnetic fields. Such systems are far from being spherically symmetrical; indeed, they are markedly anisotropic, and so the effects of the induced field at a point in the surrounding space do not average to zero for all possible orientations of the ring with respect to the applied field. Aromatic rings therefore provide a strong source of long-range shielding and deshielding. The induced field and resulting shielding zones for the benzene molecule can be pictured as in Fig. 3.5.[10] The field arising from the induced circulation of π-electrons round the ring is opposed to the applied field at the centre of the ring, but in the same direction outside, in the plane of the ring, where the benzene protons lie. These are then deshielded (negatively shielded), because there the secondary induced field reinforces the applied field. The result is that the protons come to resonance at a correspondingly lower value of the applied field than would otherwise be required. The protons of benzene have a chemical shift of τ 2·73. On the other hand, a nucleus held over the centre of the benzene ring is shielded, because in this region induced and applied fields are opposed. Thus, in the paracyclophane (IV)

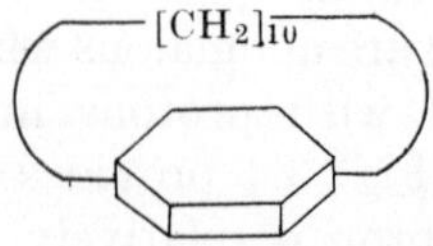

(IV)

we find the central methylene protons of the bridging chain have τ 9·20 [11] as compared with 8·75 in a normal aliphatic chain.

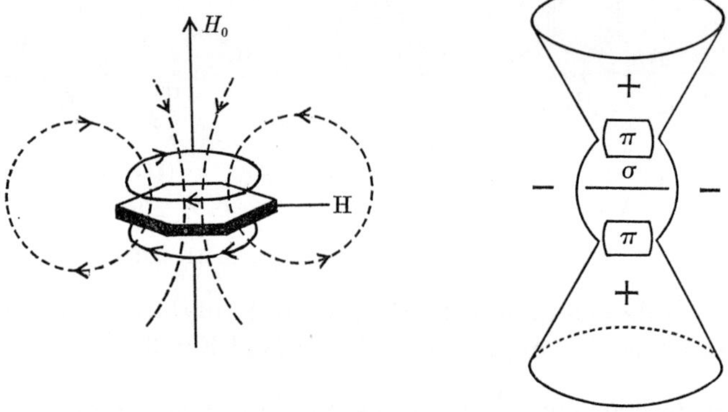

FIG. 3.5. The induced diamagnetic circulation of π-electrons round a benzene ring and the resultant shielding envelope.

Further points are that in polycyclic aromatic compounds, those ring protons nearest the largest number of rings are the most negatively shielded. So in naphthalene we find τ 2·22 for the α protons and 2·58 for those in the β positions.[12] When, in a complex molecule, an aromatic ring is not co-planar with another part of the molecule, then there will be differential long-range effects. Protons otherwise equivalent will come to resonance at slightly different fields, with the result that the signals will also show spin–spin splitting. An example is the methylenedioxy signal from dicentrine (V), which appears as an AB quartet.[13a] There is unequal deshielding of these two protons by the benzenoid ring I because the plane of this ring does not bisect the $C\langle^{H}_{H}$ group. In the Emde degradation product (VI), there *is* coplanarity and the methylenedioxy protons show a single resonance line[13b] because they are now equivalent, being equally deshielded by the aromatic rings.

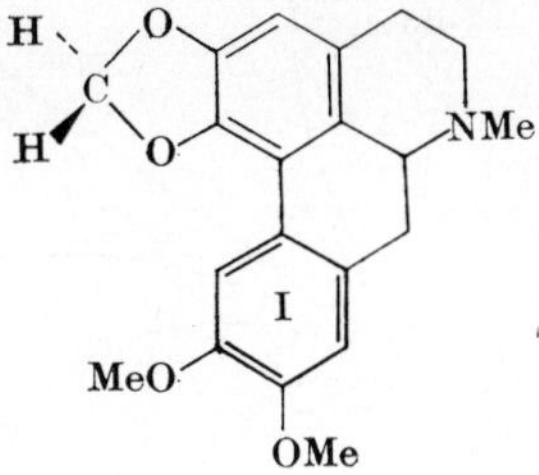

(V)

The above pictures of shielding in anisotropic groups and molecules are particularly helpful to the organic chemist and provide a means of remembering the approximate spatial geometry of the long-range shielding and deshielding zones. However, the induced magnetic field shells do not really arise from induced circulations of electrons around bonds or rings.

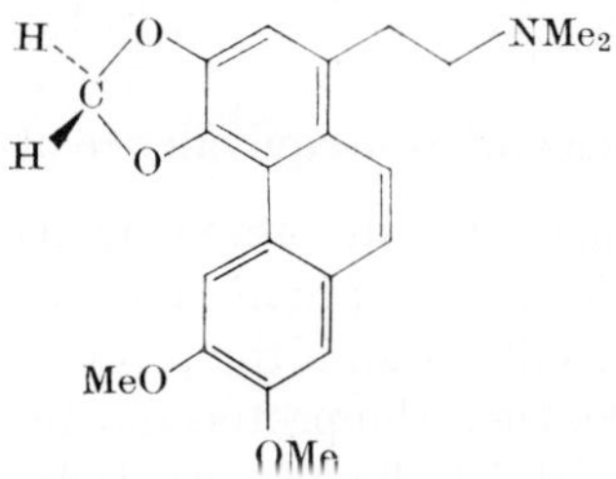

(VI)

The electrons continue to move in their normal paths but additionally perform an induced precessional motion about the lines of force of the applied field.[11] It is from the sum of their induced precessional motions that the secondary magnetic field arises. The secondary, induced field has a distribution in the surrounding space such that it could equally well have arisen from a prescribed circulation of the electrons within the bond. Thus, the induced secondary field about a benzene molecule has the same form as if the π-electrons circulated round in two superconducting loops (the π-lobes) one above and one below the plane of the σ-bond framework. The secondary field can be calculated on such a basis. In this way, the shielding and deshielding zones surrounding the benzene molecule have been computed,[15] with the result shown in Fig. 3.6. This shielding map is obviously of great help in predicting and in explaining the chemical shifts of groups held in proximity to benzene rings in the same molecule.

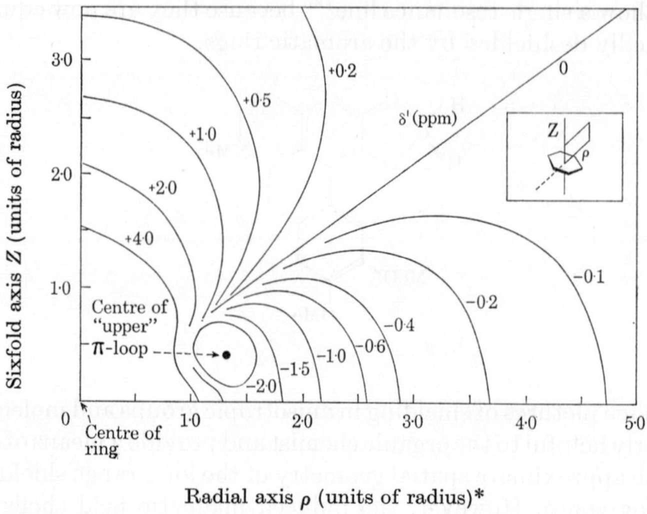

FIG. 3.6. Positive and negative shielding contours surrounding a benzene ring. The relationship of the projection to a benzene ring is shown inset.

Equivalence and Non-equivalence of Protons[16]

There are many instances in stereochemistry where protons in apparently similar environments are in fact non-equivalent. They then have different chemical shifts and, if the protons are near enough to one another in the molecule, then the proton magnetic resonance spectrum will show spin–spin multiplet signals; otherwise, separate lines will be obtained.

It is normally taken for granted that the three protons of a methyl group are equivalent, and that the two protons of a methylene group are equivalent, as suggested by the single τ or δ values given in correlation tables.

The methyl protons in a molecule CH_3—CH_2X give rise to a multiplet signal, because of coupling to the adjacent methylene protons, but a single origin position is assigned to the methyl protons; they are regarded as equivalent. This equivalence must result from the rapid rotation of the group about the single bond connecting it to the rest of the molecule and, if the rotation were made very slow (by lowering the temperature), then the methyl protons would lose their equivalence. As the three conformers

* The radius is defined as the distance from the centre of the benzene ring to the mid-point of one of the sides (see inset).

in Fig. 3.7 show, one of the protons would always be in a different environment from the other two and so would be chemically shifted from them. The conformers are effectively identical and, as the protons are mutually coupled, the methyl protons would constitute an A_2B spin–spin system; the group as a whole would be an A_2BX_2 system. Although the corresponding spectrum is unlikely ever to be encountered in practice (because the residence time of a rotamer would require to be greater than Δ_{AB}), such considerations must be borne in mind to guard against the danger of neglecting the stereochemistry of groupings.

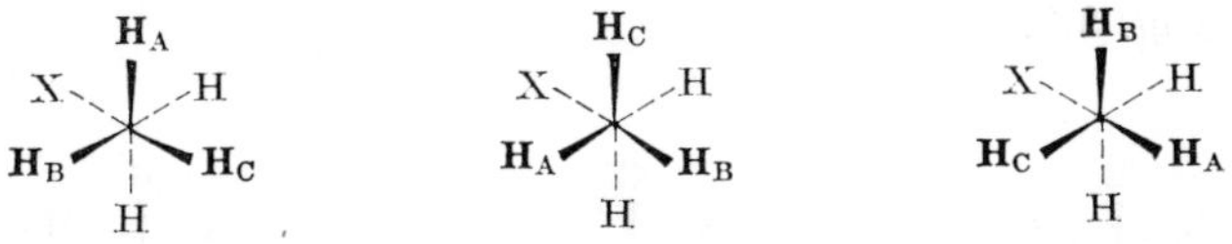

FIG. 3.7. Rotational conformations of a molecule CH_3—CH_2X.

The two protons of the methylene group in the above molecule CH_3—CH_2X remain equivalent irrespective of rotation of the methyl group. The latter is symmetrical about an axis coincident with the C—C bond and so the methylene protons are always in the same environment, as Fig. 3.7 shows. Note that if there were asymmetry in the rest of the molecule, e.g. in the group X, the two methylene protons could be non-equivalent.

When the methyl group is replaced by a group of lower symmetry, such as $-CHR_2$, so that we have the molecule CHR_2—CH_2X, then the various rotamers (i), (ii), (iii) are no longer all equivalent (Fig. 3.8).

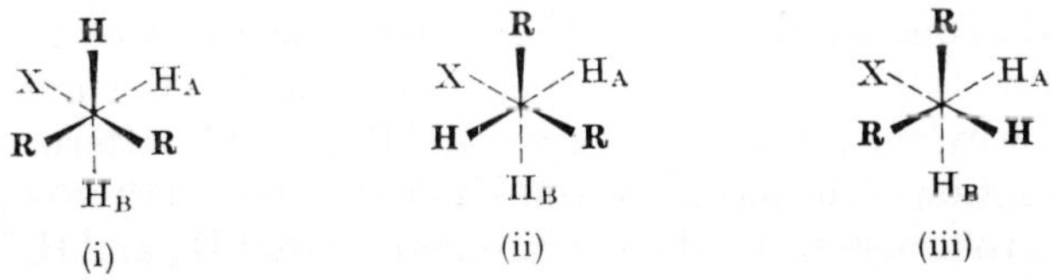

FIG. 3.8. Rotational conformations of a molecule CHR_2—CH_2X.

Although (i) and (ii) are effectively the same, (iii) is distinct. Rotation makes H_A and H_B in (i) and (ii) equivalent. In (iii), H_A and H_B are certainly equivalent to one another, but not to either of those in (i) and (ii), because the non-bonding interactions are different. Rapid rotation will give the protons H_A and H_B a single (average), origin position. Slower rotation will result in an alteration in the relative populations of the conformations, so that the averaged line position will alter; its position will be temperature dependent.

If the temperature is lowered so that rotation becomes very slow, then we may observe two spectra, one from the conformations (i) and (ii), which constitute an ABX spin-coupled system, and another from the conformation (iii), which constitutes an A_2X system. This last conformation may well not be appreciably populated, however, because it is sterically less favourable.

When the methylene group is attached to an asymmetric grouping, as in CHRR′—CH_2X, then even if this grouping is rotating rapidly, the shielding of the two methylene protons is not averaged. So these protons have different chemical shifts. This becomes clear from the rotamers, shown in Fig. 3.9.

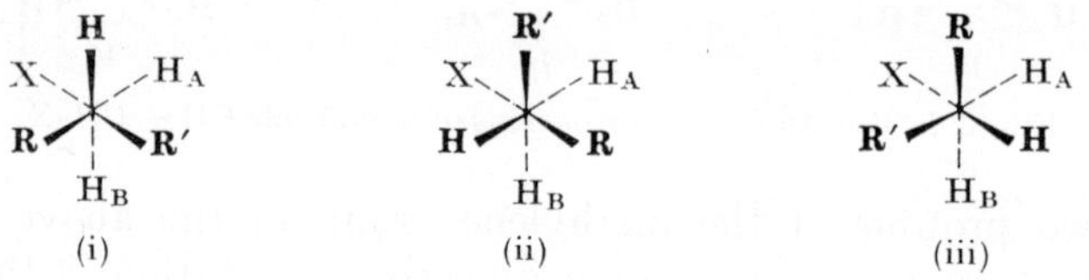

FIG. 3.9. Rotational conformations of a molecule CHRR′—CH_2X.

Each rotamer has a different set of non-bonding interactions and so the two methylene protons are always in dissimilar environments. Thus H_B in (i), between groups R and R′, is not in the same environment as H_A in (ii), also between R and R′, because in (i) R is next to X and H whereas in (ii) R is next to H and H. Similar conclusions are reached for any two pairs of the three conformations. Thus H_A and H_B are never in equivalent environments, and so will have different average chemical shifts. As they are coupled to one another and to the adjacent proton, we can expect an ABX pattern. If the terminal asymmetric group lacked the proton shown, then the methylene group would give an AB quartet pattern.

When rotation of the terminal group is slowed, the relative populations of the three rotamers will no longer be equal, so that H_A and H_B will have different average chemical shifts. In the limit at very low temperatures, we might expect to see three ABX (or three AB) patterns, one from each conformation, provided that each is still sufficiently populated.

However, it is again the room temperature result which is most important, and we must be ready to recognize that protons or groups next to an asymmetric system will not be equivalent, and so will have different chemical shifts. This is usually found in practice, though not invariably.

Thus, the sulpholane derivative (VII) shows two separate lines, from the geminal methyl groups. This is thought to arise not from the side chain having a fixed conformation, but from the $Me_2C(OH)$ group being attached to an asymmetric carbon (at the 2- position) of the ring. Consideration of the rotamers shows that, as for the two methylene

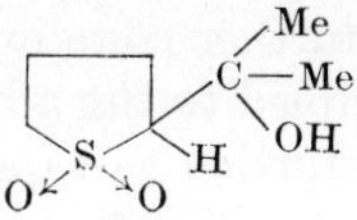

(VII)

protons in Fig. 3.9, so here the two methyl groups are always in dissimilar environments. The methyl groups therefore have different chemical shifts in spite of free rotation of the $-\underset{\displaystyle OH}{\underset{|}{C}}Me_2$ group. Non-equivalence can also be conferred by asymmetry in a part of the molecule more distant than the next carbon atom. For example, in some 21-acetoxy-20-oxosteroids, e.g. allopregnane-3,20-dione 21-acetate, the C(21)-methylene protons are in the situation $RCOCH_2OAc$, where the group R is asymmetric. The rotamers are then (i) and (ii) as shown in Fig. 3.10. H_A in

(i) (ii)

FIG. 3.10. Rotational conformations of a molecule $RCOCH_2OAc$, where the group R is asymmetric.

(i) has a similar environment to H_B in (ii), but H_B in (i) is in a different environment to H_A in (ii). So rotation does not make H_A and H_B equivalent. The C(21)-methylene protons of the steroid therefore have different chemical shifts, and thus give rise to an AB quartet pattern in the spectrum.[17]

It is not even essential that a molecule should be asymmetric as a whole.[18] Thus, it is observed in simple diethyl acetals, such as bromoacetal (VIII), that the resonance signal from the methyleneoxy protons

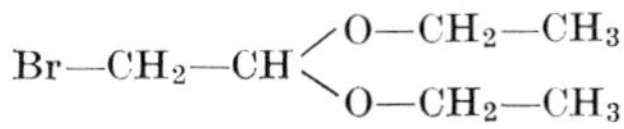

(VIII)

is complicated, there being eight lines.[19a] Evidently the two protons in a methyleneoxy group are non-equivalent, because this group is attached to a carbon atom bearing three different substituents, viz. H–, $BrCH_2$–, and $-OCH_2CH_3$. As far as the methyleneoxy protons are concerned, then, they are in an asymmetric environment and so have different chemical shifts (cf. Fig. 3.9). These protons are mutually coupled and therefore give rise, in the absence of further coupling to an AB quartet of lines. However, each proton is coupled to the adjacent methyl group, and so each of the four lines is split further as a quartet. As a result of overlap, however, eight lines are observed rather than the maximum possible sixteen. The full sixteen-line multiplet (four quartets) *is* obtained from the methyleneoxy protons of the propynal acetal (IX).[19b]

An eight-line signal similar to that from bromoacetal arises from the methylene protons of acetaldehyde diethyl acetal (X) for the same reason.[20] As soon as the internal asymmetry is removed, however, as in (XI), then the complexity of the signals disappears.[20] We must consider

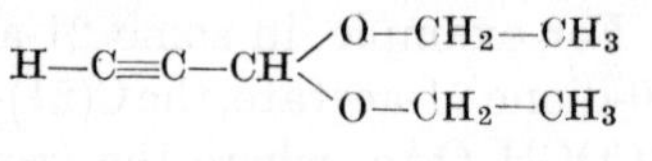

(IX)

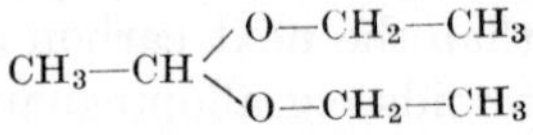

(X)

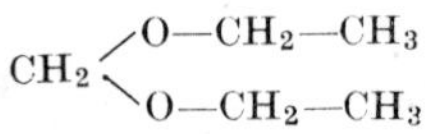

(XI)

molecules very carefully, from a new "internal" aspect, when deciding whether groups or protons will be equivalent or not.

Of course, restricted rotation can sometimes be the cause of the non-equivalence of groups. In this case, the spectrum will change with increase in temperature until, finally, the groups are rendered equivalent through free rotation in the molecule, and the originally expected spectrum is finally observed. Many substituted amides can show this behaviour. A simple case is provided by dimethylformamide, Me_2NCHO, which can be written in projection as in Fig. 3.11 to indicate the non-equivalence of the methyl groups. Rotation in this molecule is restricted by the fact that the N—CHO bond has considerable double-bond character as a result of amide resonance

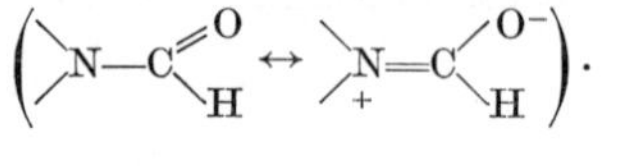

The methyl groups give two separate lines in the spectrum at room temperature, but as the temperature is raised, the energy barrier to rotation is overcome. The methyl lines then move together, and finally merge in a 6-proton singlet when free rotation has been established, at about 170–180° C.[21]

In conclusion, it must be emphasized that, even if at first sight protons or groups appear equivalent, this position should never be assumed without careful examination of the molecule. Those aspects which require detailed consideration are: (i) differential long-range effects; (ii) asymmetry of the molecule, both as a whole and from an internal viewpoint; (iii) restricted rotation.

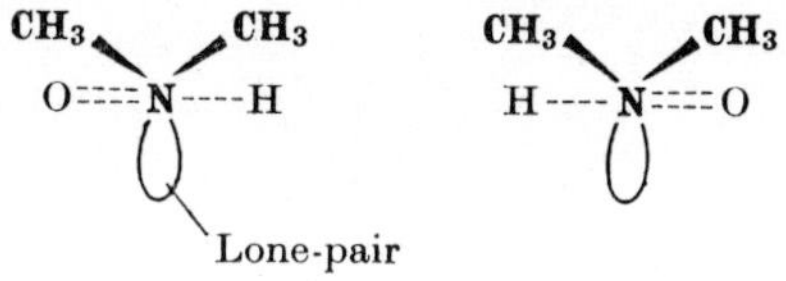

FIG. 3.11. Projection of the molecule of dimethylformamide along the N to CO bond.

References

1. P. W. Selwood, *Magnetochemistry*. Interscience, New York (1956).
2. R. U. Lemieux, R. K. Kullnig, H. J. Bernstein and W. G. Schneider, *J. Amer. chem. Soc.* **79**, 1005 (1957); **80**, 6098 (1958); K. L. Williamson and W. S. Johnson, *J. Amer. chem. Soc.* **83**, 4623 (1961).
3. L. M. Jackman, *Applications of Nuclear Magnetic Resonance Spectroscopy in Organic Chemistry*. p. 115. Pergamon Press, London (1959).
4. A. Nickon, M. A. Castle, R. Harada, C. E. Berkoff and R. O. Williams, *J. Amer. Chem. Soc.* **85**, 2185 (1963).
5. H. M. McConnell, *J. Chem. Phys.* **27**, 226 (1957).
6. J. W. ApSimon, W. G. Craig, P. V. Demarco, D. W. Mathieson, L. Saunders and W. B. Whalley, *Chem. Commun.* 359 (1966).
7. N. S. Bhacca, D. P. Hollis, L. F. Johnson and E. A. Pier, *NMR Spectra Catalog*, No. 611. Varian Associates, Palo Alto (1963).
8. Ref. 3, p. 114.
9. J. A. Elvidge and D. E. H. Jones, unpublished work.
10. Ref. 3, p. 125.
11. J. S. Waugh and R. W. Fessenden, *J. Amer. Chem. Soc.* **79**, 846 (1957).
12. R. H. Martin, *Tetrahedron* **20**, 897 (1964).
13. N. S. Bhacca, L. F. Johnson and J. N. Shoolery, *NMR Spectra Catalog*, (a) No. 342; (b) No. 349. Varian Associates, Palo Alto (1962).
14. J. H. Van Vleck, *The Theory of Electric and Magnetic Susceptibilities*, p. 22. Oxford University Press (1959).
15. C. E. Johnson and F. A. Bovey, *J. Chem. Phys.* **29**, 1012 (1958).
16. Ref. 3, p. 99.
17. J. N. Shoolery and M. T. Rogers, *J. Amer. Chem. Soc.* **80**, 5121 (1958).
18. P. M. Nair and J. D. Roberts, *J. Amer. Chem. Soc.* **79**, 4565 (1957).
19. (a) J. A. Elvidge and R. G. Foster, *J. Chem. Soc.* 981 (1964); (b) J. A. Elvidge and R. G. Foster, unpublished work.
20. J. S. Waugh and F. A. Cotton, *J. Phys. Chem.* **65**, 562 (1961).
21. W. D. Phillips, *J. Chem. Phys.* **23**, 1363 (1955); H. S. Gutowsky and C. H. Holm, *J. Chem. Phys.* **25**, 1228 (1956).

CHAPTER 4

INTRODUCTION TO SPIN–SPIN COUPLING[1–7]

J. A. ELVIDGE

When nuclear magnetic resonance spectra are determined under conditions of high resolution, not only are discrete lines or bands observed for each group of equivalent nuclei but frequently, also, fine structure. Protons or groups of equivalent protons can give rise to multiplets in certain circumstances. For example, ethyl bromide (I) shows a triplet from the methyl group and a quartet signal from the methylene. It is important to note that the total integrated intensities of these multiplet signals are still in the ratio of the numbers of protons responsible. The added multiplicities in the signals arise from spin–spin coupling between *non-equivalent* protons (or between protons and other magnetic nuclei, e.g. H and F) in the same molecule. The magnitude of the splittings observed in such signals is independent of the applied field but depends upon molecular stereochemistry, and falls off with an increasing number of bonds between the coupled nuclei.

CH_3—CH_2—Br

(I)

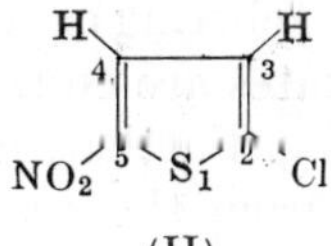

(II)

Consider the simplest case of two coupled protons. 2-Chloro-5-nitrothiophen (II), for example, has two protons in the molecule, on adjacent carbon atoms, each in a different chemical environment. We may expect, therefore, two signals in the spectrum. In fact we observe four lines, in two pairs, as indicated in Fig. 4.1. There is a low-field pair of lines from the C4-proton and a higher-field pair of lines from the C3-proton. Each

FIG. 4.1. Proton magnetic resonance lines from 2-chloro-5-nitrothiophen (II).

pair shows the same spacing; this, in c/s, is known as the coupling constant J, the magnitude of which is characteristic of that particular situation.

A simple explanation[8] of this phenomenon is that H-(3), for example, experiences not just the applied field (or rather a screened value of this), but *two* modifications of the applied field, so that it comes to resonance at two positions in the swept spectrum. These two modifications of the field could arise from the adjacent proton being at various times in its two different spin states. We may then imagine that the applied field experienced by the H-(3) proton is augmented in some way by the adjacent proton being in one spin state (↑), and then diminished (by the same small energy) by the adjacent proton being in the opposite spin state (↓), as indicated in the accompanying energy diagram (Fig. 4.2). The energy

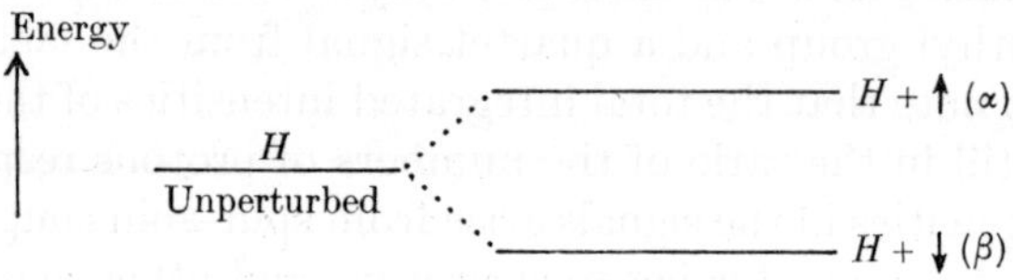

FIG. 4.2. Perturbation of the applied magnetic field H by the two spin-states of a nucleus with $I=\frac{1}{2}$.

difference between the two field modifications (which results from the effect of one change of proton-spin orientation undergone by the *adjacent* proton) is the splitting or coupling J, measured from the spectrum in c/s. Those spin states are equally probable, so the two spectral lines are of equal intensity. Further, it is apparent from the energy diagram that, if the signal from the proton H-(3) were not perturbed by spin–spin coupling, then there would be a single line mid-way between the components of the doublet. This so-called origin position gives the chemical shift of H-(3).

Now spin-coupling effects are mutual and in the same way, therefore, the signal from proton H-(4) is split as a doublet by proton H-(3). So we observe two doublets (Fig. 4.1.)

A more complicated situation is provided by the coupling between the protons of the adjacent methyl and methine groups in 1,1-dibromoethane (III).[9a] The spectrum shows two signals, as expected from our knowledge

$CHBr_2$—CH_3

(III)

of chemical shifts; there is a high-field 3-proton signal and a lower-field 1-proton signal. However, the former appears as a doublet and the latter as a quartet, as in Fig. 4.3. The spacings in the quartet are equal and this spacing occurs again in the doublet signal. This common spacing, in c/s, is the coupling constant J. Further, the components of the quartet (i.e. the individual lines) have integrated intensities in the ratio 1:3:3:1, while those of the doublet are 1 : 1 (at least approximately).

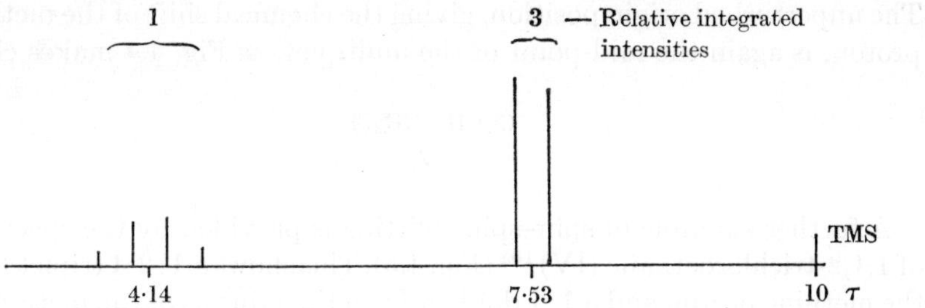

FIG. 4.3. Proton magnetic resonance spectrum of 1,1-dibromethane (III).

As in the case of the thiophen (II) above, we may imagine that the protons of the methyl group, responsible for the doublet signal, experience two modifications of the applied field (as indicated in Fig. 4.2). This is caused in some way by the adjacent proton being at various times in one and then in the other of the two spin states. These are equally probable, and so we obtain a 1 : 1 doublet from the *methyl* protons when the spectrum is swept.

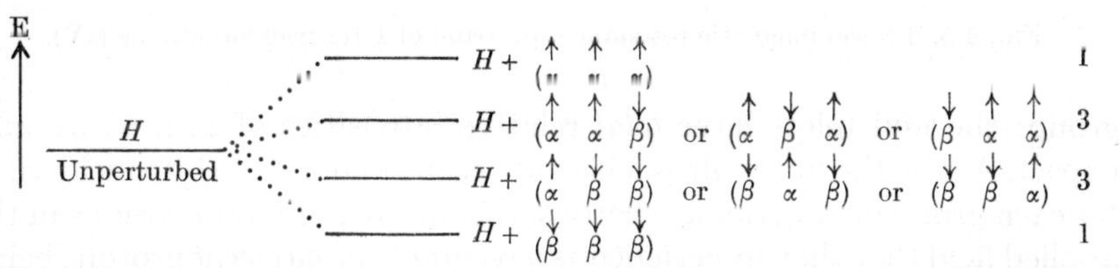

FIG. 4.4. The *four* modifications of the magnetic field H caused by the eight spin-states of a group of *three* protons.

Similarly, we consider the quartet splitting of the low-field methine signal as resulting from the effects of the spin states of the three protons of the methyl group. The spins of a group of three protons can be arranged as ↑↑↑, ↑↑↓, etc. in eight ways, but these fall into only *four* categories or energy groups. So we conclude, as indicated in the accompanying energy diagram (Fig. 4.4), that the three protons of the methyl group cause four modifications of the applied field. Each of the four modifications differs from that on an adjacent level by the effect of one change in spin orientation. Consequently, the methine proton, which experiences these field modifications, gives rise to 4 spectral lines as the field is swept. The lines are equally spaced by J (c/s), and this is the same spacing as in the methyl doublet signal. Two of the four field-modifications can be made up in three equivalent ways, as shown above; these two modifications are each three times as probable as the other two, and so the component lines of the spectral quartet have the relative intensities 1:3:3:1. The unperturbed origin position, giving the chemical shift of the methine proton, is again the mid-point of the multiplet, as Fig. 4.4 makes clear.

$$Cl_2CH{-}CH_2Cl$$

(IV)

A further example of spin–spin splitting is provided by the spectrum of 1,1,2-trichloroethane (IV)[10a] (Fig. 4.5). This shows a 1:2:1 triplet from the methine proton and a 1:1 doublet from the protons of the methylene

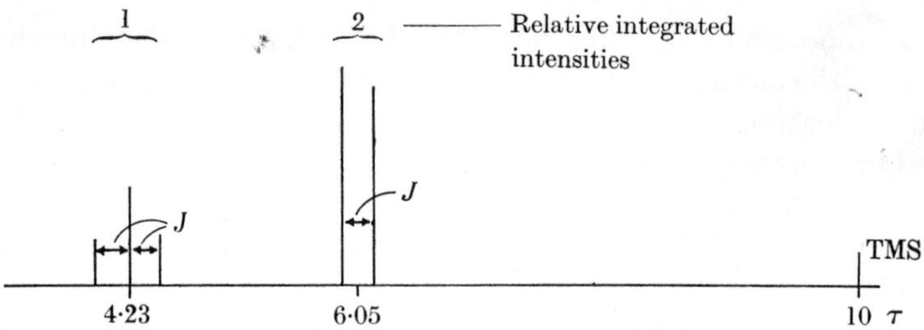

FIG. 4.5. Proton magnetic resonance spectrum of 1,1,2-trichloroethane (IV).

group; the multiplets have *total* relative intensities of 1:2, as we now expect. Again, the line multiplicities are easily understood by considering, for each group of equivalent protons, the number of modifications in the applied field that they experience as a result of the *adjacent* protons being in their various possible spin states.

Simple Splitting Rules. To appreciate the significance of the number of lines in a multiplet we must think in terms of the number of protons not on the group giving rise to the multiplet signal, but *on adjacent groups*. We may thus formulate a simple splitting rule:

the number of lines in a multiplet is $(n+1)$ where n is the number of neighbouring protons.

Stated more generally,

the number of lines in a multiplet is $(2nI+1)$ where n is the number of adjacent *nuclei* of spin I.

It must be remembered that these rules require equal coupling between all the nuclei concerned. We also find that the relative intensities of the components of a multiplet are given, ideally, by the ratios of the coefficients of a binomial expansion of the form $(1+a)^n$: where we have

for $n = 1$, 1:1
$n = 2$, 1:2:1
$n = 3$, 1:3:3:1
$n = 4$, 1:4:6:4:1 etc.

At this point it is useful to make some predictions on the basis of the foregoing ideas. The spectrum of isopropyl bromide (V) will show 2 groups

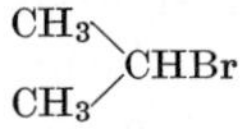

(V)

of lines. The higher field signal arises from the two equivalent methyl groups and will have a total intensity corresponding to six protons. It will be split as a doublet owing to the adjacent single proton; each component of the doublet will be of the same intensity (see Chapter 7, pp. 121–124). The single tertiary proton will, of course, give a signal of total intensity corresponding to one proton, but this signal will be split as a *septuplet* by the equal coupling to the adjacent six equivalent protons. These seven lines should have relative intensities of 1:6:15:20:15:6:1 (as indicated by the binomial mnemonic). We may note that unless the spectrum is recorded at very high gain, the outermost very weak components of the septuplet may not be visible. Further, the spacings in the septuplet are all equal and the same as the spacing in the doublet signal.

From a compound X—CH_2—CH_2—Y we expect two equally intense 1:2:1 triplet signals. From diethyl ether $(CH_3—CH_2)_2O$, we expect a 1:2:1 triplet from the methyl protons and a 1:3:3:1 quartet from the methylene protons—remembering that the splitting of a signal is caused by the *adjacent* protons. The case of 1,2-disubstituted ethanes is further discussed in Chapter 7, p. 113.

We may, of course, observe splitting of a proton signal as a result of coupling to a magnetic nucleus of another kind, such as fluorine. Fluorine has a spin $I = \frac{1}{2}$, so the simple splitting rule above, still applies. Thus 1,2-dichloro-1,1-difluoroethane (VI) [9b] gives a 1:2:1 triplet signal in the

$Cl\,CH_2—CF_2\,Cl$

(VI)

proton magnetic resonance spectrum with $J = 11$ c/s. The *triplet* from the two equivalent protons results from their equal coupling to the *two* fluorine nuclei ($I = \frac{1}{2}$) attached to the next carbon atom. A table of fluorine–proton coupling constants is given in Table 9.3, p. 156 (see also Refs. 11, 12).

Coupling between protons and nitrogen is occasionally observed.[13, 1, 3, 7] Because nitrogen has a spin $I = 1$, we must use the generalized splitting rule (p. 46) to deduce the extent of splitting of the signal from an attached proton. In general, the coupling is observed only with protonated primary amines, RNH^{+}_3 and these show a triplet, with J about 50 c/s, from the equivalent protons attached to the nitrogen atom. With more extensive substitution, as in protonated secondary and tertiary amines, the nitrogen relaxation (i.e. the nitrogen spin-state change) becomes too rapid for the attached protons to be effectively coupled, and so a single signal is observed. With the free bases and their solutions, there is generally more or less rapid proton exchange between molecules (unlike the situation in their salts). Coupling to the nitrogen is thereby prevented, and multiplet signals are not given. Depending on the rate of exchange of protons and on residence times, so the derived proton resonance signal may be sharp or broadened. In amides and imides, both proton exchange and the nitrogen-relaxation rate tend to prevent effective coupling between the nitrogen nucleus and attached proton(s). For amides, depending on the rate of the exchange process, so the signal may be broad (ordinary temperature) or show multiplet structure (raised temperature, reduced nitrogen relaxation); it becomes sharp at higher temperature or in the presence of catalysts when exchange is made rapid[1 14] (see notes in Table 5, Appendix I).

The hydroxyl proton in primary and secondary alcohols may give rise to a multiplet signal because of coupling to protons on the nearest carbon atom. For this splitting to be observed, however, the samples must be extremely pure. Minute traces of acid and/or alkali (normally present), catalyse exchange of the hydroxyl proton between molecules to such an extent that the residence time on any particular molecule is too short to allow of coupling to adjacent protons. When there *is* coupling, the simple rules predict (correctly) that the hydroxyl proton of a primary alcohol will give a triplet signal and that of a secondary alcohol a doublet. The magnitude of J will be 3–5 c/s.

Unless thiols are appreciably acidic, exchange of the thiol proton between molecules is slow, and so thiol protons are coupled usually to the proton(s) on the nearest carbon atom and give rise to multiplet signals. J will have a magnitude of 6–7 c/s. The relevant case of methyl mercaptan is discussed in Chapter 7, p 104.

So far, situations have been considered where only one coupling constant obtains, either in fact or apparently (i.e. one average coupling due to free rotation of groups about bonds). The splitting of a signal may arise, however, from two or more couplings of *different* magnitudes to several other protons or groups of protons.

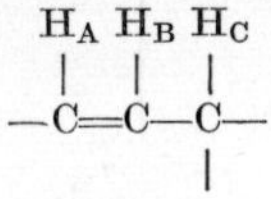

(VII)

For this situation, it is best to consider the effects of each coupling in turn, and it is often helpful to draw the splittings on graph paper. The hypothetical part structure (VII) will serve for illustration. The line

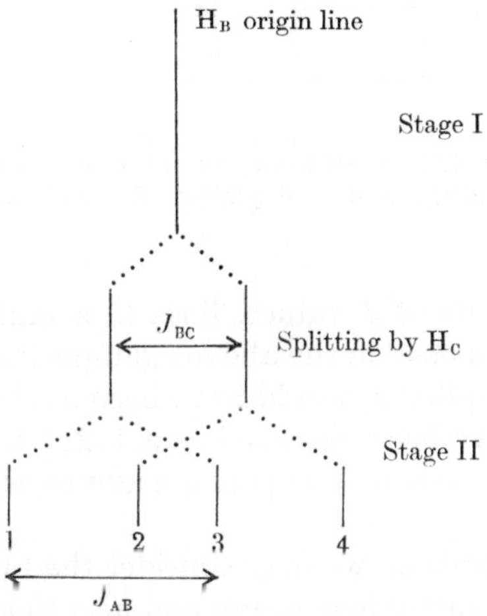

FIG. 4.6. Stage I. The splitting of the line from proton B in structure (VII) by proton C. Stage II, the further splitting of the signal from proton B in structure (VII) by proton A to give a 1:1:1:1-quartet. The "1,2"-spacings give J_{BC} and the "1,3"-spacings give the larger coupling J_{AB}.

from H_A will be split by H_B as a doublet, $J_{AB} = 15$ c/s. The line from H_C will be split by H_B as a doublet, $J_{BC} = 9$ c/s. The line from H_B will first be split by H_C as a doublet with $J_{BC} = 9$ c/s, as in Stage I, Fig. 4.6. Each component of the split signal will then be further split as a doublet by coupling to H_A (Stage II): two 1:1 doublets thus arise. Exactly the same pattern of four lines results if the effects of the splittings are considered in the reverse order: the order of treatment is immaterial. Thus, taking J_{AB} first, and then J_{BC} we obtain the result shown in Fig. 4.7. In either of the "splitting" diagrams (Figs. 4.6, 4.7), the distance between lines 1 and 2 (or 3 and 4) gives the smaller coupling constant, and that between lines 1 and 3 (or 2 and 4) the larger coupling constant.

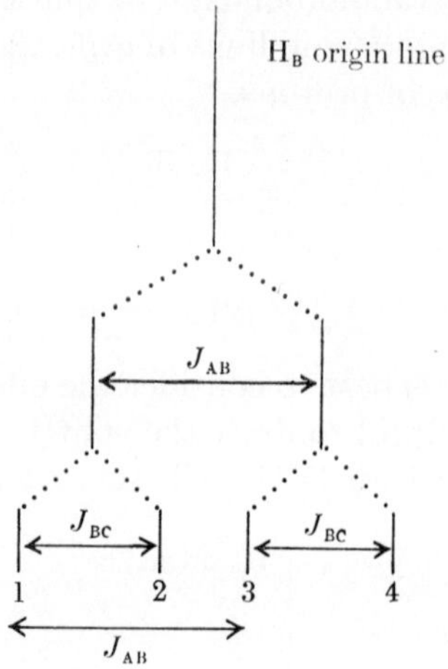

FIG. 4.7. The splitting of the signal from proton B in structure (VII), taking the effect of J_{AB} first, and then J_{BC}. The result is a quartet of lines identical to that in Fig. 3.7.

With particular sets of J values, lines in a multiplet may overlap. This would have happened in the above example if $J_{BC} = 15$ c/s, for then the second stage of splitting would have been as shown in Fig. 4.8, and the signal would have been observed as a 1:2:1 triplet. Note that the intensities of the two central components add together because of their coincidence.

As a further illustration, we may consider the hypothetical case of a proton coupled to a methylene group and also to a methyl group. The origin signal is first split into a 1:2:1 triplet by the methylene protons, with a particular J value. *Each* of these three lines is then split by the methyl protons into 1:3:3:1 quartets, with a second value of J. Only by drawing the pattern in stages can we easily predict the form of the resultant multiplet, i.e. the number of component lines and their relative intensities.

The foregoing picture of spin–spin coupling is much simplified, and actual spectra can be considerably more complicated than would be expected on the basis of such simple rules.[1, 3, 15, 16] These rules apply *only* when the chemical shift Δ c/s between the coupled protons is large compared to the coupling constant J c/s so that $\Delta/J = 10$ or more. The multiplet patterns are then said to be first order patterns.

When the chemical shift is not much greater than J, say, $\Delta/J = 2$ to 5, then the component lines in each multiplet no longer have the ideal intensities. There is a distortion within each multiplet such that the innermost lines (those nearest the other multiplet) increase in intensity at the expense of the outermost lines (those furthest away from the other multiplet). This has qualitative usefulness in that we can at once deduce on which side of a multiplet signal the related multiplet must be. The effect may be discerned in Fig. 4.5.

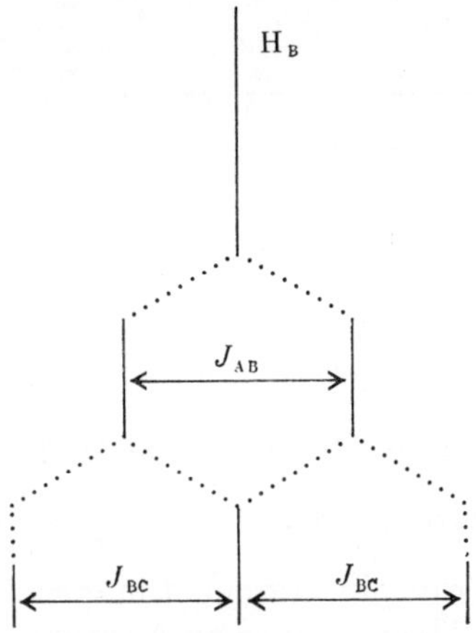

FIG. 4.8. A 1:2:1-triplet signal arising from two equal couplings ($J_{AB}=J_{BC}$).

When the chemical shift between coupled protons is comparable to or smaller than J, then the splitting patterns are no longer "first order". The number of lines observed may be increased and either or both of the parameters Δ and J are no longer directly extractable from the multiplet patterns (cf. Chapters 5–7). Some of these more complicated situations can still be examined and analysed according to rules summarized in the section from p. 195 onwards. Others require treatment from first principles by quantum mechanics.[3, 15, 16]

In order to designate spin-coupled nuclei, letters of the alphabet are used.[17] Thus a first-order two-spin system is denoted as AX and, as already shown above with compound (II) we obtain two equal doublets. The coupling J_{AX}, and the origin positions of the nuclei, ν_A and ν_X, are directly obtainable from the spectrum as in Fig. 4.9.

When in a two-spin system the magnitude of the chemical shift Δ between the nuclei becomes similar to or smaller than J, then we designate it an AB system. Two pairs of lines are again observed, each separated by J_{AB} c/s, but the origin positions are no longer the mid-points of the doublets, and the lines have distorted intensities as shown in Fig. 4.10. Such a pattern is given by the olefinic protons of cinnamic acid.[10b] The pattern is described as an AB quartet: it is no longer first-order. If m is the distance between the outer lines and n that between the inner lines, then the chemical shift between the nuclei A and B is given by $\Delta_{AB} = \sqrt{m.n}$.

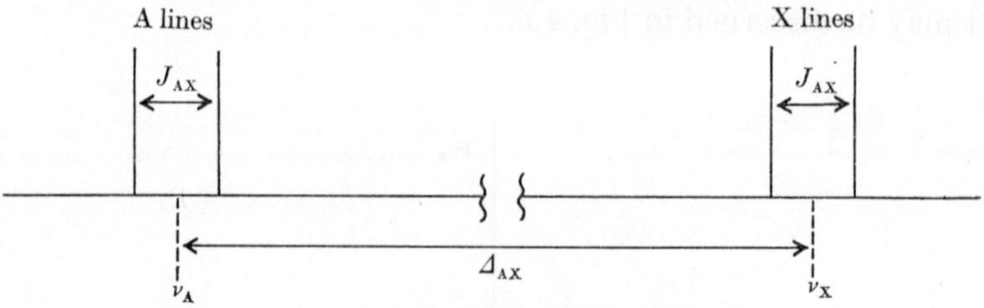

FIG. 4.9. First-order AX quartet.

If the mid-point of the quartet of lines is ν_0 c/s (from TMS) (obtainable by inspection), then the origin position of nucleus A,

$\nu_A = \nu_0 + \frac{1}{2}\sqrt{m.n}$ c/s from TMS, and the origin position of nucleus B is,

$\nu_B = \nu_0 - \frac{1}{2}\sqrt{m.n}$ c/s from TMS.

In the limit, in a two-spin system, the chemical shift between the coupled nuclei is zero. Then we have an A_2 group and, of course, a single

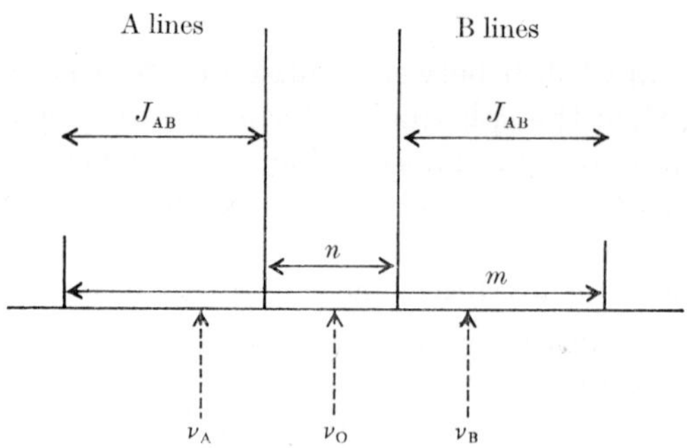

FIG. 4.10. Non-first-order AB quartet.

line in the spectrum. We can imagine the inner two lines of the AB quartet increasing in intensity and finally merging together as the outermost lines dwindle to zero, as $\Delta_{AB} \to 0$.

An important point which the above simple treatment of spin–spin coupling does not explain is the fact that groups of equivalent protons give rise to single lines. We might expect one proton in a group to experience modifications in the applied field as a result of the other protons of the group undergoing spin-state changes, but this appears not to be so.

In actual fact there *is* coupling, but it happens that the quantum mechanical selection rules forbid any observable effect from the spin–spin coupling of equivalent nuclei. Thus, the transitions represented by the outermost lines in an AB spectrum become forbidden in the limiting A_2 case.

The sensing of the spin states of one group of nuclei by another appears to occur through the intervening bonds rather than across space.* This is suggested by the dependence of the magnitude of the coupling $|J_{AB}|$ upon

(i) the number of intervening bonds

e.g. H_A—C—H_B (methanes) 10–18 c/s.[18]
H_A—C—C—H_B (ethanes; free rotation) 6–8 c/s.[19]
H_A—C—C—C—H_B (propanes and higher; free rotation) 0 c/s.[1,2]

(ii) the bond angles[5,21]

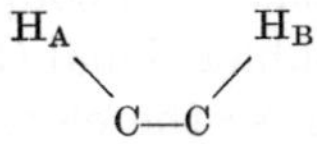

(fixed conformation)

$\angle\phi_{AB}$	J_{AB}
0°	8 c/s
85°	0 c/s
180°	10–12 c/s

and (iii) bond hybridization[1,2,3]

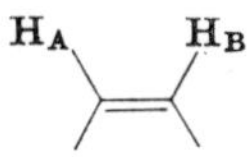

(*cis*-olefin) — J_{AB} 12–14 c/s

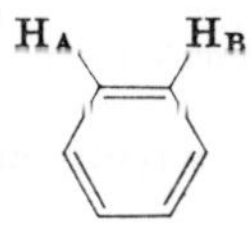

(*ortho*-protons on benzene) — usually 8–10 c/s

} For the same bond angles

The transmission of spin state information between coupled nuclei may be pictured in the following terms. A proton in its particular spin state produces a small magnetic polarization of the adjacent electrons, so that the spins tend to become antiparallel: H↑ *e*↓ *e*↑. This effect really results from there being a small probability of the electron being *at* the nucleus itself: of necessity, spins would then *have* to be paired. This correlation effect is transmitted through further bonding electrons to neighbouring protons. Any two electrons in one bonding orbital have their spins antiparallel by the Pauli exclusion principle. Neglecting for the moment how the correlation occurs between bonds on opposite sides of a carbon nucleus, we arrive at the picture shown in Fig. 4.11, and here the effective field at nucleus H_X would be *increased* to (H_0↑ + ↑*e*). Equally probable

* For F—F and F—H couplings, interactions across space may be significant.[20]

would be the situation in which nucleus H_A had the opposite spin. The electron spins would then tend to correlate in the opposite sense, as in Fig. 4.12, and the effective field at nucleus H_X would be *diminished* to $(H_0\uparrow + \downarrow e)$. It follows that nucleus H_X experiences two modifications of the applied field, dependent on the spin state of the first nucleus H_A.

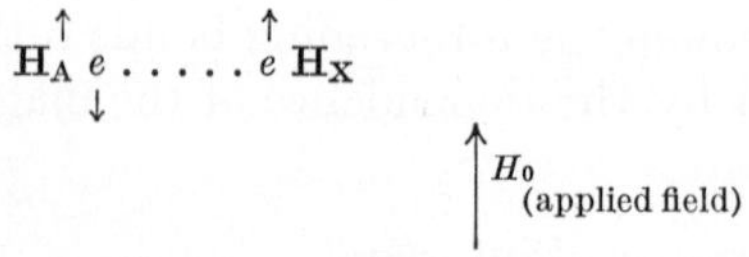

FIG. 4.11. The applied field H_0 at a proton H_X increased by a contribution from the spin of the adjacent bonding electron.

So nucleus H_X comes to resonance at two values of the applied field, one on each side of the value which would be required for the unperturbed nucleus (Fig. 4.2). The effect is mutual of course, so that two coupled protons perturb each other, and each gives rise to a doublet signal (Fig. 4.9). The coupling J can have a range of values because the electron correlation of nuclear spins will be more or less efficient, depending on the number of intervening bonds (i.e. electrons), the bond angles (i.e. degree of overlap of electron orbitals), and the nature of the orbitals involved (cf. chapter 8).

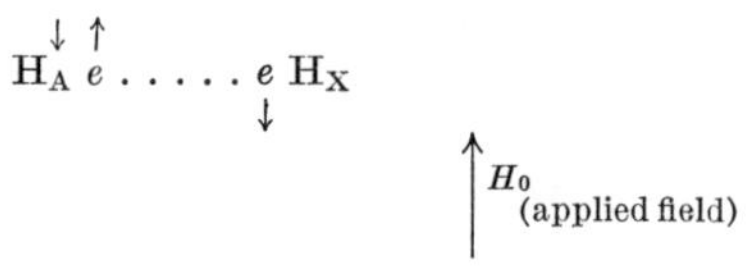

FIG. 4.12. The applied field H_0 at a proton H_X decreased by a contribution from the spin of an adjacent electron.

From this qualitative picture, it may thus be appreciated that the magnitudes of the spin–spin couplings (i.e. spectral line splittings) can be correlated with stereochemistry, and so can provide valuable structural information. Appendix I, Table 8, p. 186 lists their values and some structural correlations.

It will also be appreciated that couplings are *independent* of the applied magnetic field strength, in contrast to chemical shifts (τ and δ values). There is nothing in the picture of electron correlation of nuclear spins to suggest that the strength of the applied field plays any part.[3] This is experimentally verified: multiplet patterns obtained with different spectrometer operating frequencies (and hence different field strengths) show the same splittings, and only the chemical shifts between multiplets are altered. We note, however, that a first order spectrum obtained at a high frequency, say, 100 Mc/s, may become complex at 40 Mc/s for example, because Δ/J will diminish. The whole appearance of the spectrum might then change radically (cf. the change from A_2X to A_2B shown in Fig. 5.6 (p. 79).

A further point is that couplings can be either positive or negative. J is positive or negative depending on whether the energetically more stable orientation of coupled nuclear spins is antiparallel or parallel, respectively[19] (Fig. 4.13). The absolute signs of coupling constants are

(i) When $H_A{\uparrow}$ ∿∿∿ $H_X{\downarrow}$ and $H_A{\downarrow}$ ∿∿∿ $H_X{\uparrow}$ are relatively *more* stable and $H_A{\uparrow}$ ∿∿∿ $H_X{\uparrow}$ and $H_A{\downarrow}$ ∿∿∿ $H_X{\downarrow}$ are relatively *less* stable than the corresponding situations in the absence of coupling, then J is positive.

(ii) When $H_A{\uparrow}$ ∿∿∿ $H_X{\uparrow}$ and $H_A{\downarrow}$ ∿∿∿ $H_X{\downarrow}$ are relatively *more* stable and $H_A{\uparrow}$ ∿∿∿ $H_X{\downarrow}$ and $H_A{\downarrow}$ ∿∿∿ $H_X{\uparrow}$ are relatively *less* stable than the corresponding situations in the absence of coupling, then J is negative.

FIG. 4.13. The meanings of the signs of a coupling constant.

not directly obtainable from normal spectra, although the relative signs of a set of J values can sometimes be deduced from the form of a spectrum[3, 15] (see, for example, the ABX case described, in Chapter 6, p. 85).

Finally, we may picture the AX and AB cases in terms of the total energies of these two-spin systems.[3, 15] We write each nuclear spin as α (for the upper state) and β (for the lower state)—these are really the spin wave functions—and assume that the energy associated with nucleus A undergoing a spin transition, is larger than for nucleus X. This is not strictly true, but it *appears* to be so, to the external observer. Because of the differential screening, there is a chemical shift between the two nuclei. They come to resonance at different values of the applied field. For the situation when there is no coupling ($J = 0$), we have an energy diagram as in section (i) of Fig. 4.14. With the proviso that only one nucleus is likely to undergo spin change at a time, then we find there are 4 allowed transitions—two equal A transitions $4 \rightarrow 2$ and $3 \rightarrow 1$, and two X transitions $4 \rightarrow 3$ and $2 \rightarrow 1$ equal in energy to one another. In consequence, only two absorption lines appear in the spectrum; the lines correspond to the two different transition energies and hence appear at different fields, one from nucleus A at ν_A and one from nucleus X at ν_X (in c/s from TMS) (Fig. 4.15.)

For the situation when there *is* coupling (with $J \neq 0$) then we find that a new set of energy levels applies as in section (ii) of Fig. 4.14. We can say

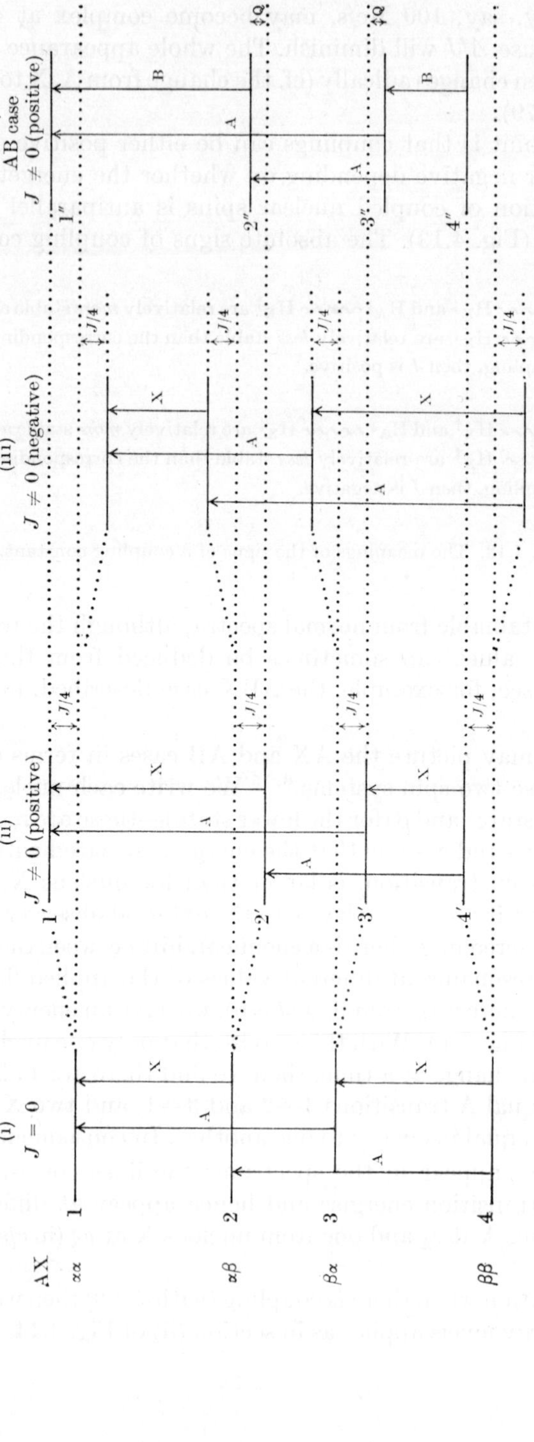

FIG. 4.14. Energy levels and allowed transitions for the two-spin systems (i) AX, $J = 0$; (ii) AX, $J \neq 0$ (J positive); (iii) AX, $J \neq 0$ (J negative); (iv) AB, $J \neq 0$ (J positive).

that the electron correlation of nuclear spins will *increase* the energies of the $\alpha\alpha$ and $\beta\beta$ states, equally. This becomes clear if we draw the spins diagrammatically: for the $\alpha\alpha$ state we have ($H_A\uparrow$ $e\downarrow \ldots \uparrow e$ $\downarrow H_X$) or ($H_A\uparrow$ $e\uparrow \ldots \downarrow e$ $\uparrow H_X$) and for the $\beta\beta$ state, ($H_A\downarrow$ $e\uparrow \ldots \downarrow e$ $\downarrow H_X$) or ($H_A\downarrow$ $e\downarrow \ldots \uparrow e$ $\downarrow H_X$). Both the $\alpha\alpha$ and $\beta\beta$ states are destabilized, i.e. increased in energy by the lack of full spin-correlation. When the nuclear spins are $\alpha\beta$ and $\beta\alpha$, then there can be full correlation of spins, as in ($H_A\uparrow$ $e\downarrow \ldots \uparrow e$ $\downarrow H_X$) and ($H_A\downarrow$ $e\uparrow \ldots \downarrow e$ $\uparrow H_X$) and so in these two states the energies of the system are *reduced* by the same small amount.

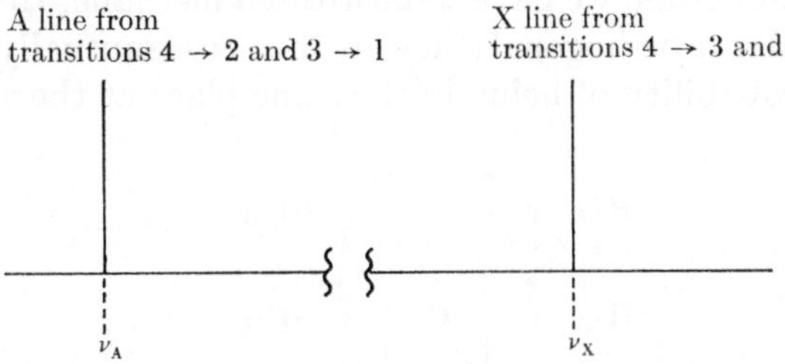

FIG. 4.15. Spectrum from a two-spin AX system, with $J=0$.

Hence levels 2 and 3 move down, whereas levels 1 and 4 move up by the same amount. If the energy change is assigned the value $J/4$ c/s, Fig. 4.14 (ii) shows that levels 4 and 3 move together by $J/2$, and levels 2 and 1 move apart by $J/2$, whilst levels 3 and 2 stay the same distance apart. The new energy level system (ii) now shows two unequal A transitions $4' \sim 2'$ and $3' \sim 1'$, which are respectively diminished and increased (over the previous value in (i)) by $J/2$. So the spectrum now shows two A lines, one each side of the original position ν_A, spaced from it $+J/2$ and $-J/2$. The A doublet lines are therefore J c/s apart. The X transitions change in a similar fashion and a four-line AX spectrum (Fig. 4.9) is observed.

A further comment may now be made about the signs of coupling constants. We began the above discussion by assuming that lack of nuclear spin-correlation resulted in an increase in the energy of the system ("destabilization"): this corresponded to assigning J a positive sign. Had we supposed that *lack* of nuclear spin-correlation results in a decrease in energy (*stabilization* of the system), corresponding to J being negative, then we should have drawn the energy diagram (ii) in the reverse sense, as in Fig. 4.14 (iii). The states $\alpha\alpha$ and $\beta\beta$ would now be lowered in energy, and the states $\alpha\beta$ and $\beta\alpha$ raised in energy. There would still result two unequal A transitions and two unequal X transitions, and these would have the same four energies previously found. The resulting spectrum would have exactly the same appearance as before (see Fig. 4.9). So, in the AX spin-coupled case, the absolute sign of J has no *observable* effect on the spectrum.

It might be objected that it is difficult to appreciate how a lack of nuclear spin correlation could result in stabilization of a system, i.e. how the state with parallel nuclear spins could have the lower energy. It depends on how the nuclear spins are correlated by the intervening electrons of the connecting bonds. For directly linked protons it should be obvious that the state of paired (antiparallel) spins should be more stable, i.e. have a lower energy content, than a state with unpaired proton spins. Thus, for H_2, the coupling constant is positive. With two protons attached to the same carbon atom, we have a substituted methane. Here, the two C–H bond orbitals are orthogonal (they are in a sense mutually exclusive; they have no probability of being in the same place at the same time).

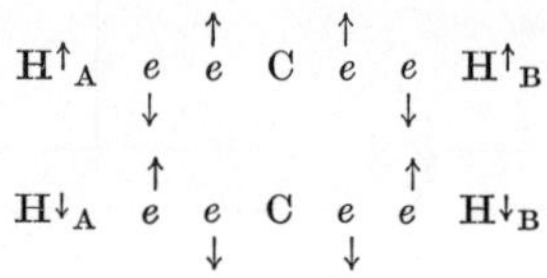

FIG. 4.16. The preferred situation in a substituted methane 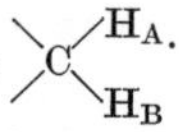 These states are relatively stabilized as compared with those having antiparallel proton spins.

The two electrons nearest to the carbon atom, one in each orbital, do not tend to be paired and they will have parallel spins (cf. the filling of the d-orbitals of a transition metal according to Hund's rule). Hence the preferred, lower-energy state of the $-CH_2-$ system will be as in Fig. 4.16 with unpaired, parallel proton-spins. This corresponds to J being negative. If we follow the reasoning further, we shall find that the coupling is positive between protons on vicinal carbon atoms, negative between protons on carbon atoms 1 and 3 of a chain, and so on.[19] It must be emphasized that this picture is much simplified and that proton–proton couplings can sometimes change sign with change in bond hydridization and change in substitution[18, 22, 19] (see Chapter 7 and Table 8 (coupling constants) p. 186).

When the chemical shift between two coupled protons is small compared to the coupling constant, this constitutes the AB system and at this stage we cannot necessarily arrive at a correct prediction of the spectrum by intuition. Because the nuclei are now similar (they have nearly the same chemical shift), it is no longer accurate to describe the two states with antiparallel nuclear spins as $\alpha\beta$ and $\beta\alpha$. The nuclei become too difficult to distinguish. So mixed spin functions are employed, of the form $\Psi_2 = (a\alpha\beta + b\beta\alpha)$ and $\Psi_3 = (a\beta\alpha - b\alpha\beta)$ where a and b are the so-called mixing coefficients which have to satisfy certain conditions. A detailed description is given on p. 66. The mixing causes the levels 2′ and 3′ in Fig. 4.14 to move apart slightly in energy as shown under (iv). There are still two different A transitions and two different B transitions, so four lines are observed in the spectrum. However, the A transitions 4″, 2″ and 3″, 1″ are both increased in energy as a result of the mixing, and the B transitions are reduced in energy by the same amount Q. The effect observed in the spectrum is a shifting of the pairs of doublets apart so that neither pair is any longer centred upon the respective unperturbed origin position, as shown in Fig. 4.10. A further effect of the quantum-mechanical mixing of states is upon the relative transition probabilities: these probabilities increase for the inner pair of lines and decrease for the outer pair.

Thus, for calculation of the origin positions of the protons A and B, the simple formula on p. 52 has to be used. The more complex the system, the greater the difficulties, and we must either follow through with quantum mechanics,[3, 15, 23] use rules derived from the full treatment (cf. p. 195), or try to match the spectrum with one in a compilation of calculated spectra.[16] Fortunately, rules for many of the more common spin–spin situations can readily be used by the organic chemist to derive the parameters required for structural correlation and no more than an elementary knowledge of algebra is required.

A more detailed discussion of spin-coupled systems follows in Chapters 5, 6 and 7.

References

1. L. M. Jackman, *Applications of Nuclear Magnetic Resonance Spectroscopy in Organic Chemistry*. Pergamon Press, London (1959).
2. J. D. Roberts, *Nuclear Magnetic Resonance. Applications to Problems in Organic Chemistry*. McGraw-Hill, New York (1959).
3. J. A. Pople, W. G. Schneider and H. J. Bernstein, *High-resolution Nuclear Magnetic Resonance*. McGraw-Hill, New York (1959).
4. J. B. Stothers, "Applications of Nuclear Magnetic Resonance Spectroscopy", Chap. IV of Part I, Vol. XI in the series *Technique of Organic Chemistry*, ed. by A. Weissberger. Interscience, New York (1963).
5. H. Conroy, "Nuclear Magnetic Resonance in Organic Structural Determination", *in Advances in Organic Chemistry*, ed. by R. A. Raphael, E. C. Taylor, and H. Wynberg, Vol. 2, p. 265. Interscience, New York (1960).
6. R. F. M. White, "Nuclear Magnetic Resonance Spectra", Chap. 9 of *Physical Methods in Heterocyclic Chemistry*, Part II, ed. by A. R. Katritzky. Academic Press, New York (1963).
7. R. J. Gillespie and R. F. M. White, "Nuclear Magnetic Resonance and Stereochemistry", Chap. 2 of *Progress in Stereochemistry*, Vol. 3, ed. by P. B. D. de la Mare and W. Klyne, Butterworths, London (1962).

8. R. E. Richards, *Nuclear Resonance Spectra of Liquids*, in Special Publication No. 12, pp. 171–176. Chemical Society, London (1958).
9. N. S. Bhacca, D. P. Hollis, L. F. Johnson and E. A. Pier, *NMR Spectra Catalog*, Vol. 2 (a) No. 370, (b) 369. Varian Associates, Palo Alto (1963).
10. N. S. Bhacca, L. F. Johnson and J. N. Shoolery, *NMR Spectra Catalog*, (a) No. 2, (b) No. 230. Varian Associates, Palo Alto (1962).
11. Ref. 1, p. 86.
12. W. D. Phillips, "High Resolution ^{1}H and ^{19}F NMR Spectra", *in Determination of Organic Structures by Physical Methods*, by W. Nachod and W. D. Phillips, Vol. 2, p. 401. Academic Press, New York (1962). N. Muller and D. T. Carr, *J. phys. Chem.* **67**, 112 (1963).
13. R. A. Ogg and J. D. Ray, *J. Chem. Phys.* **26**, 1339 (1957).
14. J. D. Roberts, *J. Amer. Chem. Soc.* **78**, 4495 (1956).
15. J. D. Roberts, "An Introduction to the Analysis of Spin–spin Splitting", *in High Resolution Nuclear Magnetic Resonance Spectra*. Benjamin, New York (1961).
16. K. B. Wiberg and B. J. Nist, *Interpretation of NMR Spectra*. Benjamin, New York (1962).
17. Ref. 1, p. 89.
18. H. J. Bernstein and N. Sheppard, *J. Chem. Phys.* **37**, 3012 (1962).
19. C. N. Banwell and N. Sheppard, *Discuss. Faraday Soc.* **34**, 115 (1962) and references there cited.
20. L. Petrakis and C. H. Sederholm, *J. Chem. Phys.* **35**, 1243 (1961); D. R. Davis, R. P. Lutz and J. D. Roberts, *J. Amer. Chem. Soc.* **83**, 246 (1961).
21. M. Karplus, *J. Chem. Phys.* **30**, 14 (1959).
22. C. N. Banwell and N. Sheppard, *Mol. Phys.* **3**, 351 (1960).
23. P. L. Corio, *Chem. Rev.* **60**, 363 (1960).

CHAPTER 5

BASIC PRINCIPLES OF SPECTRAL ANALYSIS: SYSTEMS CONTAINING UP TO THREE SPIN-COUPLED NUCLEI

E. O. Bishop

Comparison of First Order and Non-first Order Spectra

The spectral complexity and ease of analysis for a given number of spin-coupled nuclei depends very largely on whether the system is weakly or strongly coupled. By weakly coupled, we mean that, for all pairs of nuclei i, j* in *different* chemical environments, the ratio $|J_{ij}|$ to $|\nu_i - \nu_j|$ is small. With certain reservations (see later), coupling between magnetically equivalent nuclei has no effect on the spectrum, and the values of J_{ii}, J_{jj} are therefore irrelevant in this context. Since chemical shift differences are proportional to the applied magnetic field (H_0), while coupling constants are not, the strength of coupling within a given molecule can be varied within certain practical limits by changing H_0. The higher the applied field, the less strongly coupled is the system. It then approaches the "high field limiting condition" in which all ratios $|J_{ij}|$ to $|\nu_i - \nu_j|$ are vanishingly small and the spectrum is at its simplest. With finite values of J_{ij} this condition would in theory be attained only at infinite H_0, but in practice a spectrum is usually indistinguishable from its appearance in the high field limit if none of the above ratios exceeds about 1/10. Except in a few special cases involving magnetic equivalence without spin-coupling equivalence (see the end of this chapter, and AA′XX′ in Chapter 7), the high field limiting appearance of a spectrum is a very striking and simple pattern termed the "first order" condition. Rules for predicting and interpreting first order spectra are given in Chapter 4—and for the present we will take the description "high field" and "first order" as synonymous.

For most purposes it is an advantage to work with spectra which are as nearly first order as possible. In cases of more than two spin-coupled nuclei per molecule, more information can in principle be extracted from

* The letters i and j are used in this chapter to denote the general case of any two nuclei with different chemical shifts.

spectra which are not first order, but the analysis is invariably more difficult. The main differences may be summarized thus.

First Order Spectra	*Non-first Order Spectra*
Symmetrical multiplets	Intensity pattern distorted
Independent of the signs of the coupling constants	Usually depend on relative signs of the coupling constants
"Fundamental" lines only	Additional "combination" lines
Superposition of transitions if some nuclei have the same shift	Previously coincident transitions are split into separate lines
Analysis (for ν and $\|J\|$) is easy	Analysis usually difficult

In this chapter, the simplest spin systems of one, two and three nuclei are considered in some detail, firstly, to outline the theoretical principles of spectral analysis, and secondly, to underline the reasons for the complexities which arise in non-first order spectra. This treatment is intended as a qualitative description, and the reader who wishes to learn details of the calculations involved, is advised to study Ref. 1. The present chapter is used as a basis for a further discussion of systems containing more than 3 nuclei, in Chapter 7.

The One-spin System

The simplest case of an n.m.r. spin system is that of a single magnetically active nucleus, of spin quantum number $(I) = \frac{1}{2}$. The proton resonance of chloroform is effectively an example of this, since the magnetic moment of the chlorine nuclei has no detectable effect on the ^1H spectrum. The fact that the spectrum consists of a single sharp line tells us that we are observing a transition between discrete energy states of the nucleus in the applied magnetic field. This must arise from restrictions imposed by quantum mechanics, since classical Newtonian mechanics (which one would apply to a large-scale magnet) would have predicted an infinite series of energy values and a continuous spectrum. It is convenient therefore to use some of the terminology of quantum mechanics, even in a simple qualitative description.

It was seen in Chapters 1 and 2 that the nucleus, considered as a small magnet, can adopt only $(2I+1)$ orientations in an externally applied magnetic field H_0. For $I = \frac{1}{2}$ there are thus two possible orientations, that in which the nuclear magnetic moment is commonly (but inexactly) said to be "parallel" and that designated "antiparallel" to the direction of H_0. More exactly, the two possible orientations are those in which the geometrical projection of the nuclear magnetic moment (a vector quantity) upon the direction of the lines of force of H_0 is $+\frac{1}{2}$ or $-\frac{1}{2}$ in the appropriate units. This projection is denoted by the symbol I_z, since the direction of the lines of force of H_0 is taken to define the z-axis in Cartesian co-ordinates—i.e. the two possible orientations have $I_z = +\frac{1}{2}$ and $-\frac{1}{2}$. That orientation aligned *against** the field ($I_z = +\frac{1}{2}$) is said to be "parallel" and is of higher energy than the corresponding "antiparallel" orientation ($I_z = -\frac{1}{2}$). Relative to $H_0 = 0$ taken as zero, the energies are given by

$$E = \pm\tfrac{1}{2}\eta H_0(1-\sigma) \tag{1}$$

where η is a constant of proportionality and σ is the screening constant.

In the terminology of quantum mechanics, these are said to be *stationary states* of the nuclear spin system, and will be numbered (1) and (2) respectively. Each stationary state is associated with a characteristic *energy level*, here E_1 or E_2, and a "shorthand description" or *wave function* ψ_1 or ψ_2. The description need tell us only whether the nuclear moment is parallel or antiparallel to H_0, for which the two symbols α and β respectively will suffice. Thus the nuclear spin stationary states (1) and (2) have energies E_1 and E_2, and wave functions $\psi_1 = \alpha$, $\psi_2 = \beta$ respectively†.

The single-line nuclear resonance spectrum arises from transitions between states (1) and (2) at a frequency given by Planck's equation as $\Delta E/\mathbf{h}$. Such transitions can be brought about by applying at right angles to H_0, a magnetic field H_1 oscillating at the correct frequency (this is in the radio frequency region). From the point of view of a single nucleus, the transitions (1)→(2) and (2)→(1) are induced with equal probability, so that the observation of a net absorption signal depends on the presence of more nuclei in state (2) than (1) by the Boltzmann distribution. For a given H_1 field strength (r.f. power level) at the correct frequency $\Delta E/\mathbf{h}$, the intensity of transition between two states in any spin system is decided by the accompanying *change in wave function*. Here the change must be $\beta \rightarrow \alpha$ in absorption, or $\alpha \rightarrow \beta$ in emission—i.e. a "spin-flip" of the nucleus.

From equation (1), the absolute frequency of the resonance signal is $\dfrac{\eta H_0}{\mathbf{h}}(1-\sigma)$. In practice, line positions are always measured relative to some convenient reference signal (e.g. that of TMS for proton spectra).

* In terms of the classical bar magnet this represents a N to N orientation of two magnetic poles.

† E_1, ψ_1, and E_2, ψ_2 are the two valid solutions to the *wave equation* $\mathcal{H}\psi = E\psi$ where, $\mathcal{H}$ is the Hamiltonian operator, the nature of which need not be considered here.

It is a convenient shorthand device to use ν with a suitable subscript to denote not the *absolute* frequency but the separation in c/s of a spectral line or a chemical shift from such a reference (e.g. see Tables 5.1 and 5.3). Sometimes, $\Delta\nu$ is used, without subscripts if there is no ambiguity, to denote the separation between two chemical shifts.

The Two-spin System

For two nuclei A and B of a two-spin system, the spectral parameters are the two chemical shifts ν_A, ν_B and the spin coupling constant J_{AB}. The limiting case of coincident chemical shifts is denoted "A_2", the proton resonance of methylene dichloride CH_2Cl_2 being an example of this kind. At the other extreme, the "first order" limit is denoted "AX", and here $|J_{AB}| \ll |\nu_A - \nu_B|$.

1. AX case: zero coupling

In the case in which the coupling constant is zero, this satisfies the requirement for a first order spectrum, and we should letter the nuclei A, X rather than A, B. As we have already seen, the spectrum consists of two lines only, one at the chemical shift position ν_A, due to a spin-flip of A, and the other at ν_X due to a spin-flip of X.

This can be expressed easily in more precise quantum-mechanical terms, the advantages of which will become apparent in later sections with more complex systems. To do so, we must consider the system of two nuclear spins taken as a pair. It is clear that the *total* nuclear spin system has four stationary states. The first, of highest energy, is that in which both nuclear moments are "parallel" to H_0. The total nuclear spin wave function ψ_1 is conveniently written

$$\psi_1 = \alpha(A)\alpha(X) \text{ (or simply } \alpha\alpha)$$

and it follows from equation (1) that the corresponding energy

$$E = E(A) + E(X)$$

$$= \tfrac{1}{2}\eta H_0(2 - \sigma_A - \sigma_X)$$

The remaining three spin functions are $\alpha\beta$, $\beta\alpha$ and $\beta\beta$, and the total "energy level diagram" is shown in Fig. 5.1. Two transitions between states on this diagram correspond physically to a spin-flip of nucleus X, namely, (2)→(1) and (4)→(3), the arrow being drawn in the sense of absorption of energy, from the lower to the upper level. In the first of these, the total spin function changes from $\alpha\beta$ to $\alpha\alpha$, i.e. this corresponds to a spin-flip of X in a molecule in which the spin of A is α, whilst the second refers to one in which the spin of A is β. From the symmetry of the diagram it can readily be seen that these transitions are associated with the same change in energy, $\Delta E = \eta H_0(1-\sigma_X)$, so that they *coincide* in the spectrum at the chemical shift of X. Similarly, transitions (3)→(1) and (4)→(2) correspond to a spin-flip of A only, and so coincide at the chemical shift position of A. It should be noted that energy level diagrams as in Fig. 5.1 *are never drawn to scale*. For protons at an H_0 value of 14,000 gauss, the separation between levels (1) and (2),

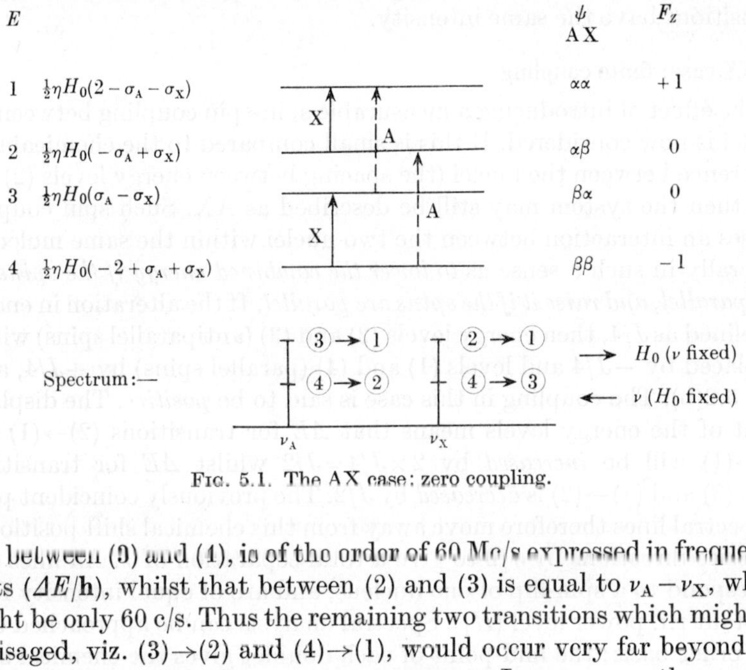

FIG. 5.1. The AX case: zero coupling.

also between (3) and (4), is of the order of 60 Mc/s expressed in frequency units ($\Delta E/\mathrm{h}$), whilst that between (2) and (3) is equal to $\nu_A - \nu_X$, which might be only 60 c/s. Thus the remaining two transitions which might be envisaged, viz. (3)→(2) and (4)→(1), would occur very far beyond the range of the normal n.m.r. proton spectrum. In any case, they would require a simultaneous spin-flip of both nuclei, and this is not possible under the usual experimental conditions.

It is convenient to define a new quantity

$$F_z = \Sigma\, I_z \text{ for all nuclei in the spin system.}$$

The numerical values of F_z for the AX stationary states are shown in Fig. 5.1, and the above results can then be summarized in terms of a *selection rule:*

$$\Delta F_z = \pm 1^*$$

* A second selection rule is given on p. 82.

In transitions (4)→(2), (3)→(1) and in (4)→(3), (2)→(1), F_z changes by 1 unit and this is a necessary condition for an allowed transition. It follows also that transitions (3)→(2) and (4)→(1) are *forbidden* by this rule. Since the allowed transitions correspond to a spin-flip of a single nucleus, each is induced with equal probability for a given field strength H_1. Furthermore, the scale of the energy level diagram ensures that the Boltzmann distribution factor (population ratio of upper to lower states) favouring absorption over emission, is effectively the same for *any* transition between states differing by unity in F_z. Hence all four allowed transitions have the same intensity.

2. AX case: finite coupling

The effect of introducing a measurable spin-spin coupling between the nuclei is now considered. If this is small compared to the chemical shift difference between the nuclei (the spacing between energy levels (2) and (3)) then the system may still be described as AX. Such spin coupling causes an interaction between the two nuclei within the same molecule, generally in such a sense as *to lower the combined energy if the spins are antiparallel, and raise it if the spins are parallel*. If the alteration in energy is defined as $J/4$, then energy levels (2) and (3) (antiparallel spins) will be displaced by $-J/4$ and levels (1) and (4) (parallel spins) by $+J/4$, as in Fig. 5.2(b). The coupling in this case is said to be *positive*. The displacement of the energy levels means that ΔE for transitions (2)→(1) and (3)→(1) will be *increased* by $2 \times J/4 = J/2$ whilst ΔE for transitions (4)→(3) and (4)→(2) is *decreased* by $J/2$. The previously coincident pairs of spectral lines therefore move away from this chemical shift position in opposite directions by $J/2$ to give a total separation of J. All lines still correspond to a spin-flip of one nucleus, and are of equal intensity for a constant r.f. power level (H_1). Spectral analysis can be approached on a first order basis; the mid-point of each doublet gives the chemical shift of the corresponding nucleus, and the component separation within each doublet gives the *magnitude* of J (Fig. 5.2(b)). If J were negative, the energy level displacements in Fig. 5.2 (b) would be reversed, as would the assignment of lines within each doublet. However, the appearance of the spectrum would not be altered, so the *sign* of J cannot be determined.

3. The general AB case

A second effect of the nuclear spin-coupling comes into play if the chemical shifts of an AX case lie closer together such that $J/\Delta\nu \approx 1$. This may be the result of a molecular situation *per se* or it may be caused by lowering H_0 until an AX case assumes the AB condition. The effect may be likened to quantum mechanical "resonance" in the theory of molecular structure, whereby, subject to certain conditions, two formal electronic states of similar energy may be considered as replaced by two new states, one of lower energy (bonding) and the other of higher energy (antibonding). So here, if two nuclear spin states approach closely in energy on lowering H_0, there operates what amounts to a "repulsion" between them—i.e. the state of higher energy E is displaced to a still higher value $(E+Q)$ whilst the lower one is displaced to $(E-Q)$. Moreover, the wave functions of these displaced (perturbed) states will be a *mixture* of those of the original states.

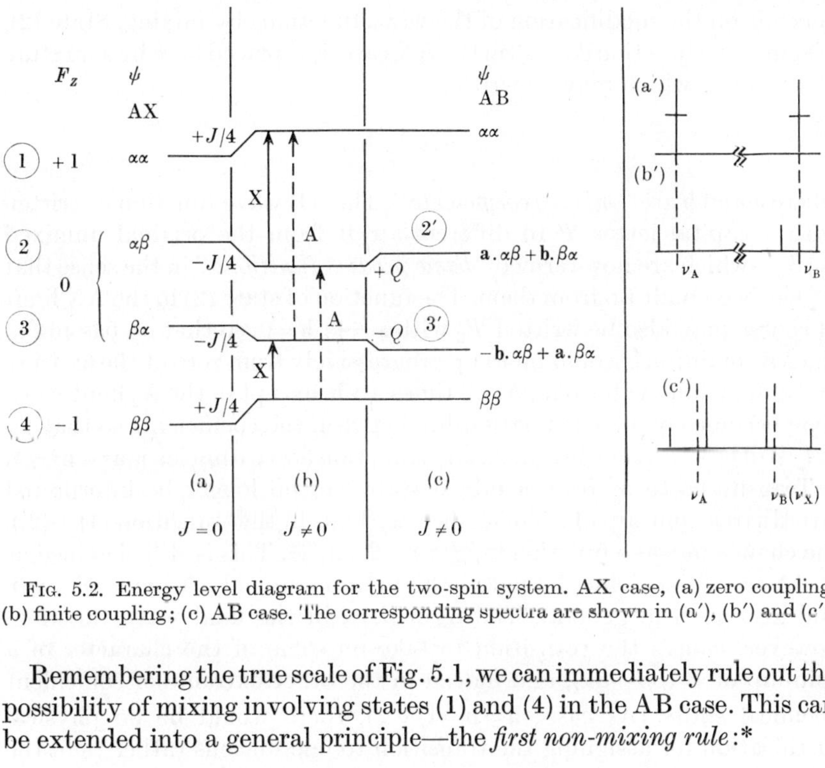

FIG. 5.2. Energy level diagram for the two-spin system. AX case, (a) zero coupling. (b) finite coupling; (c) AB case. The corresponding spectra are shown in (a′), (b′) and (c′).

Remembering the true scale of Fig. 5.1, we can immediately rule out the possibility of mixing involving states (1) and (4) in the AB case. This can be extended into a general principle—the *first non-mixing rule*:*

states of different F_z values cannot mix.

The only possible mixing is thus between (2) and (3). The mechanism for mixing is available through the finite coupling J_{AB}, and its effect upon the energy level diagram is shown in Fig. 5.2 (c).

The repulsion effect Q depends on the ratio $|J_{AB}/(\nu_A-\nu_B)|$ and as the magnetic field H_0 is decreased it becomes progressively larger, for a

* cf. p. 81 for second non-mixing rule.

constant value of J_{AB}. The spacing within each spectral doublet remains equal to J_{AB} so that $|J_{AB}|$ can still be found by inspection. However, the mid-point of each doublet no longer coincides with the appropriate chemical shift, but is displaced outwards from it by the quantity "Q" (see Fig. 5.2 (c′)). Analysis of the spectrum, whilst simple, is no longer first order. The energies and intensities for the lines of an AB spectrum are shown in Table 5.1; it should be noted, however, that the quantity Q is not explicitly stated but is incorporated in terms of J and ν.

The AB case serves also to illustrate a second feature of all non-first order spectra, namely, that the lines become *unequal* in intensity. This depends on the modification of the wave functions by mixing. State (2), of spin wave function $\psi_2 = \alpha\beta$ in the AX case, is replaced here by a mixture of ψ_2 and ψ_3 which may be written

$$\Psi_2' = \mathbf{a}\,.\,\alpha\beta + \mathbf{b}\,.\,\beta\alpha$$

where **a** and **b** are "*mixing coefficients*". The AB wave function is written with a capital letter Ψ to differentiate it from the original unmixed ψ_2, ψ_3—which are now termed "*basic product functions*" in the sense that Ψ_2' has been built up from them. The function of state (2) in the AX limit of course may also be written Ψ_2 with $\mathbf{a} = 1$: $\mathbf{b} = 0$, so that on passing to the AB condition **b** would build up progressively from zero at the expense of a diminishing value of **a**. At all times $\mathbf{a} > \mathbf{b}$, except in the A_2 limit when they become equal. The relationship between the coefficients so that Ψ_2' may still be a correct ("normalized") function for two nuclei, is $\mathbf{a}^2 + \mathbf{b}^2 = 1$.

Transitions to or from a mixed state can no longer be interpreted strictly as a spin-flip of a single nucleus. Thus in the transition (4)→(2′), the change in wave function is $\beta\beta \rightarrow \mathbf{a}\,.\,\alpha\beta + \mathbf{b}\,.\,\beta\alpha$. This is still classified as an "A-transition" since it tends to this in the high-field AX limit ($\beta\beta \rightarrow \alpha\beta$), and also $\mathbf{a} > \mathbf{b}$ in general. The term $\mathbf{b}\,.\,\beta\alpha$ in the final wave function, however, causes the transition to take on some of the character of a spin-flip of B ($\beta\beta \rightarrow \beta\alpha$), and indeed in the other extreme of coincident chemical shifts (A_2 case; $\mathbf{a} = \mathbf{b} = 1/\sqrt{2}$), there would be no physical justification for assigning the transition to one nucleus rather than the other. It has been stated that transition intensity is governed by the wave functions of the initial and final states. Both terms $\mathbf{a}\,.\,\alpha\beta$ and $\mathbf{b}\,.\,\beta\alpha$ in Ψ_2' here make a contribution, since each differs from the initial function $\psi_4 = \beta\beta$ in the spin of one nucleus. The intensity, relative to that of the first-order lines as unity is in fact (sum of coefficients)$^2 = (\mathbf{a} + \mathbf{b})^2 = 1 + 2\mathbf{ab}$ since $\mathbf{a}^2 + \mathbf{b}^2 = 1$. This applies also to the inner B transition (2′)→(1), and, since **b** and **a** are here of the same sign, both inner lines of the quartet have increased in intensity.

Ψ_3' is also a mixture of $\alpha\beta$ and $\beta\alpha$, in such proportions as to be complementary to Ψ_2'—thus, since the latter consists mainly of $\alpha\beta$ so Ψ_3' must consist mainly of $\beta\alpha$. In quantum-mechanical terms, Ψ_3' must be *orthogonal* to Ψ_2', and this is effected by writing $\Psi_3' = -\mathbf{b}.\alpha\beta + \mathbf{a}.\beta\alpha$. It is not easy to see why this requires changing one of the signs as well as interchanging the coefficients, but this must be so to explain the observed AB intensity pattern. The transitions between (3′) and the unmixed states (1) and (4) are the outer lines of the AB quartet. With Ψ_3' as above, similar reasoning gives their intensities as $(\mathbf{a}-\mathbf{b})^2 = 1-2\mathbf{ab}$, so that they lose intensity to the same extent as it is gained by the inner lines. In the A_2 case, their intensity drops to zero whilst the inner lines, which now coincide at the chemical shift $\nu_A = \nu_B$, each have an intensity of 2 units.

If J_{AB} is negative, the assignment of lines is reversed in pairs as given above, and it is now Ψ_2' which has the coefficients with opposite signs, $= \mathbf{a}.\alpha\beta - \mathbf{b}.\beta\alpha$. The *appearance* of the spectrum is thus completely unchanged and the sign of J_{AB} cannot be found.

4. Analysis of the AB case

To relate the line positions to parameters J_{AB}, ν_A and ν_B, the quantity Q must be expressed in terms of these parameters. To confirm the analysis by calculation of relative line intensities, mixing coefficients **a** and **b** must also be expressed in terms of J_{AB}, ν_A and ν_B. We have thus three "unknowns" Q, **a** and **b**, which can immediately be reduced to two by using the relation $\mathbf{a}^2 + \mathbf{b}^2 = 1$. This is summarized conveniently by making the trigonometric substitutions $\mathbf{a} = \cos\theta$, $\mathbf{b} = \sin\theta$, which explains the common use of such functions in expressions for line intensities. We know that the mixed states (2′) and (3′) must be solutions of the "wave equations" (see footnote, p. 63), and this knowledge enables us to write down two quadratic equations in Q and θ, which can be solved to give Q and θ separately in terms of J_{AB}, ν_A and ν_B. Hence all energy values, wave functions, line positions and relative intensities in the AB case can be related by simple algebraic formulae to the spin-coupling and chemical shift parameters, and this conclusion will apply to systems of any number of nuclei *provided that mixing is restricted to two states at a time.*

It will suffice to quote the results for the AB case in Table 5.1, in which $J = J_{AB}$. This shows the line positions relative to the centre of the spectrum. In the column of relative intensities,

$$\sin 2\theta = J/\sqrt{J^2 + (\nu_A - \nu_B)^2}.$$

TABLE 5.1

Transition	Assignment	Position	Rel. intensity
$3' \to 1$	A	$\frac{1}{2}\sqrt{J^2+(\nu_A-\nu_B)^2}+\frac{1}{2}J$	$1-\sin 2\theta$
$4 \to 2'$	A	$\frac{1}{2}\sqrt{J^2+(\nu_A-\nu_B)^2}-\frac{1}{2}J$	$1+\sin 2\theta$
$2' \to 1$	B	$-\frac{1}{2}\sqrt{J^2+(\nu_A-\nu_B)^2}+\frac{1}{2}J$	$1+\sin 2\theta$
$4 \to 3'$	B	$-\frac{1}{2}\sqrt{J^2+(\nu_A-\nu_B)^2}-\frac{1}{2}J$	$1-\sin 2\theta$

The chemical shift difference

$$|\nu_A-\nu_B| = \sqrt{\text{separation of outer lines} \times \text{separation of inner lines}}\dagger$$

and in most cases it is sufficient to use this formula. For doubtful cases, however, the reader is advised to check the relative intensities of the lines using the data given in the last column of the above.

It should be noted that the chemical shift of nucleus A occurs at the "centre of gravity" of the two A lines, and similarly for B. Notice also that the appearance of the spectrum, apart from a scaling factor, depends *only* on the ratio $|J_{AB}/\nu_A-\nu_B|$ so that the changes from AX through to A_2 can be summarized by a sequence of spectra calculated for a range of values of this ratio—see Fig. 5.2 (a′, b′, c′).

The examples of the olefinic protons in ethyl cinnamate (Fig. 5.3 (a)) and ω-nitrostyrene (Fig. 5.3 (b)) are instructive. The former case is more nearly first order than the latter and a comparison of "first-order" and exact analysis for each, is given in Table 5.2. Agreement is less good in the latter case.

TABLE 5.2

		ν^*_A	ν^*_B	J (c/s)
Ethyl cinnamate	First order	469	391	16
	Exact analysis	468·5	391·8	16
ω-Nitrostyrene	First order	133·5	105·5	14
	Exact analysis	131·6	107·4	14

* c/s from TMS at 60 Mc/s.

† Cf. Chapter 4.

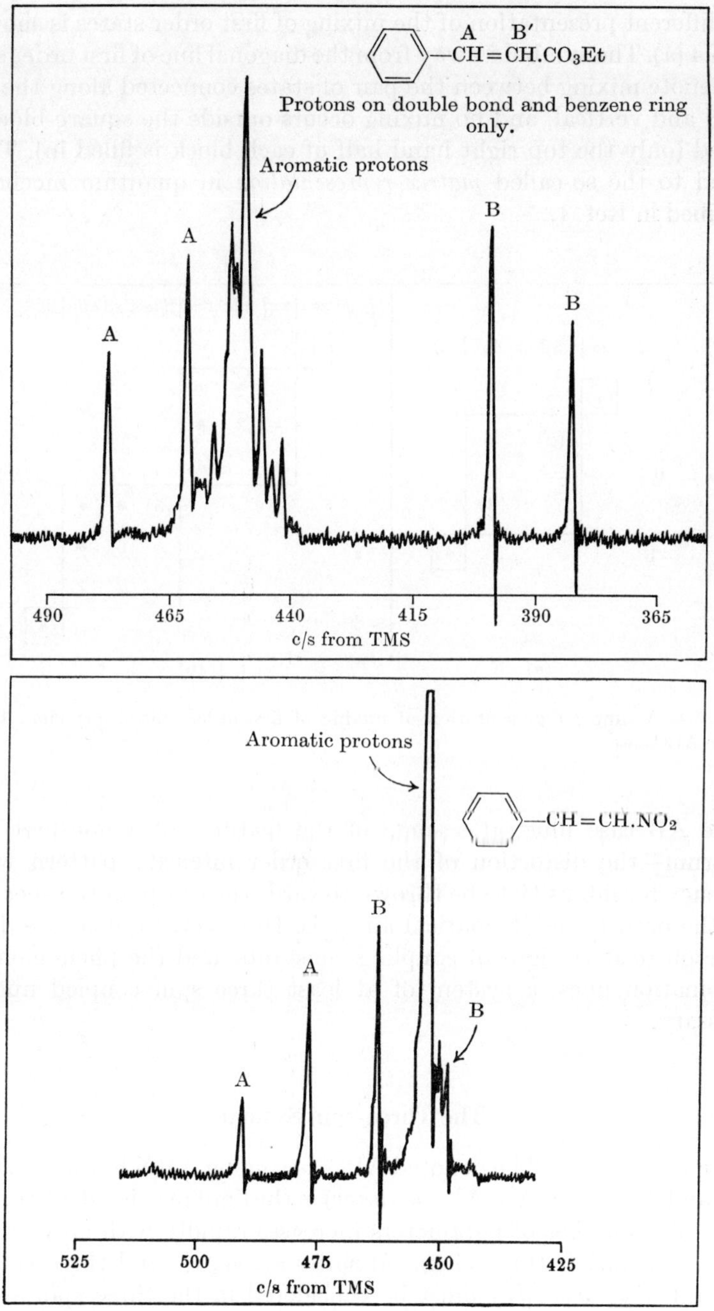

FIG. 5.3 Olefinic proton resonance spectra of (a) ethylcinnamate, (b) ω-nitrostyrene at 60Mc/s.

A different presentation of the mixing of first order states is shown in Fig. 5.4 (a). The symbol * away from the diagonal line of first order states will denote mixing between the pair of states connected along the horizontal and vertical, and no mixing occurs outside the square blocks so formed (only the top right-hand half of each block is filled in). This is related to the so-called *matrix representation* in quantum mechanics, described in Ref. 1.

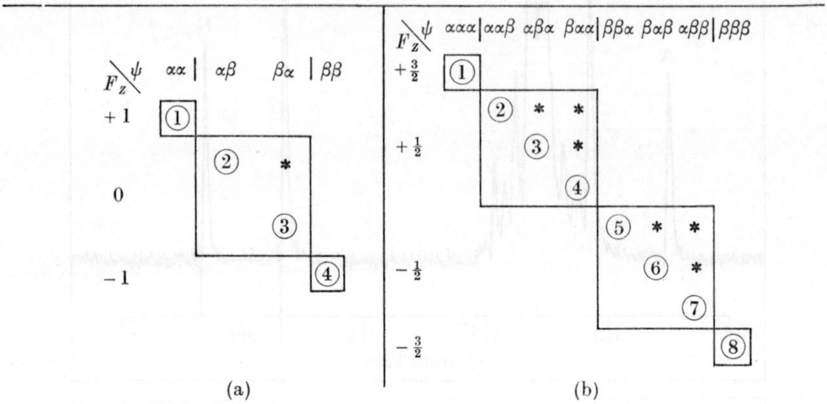

FIG. 5.4. A simple representation of mixing of first-order states. (a) The AB case; (b) The ABC case.

The AB case illustrates some of the features of a non-first order spectrum—the distortion of the first order intensity pattern with a tendency for intensity to be thrown towards the centre of the spectrum, and the need for mathematical analysis. However, to illustrate dependence on relative signs of coupling constants, and the phenomenon of combination lines, a system of at least three spin coupled nuclei is necessary.

The Three-spin System

In general this will be denoted ABC, becoming AMX in the first order limit and A_3 (better, AA′ A″—see later) in the limit of coincident chemical shifts. The number of parameters increases rapidly with an increasing number of nuclei; three chemical shifts ν_A, ν_B, ν_C, and three coupling constants J_{AB}, J_{BC}, J_{CA} must be considered in the three spin system. Spectral analysis for ABC will thus be much more difficult than for AB.

1. The AMX case

The appearance of the spectrum is shown in Example 13 of the problems. The 12 lines, of equal intensity, are grouped in three well-separated, symmetrical quartets, centred upon the chemical shift positions of the three nuclei. The spacing J_{AX} is repeated twice in the A quartet and twice in the X quartet, and similarly for the other nuclei taken in pairs. Analysis of the spectrum is first-order, and it should be noted that the signs of the coupling constants cannot be found.

A quantum-mechanical interpretation can be given as for the AX case with finite coupling. For n nuclei there will be 2^n stationary states; here, $2^3 = 8$. Their wave functions are as follows:

AMX	AMX	AMX	AMX	AMX	AMX	AMX	AMX
$\alpha\alpha\alpha$	$\alpha\alpha\beta$	$\alpha\beta\alpha$	$\beta\alpha\alpha$	$\beta\beta\alpha$	$\beta\alpha\beta$	$\alpha\beta\beta$	$\beta\beta\beta$
(1)	(2)	(3)	(4)	(5)	(6)	(7)	(8)
$F_z = +3/2$		$+\frac{1}{2}$			$-\frac{1}{2}$		$-3/2$

In each of the twelve allowed transitions between these, the spin of one nucleus is reversed, and all are subject to the selection rule $\Delta F_z = \pm 1$. These criteria, however, are not identical as they are in the AX case. Thus although the transition $\beta\beta\alpha \rightarrow \alpha\alpha\beta$ does not violate the ΔF_z selection rule, it is precluded by the condition that the spin of one nucleus only may be reversed.

2. The general ABC case

There are three states each of F_z value $+\frac{1}{2}$ and $-\frac{1}{2}$. Consequently, if all the coupling constants have finite values, mixing of these states *in sets of three* will occur on lowering H_0 to the ABC condition. This is shown in Fig. 5.4 (b)†. To express the energies and wave functions of the mixed states ($F_z = +\frac{1}{2}, -\frac{1}{2}$) in terms of the various values of J and ν, cubic equations must be solved for each of the interaction energies and mixing coefficients.

Algebraic expressions for these states in terms of the spectral parameters would thus be very cumbersome, and since all allowed transitions involve one or both of the $F_z = +\frac{1}{2}, -\frac{1}{2}$ "blocks" *none* of the transition energies or intensities can be expressed in simple algebraic form. Consequently, there can be no simple set of rules for the reverse procedure of finding the coupling constants and chemical shifts from line positions. Certain regularities nevertheless persist in the spectrum. On passing from the AMX to the ABC condition, the intensity pattern is distorted, but the

† The condition for putting a * connecting two first-order states of the same F_z value is that the wave functions shall differ in the spin of two nuclei only, and that the coupling constant between these two nuclei shall be non zero.

spacings of the four lines associated with each nucleus remain those of a symmetrical quartet. A spacing S_{AC} is again common to the A and C quartets, being repeated twice in each, but this is no longer equal to the coupling constant J_{AC}; similarly for the other nuclei taken in pairs. Although S_{AB}, S_{BC}, S_{CA}, are not equal to the respective coupling constants J_{AB}, J_{BC}, J_{CA}, yet

$$S_{AB}+S_{BC}+S_{CA} = J_{AB}+J_{BC}+J_{CA}.$$

A further complication is the appearance of three weak "combination lines", so that the ABC spectrum may contain 15 lines instead of the original 12 (AMX). For example, state (2) ($\alpha\alpha\beta$) of the AMX case, is replaced in the ABC case by the mixed state (2′) of wave function $\Psi_2' = \underline{a.\alpha\alpha\beta} + b.\alpha\beta\alpha + c.\beta\alpha\alpha$, and state (5) ($\beta\beta\alpha$) is similarly replaced by (5′) of wave function $\Psi_5' = \underline{p.\beta\beta\alpha} + q.\beta\alpha\beta + r.\alpha\beta\beta$ where a, b, c, p, q and r are constants. If the system is not too far from first order, the degree of mixing is small and the terms underlined are dominant. Transition (5)→(2) was forbidden in the AMX case ($a = p = 1$) since it would involve the simultaneous spin-flip of all three nuclei. However, in the ABC case, admixture with small amounts of the other basic product functions enables us to write down six "cross terms" which give a small but finite intensity to (5′)→(2′). These are the products of coefficients of those basic functions which differ in the spin of only one nucleus at a time: they appear in the form $\{a(q+r)+b(p+r)+c(p+q)\}^2$. Similarly, transitions (6′)→(3′) and (7′)→(4′) become allowed as combination lines. To differentiate between these and the lines which were allowed also in the first order condition, the latter are sometimes called "fundamental" lines; these terms being borrowed from electronic spectroscopy.

From the point of view of analysis, we will arbitrarily divide ABC spectra into two categories, although the distinction is *not* sharply drawn in practice.

3. Weakly coupled ABC spectra

The first category is that in which all ratios $|J_{ij}/\nu_i - \nu_j|$ are small. The degree of mixing is small, and the spectrum bears a recognizable resemblance to AMX. In particular, the combination lines are much weaker than the fundamental lines, so there is no confusion between them. The A, B and C quartets can be picked out, and reasonable approximations to the chemical shifts and magnitudes (*not* the sign) of the coupling constants, can still be obtained by treating the spectrum as if it were first order. The closer the spectrum is to first order, of course, the better is this approximation.

Exact numerical analysis, preferably by computer, would give accurate values for the chemical shifts and both the magnitude *and* the relative signs of all the coupling constants. Line positions in the AMX case could be "matched" by taking any of the four relative sign permutations of the coupling constants J_{AB}, J_{BC}, J_{CA} $(+++, ++-;$ $+-+; -++;$ complete reversal of all signs is not considered separately). In a similar way the line *positions* in a weakly coupled ABC spectrum can generally be matched by four sets of parameters. These are given *approximately* by the chemical shifts and the four relative sign

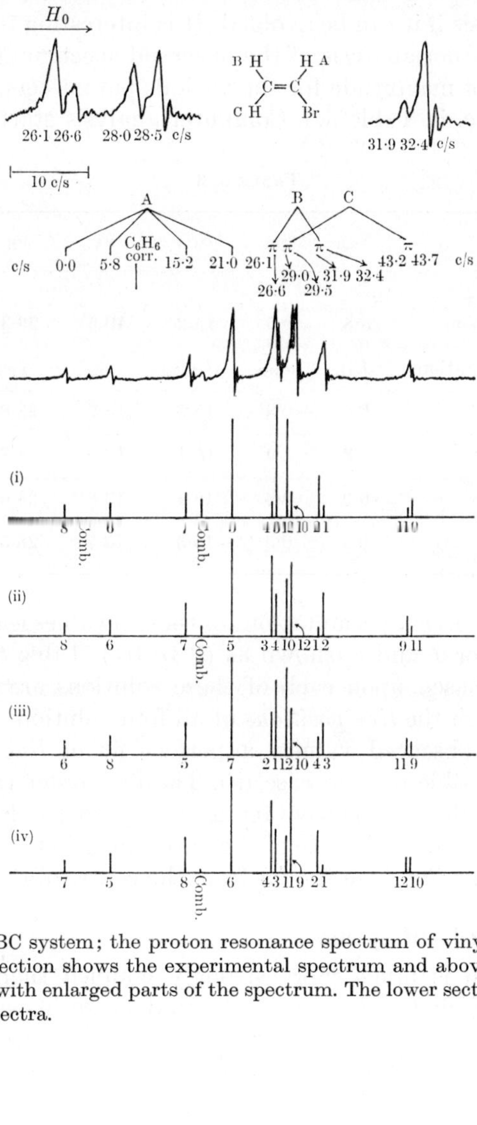

FIG. 5.5. An ABC system; the proton resonance spectrum of vinyl bromide at 29·92 Mc/s. The upper section shows the experimental spectrum and above it, the first order splitting diagram with enlarged parts of the spectrum. The lower section (i) to (iv) shows calculated ABC spectra.

permutations of S_{AB}, S_{BC}, S_{CA} obtained by the first order treatment. Whilst there is no means of deciding between these in the AMX case, where all lines are of equal intensity, in a weakly coupled ABC spectrum the four sets of parameters lead to slightly different *intensity* patterns, of which only one will agree *in all respects* with the experimental spectrum. Figure 5.5 illustrates this for the proton spectrum of vinyl bromide at 29·92 Mc/s,[2] in which two combination lines are visible. It should be noted that this spectrum departs quite considerably from the AMX condition (there is even some overlap between the B and C quartets). and that with coupling of this strength one should not be content with a first order analysis if it can be avoided. It is interesting to see the errors involved. First order analysis of the observed spectrum gives values of the right order of magnitude for the various parameters, and these are shown as S_{AB} etc. in Table 5.3. Computer analysis starting from each

TABLE 5. 3

	S_{AB}	S_{BC}	S_{CA}	ν_A	ν_B	ν_C
First order approximation	5·8	0·5	15·2	10·5	29·3	35·8
Computer solutions	J_{AB}	J_{BC}	J_{CA}	ν_A	ν_B	ν_C
ABC (i)	6·8	−0·8	15·5	14·0	28·8	32·8
(ii)	7·3	−1·9	15·1	13·9	28·8	32·9
(iii)	−6·3	0·9	15·4	13·8	28·6	33·1
(iv)	5·9	0·5	−15·3	13·7	28·5	33·4

of the four relative sign permutations of these S values leads to four possible solutions for J and ν (shown as (i) to (iv), Table 5.3). The theoretical spectra based upon each of these solutions are drawn in Fig. 5.5., and although the *line positions* of all four solutions agree with the experimentally observed values, a perfect fit of *line positions and intensities* is possible only in case (ii). The first order treatment gives of course much the best percentage accuracy for the largest coupling constant. A first order estimate of the outermost chemical shifts in any spectrum always puts them too far from the centre of the spectrum.

4. Strongly coupled ABC spectra

If a spectrum is much more strongly coupled than the previous example, then a first order analysis would be so inaccurate as to be grossly misleading. The obvious remedy is to re-examine the spectrum at a higher magnetic field, if this is practicable. Several other possibilities are discussed in Chapter 7. If the first order AMX or the ABX approximations are insufficiently accurate at the highest available magnetic field, then, failing access to a computer and a suitable programme, the organic chemist may attempt an exact ABC analysis by algebraic methods, a task more difficult in fact than computer analysis. It would be advisable to undertake such an analysis by hand only if the three fundamental quartets can be picked out unambiguously, i.e. provided that the combination lines are not so strong as to be confused with the fundamental ones. Relative to a suitable zero a unique set of (eight) energy levels can then be obtained from the line positions; this is most easily done by a graphical method.[3] Several algebraic procedures for exact ABC solutions have been published, although it will be appreciated that they will be substantially longer than cases in which mixing of states in pairs only is involved. The methods, moreover, are too lengthy to be reproduced here. One procedure[3c] is a simplification of an earlier method[3b] with the organic chemist in mind. It involves guessing plausible values for $(\nu_A^2+\nu_B^2+\nu_C^2)$ and then using experimental energy values to deduce $(J_{AB}^2+J_{BC}^2+J_{CA}^2)$ by two different routes. The correct ABC solution corresponds to the state of internal consistency, i.e. when these two calculated values are identical, and this may be found after a few guesses by a graphical intersection method. An approximate method intermediate between ABC and ABX has also been published[3a].

In *some* spin systems (e.g. AB_2, see below), the combination lines always remain weak at any strength of coupling, and cause no difficulty in analysis. In most cases however, (including ABC), as the strength of coupling increases one or more combination lines move(s) towards the centre of the spectrum where it is in a favourable position for progressive gain in intensity at the expense of outlying fundamental ones. One may then have a strongly coupled spectrum in which it is not possible to differentiate between fundamental and combination lines at sight. If an ABC system is so strongly coupled that this situation arises, then the grouping of lines into three quartets and three combination lines becomes ambiguous. There are then fifteen possibilities which would fulfil the necessary properties of repeated spacing, quite apart from sign permutations of the coupling constants, although many of these may in practice be eliminated[3b, 4]. Exact solution without use of a computer is then not advisable.

If the chemical shifts become nearly the same, then the outermost lines will fade into the baseline as the intensity piles up into a few transitions near the middle of the spectrum. These will coalesce ultimately into a single line in the limit of coincident shifts. Five transitions would generally remain of finite intensity in this limit, and dominate the spectrum in a very strongly coupled ABC case as a close quintet of intensity ratios 1:3:4:3:1.[4]

Apart from the first order AMX condition, there are two special cases of three spin-coupled nuclei for which all line positions and intensities can be related to the spectral parameters by simple algebraic equations. These must be situations in which we can avoid the mixing in 3×3 blocks, and reduce the problem to one in which mixing does not involve more than two first order states at a time:—

5. The ABX case

In this, *some*, but not all, of the interactions are treated as first order. A detailed discussion of such an example is provided in Chapter 6.

6. The AB_2 case

This is now considered together with its first order limiting condition, AX_2. Two of the nuclei are here *completely* equivalent, having the same chemical shift ν_B (or ν_X) and the same coupling J_{AB} (or J_{AX}) to the A-nucleus. Consequently there are four spectral parameters (ν_A, ν_B, J_{AB}, J_{BB}) instead of six in ABC, and at the outset we may expect the analysis to be easier. The case is fairly often encountered in practice, for example in the following types of molecule.

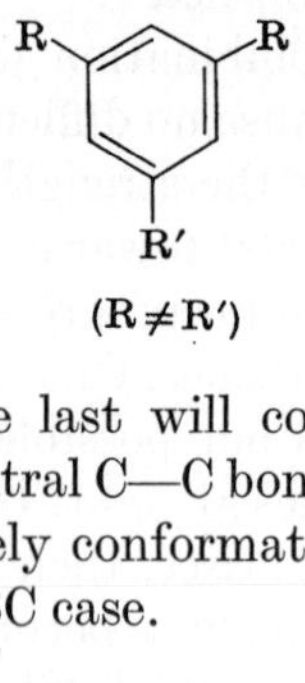

(R≠R′)

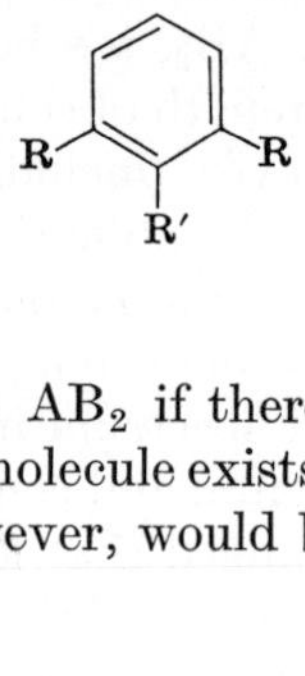

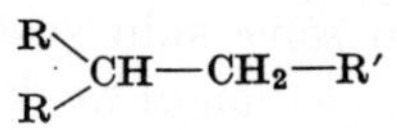

(R=R′ or ≠R′)

The last will correspond to AB_2 if there is rapid rotation about the central C—C bond, or if the molecule exists as conformation (a). The less likely conformation (b), however, would be an example of the general ABC case.

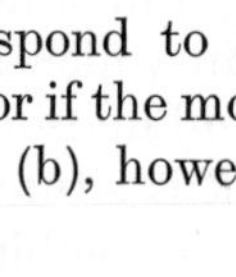

(a)

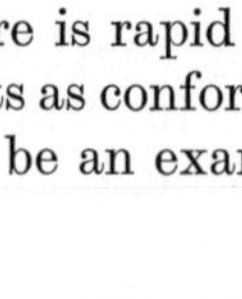

(b)

In the limiting case AX_2, rules for the appearance of first order spectra (Chapter 4) predict a 1:2:1 triplet for the A nucleus and a 1:1 doublet, of twice the total intensity, for X_2 (see Fig. 5·6 (a)). The multiplet splitting is equal to J_{AX}, and the spectrum is independent of J_{XX}. Analysis to give J_{AX}, ν_A and ν_X is first order and it should be noted that there are five lines only, instead of twelve as in the AMX case. As may be seen from the splitting diagram however, the central line of the A triplet must be a superposition of two transitions (Fig. 5.6 (a)), and we may suspect that the strong X_2 lines are also superpositions.

(a)
$J/\Delta\nu \to 0$ (AX_2)

(b)
$J/\Delta\nu = 0{\cdot}1$

(c)
$J/\Delta\nu = 0{\cdot}5$

(d)
$J/\Delta\nu = 2{\cdot}0$

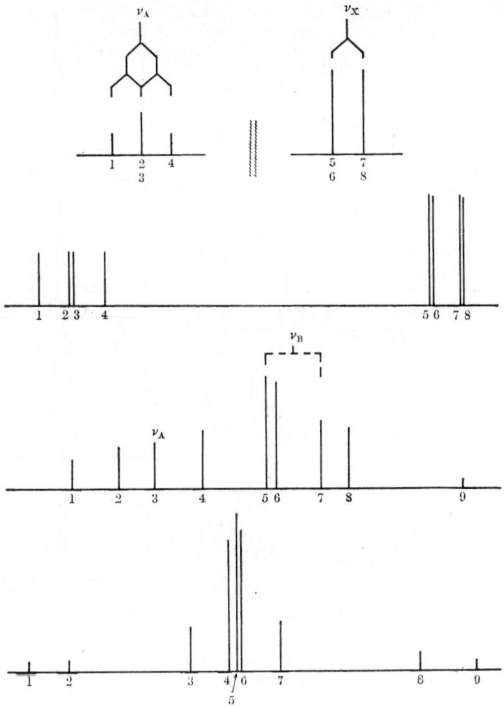

FIG. 5.6. Calculated AB_2 spectra, as a function of $J/\Delta\nu$.

Changes in the spectrum on progressively decreasing the chemical shift difference ($AX_2 \to AB_2$) are shown in Fig. 5.6. The first noticeable effect is the splitting of previously coincident transitions, giving an eight-line spectrum. In a nearly first order spectrum this effect is termed *second-order* splitting*, being readily differentiated from the "first-order* splitting", into multiplet components of spacing J_{AB}, by the fact that it increases on lowering the applied field. Second order splitting is a useful

* The descriptions first and second order are derived from a particular mathematical approach known as perturbation theory.

concept for describing weakly coupled spectra (cf. the ethyl group, Chapter 7, p. 119) although it becomes meaningless in strongly coupled systems. In the latter, a single combination line appears beyond the B_2 group of lines but always remains weak. Together with the other outlying lines, it fades to zero intensity in the limiting case where $\nu_A = \nu_B$, when the five strong central lines (cf. ABC) coincide at the chemical shift position. As in the AX_2 limit, the AB_2 spectrum is completely independent of J_{BB},

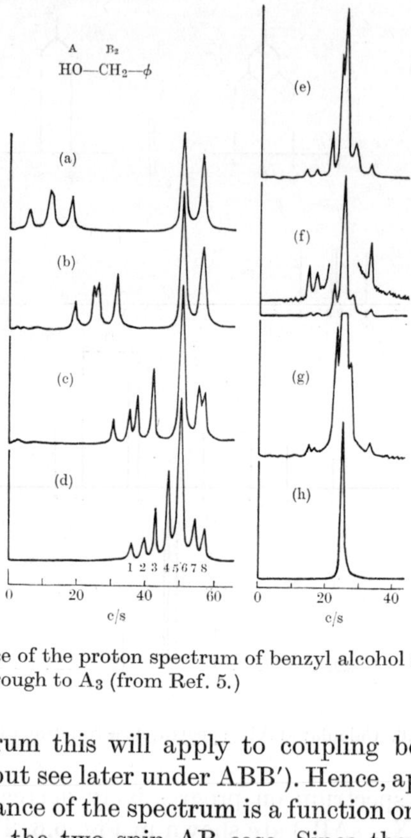

FIG. 5.7. Appearance of the proton spectrum of benzyl alcohol showing the full range of appearance AX_2 through to A_3 (from Ref. 5.)

and in *any* spectrum this will apply to coupling between *completely equivalent nuclei* (but see later under ABB′). Hence, apart from a scaling factor, the appearance of the spectrum is a function only of the quantity $|J_{AB}/\nu_A - \nu_B|$ as in the two spin AB case. Since the absolute signs of coupling constants cannot be found from an n.m.r. spectrum, that for AB_2 must be independent of the sign of J_{AB}.

For the purposes of analysis (see next section) the lines are numbered consecutively 1–9 (Fig. 5.6). The numbering starts from the A proton, and although this may lie to the high or low field end of the spectrum, its position is easily determined by comparison with the patterns shown in Fig. 5.6. Alternatively an integral trace would allow of choice. It must be remembered that in practice, close transitions may not be resolved (cf. problem 19 and Chapter 6, p. 89). An interesting demonstration is provided by the spectra of the hydroxyl and methylene protons of benzyl alcohol (Fig. 5.7). In this the total range of appearance of AB_2 spectra from nearly first order to coincident chemical shifts is exhibited within a single compound. This unusual result was obtained by McGreer and Mocek[5] by progressive addition of acetone, causing a large displacement of the hydroxyl proton shift. Again we have to note that transitions 5 and 6 are not resolved in any of the spectra a to h.

7. Analysis of the AB_2 case; use of symmetry

In the systems previously considered, the F_z value has sufficed to tell us which of the first order states will mix in non-first order cases*. There is, however, a further condition which must now be considered: this is the *second non-mixing rule**:

states of different symmetry cannot mix.

The great simplification in the AB_2 case relative to ABC is due to the operation of this principle. The nuclear spin system (although not necessarily the whole molecule) has an important element of symmetry which was lacking in ABC, namely, a plane of symmetry as shown here:—

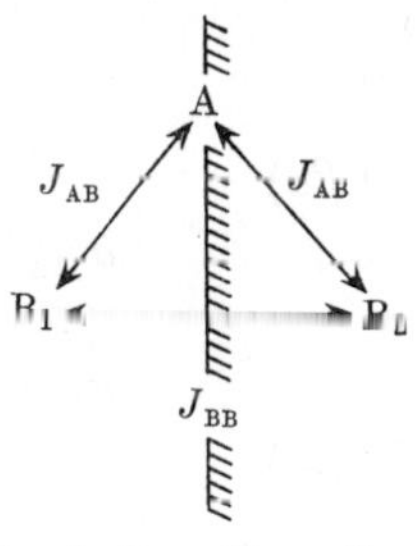

If first we write the eight basic functions for the AMX case, we must then consider in the AB_2 case how each would be modified by "reflection" about the plane indicated.

AB_1B_2	AB_1B_2	AB_1B_2	AB_1B_2	AB_1B_2	AB_1B_2	AB_1B_2	AB_1B_2
$\alpha\alpha\alpha$	$\alpha\alpha\beta$	$\alpha\beta\alpha$	$\beta\alpha\alpha$	$\beta\beta\alpha$	$\beta\alpha\beta$	$\alpha\beta\beta$	$\beta\beta\beta$
1	2	3	4	5	6	7	8
S	—	—	*S*	—	—	*S*	*S*

Since A lies in the plane of symmetry its spin remains unaltered: interchange of B_1 and B_2, however, means that the spin of nucleus B_1 becomes

* See p. 67 for first non-mixing rule.

that of B_2 and vice versa. Of the eight basic functions above, there are four (marked S) which do not alter as a result of this operation: they are said to be *symmetric*.

The remaining functions, however, are changed to different functions (e.g. $\alpha\alpha\beta$ to $\alpha\beta\alpha$), and do not have any symmetry with respect to the reflection. Thus if we take the $F_z = +\frac{1}{2}$ block in (Fig. 5.4 (b)), we have so far singled out $\beta\alpha\alpha$, and are left with $\alpha\alpha\beta$ and $\alpha\beta\alpha$. Whilst these do not have any symmetry properties with respect to the reflection, it is clear that their *sum* and *difference* do. Thus if we replace $\alpha\alpha\beta$ and $\alpha\beta\alpha$ by $\sqrt{\frac{1}{2}}(\alpha\alpha\beta \pm \alpha\beta\alpha)$, i.e. by $\sqrt{\frac{1}{2}}\,\alpha(\alpha\beta+\beta\alpha)$ and $\sqrt{\frac{1}{2}}\,\alpha(\alpha\beta-\beta\alpha)$ respectively, we find that the former is again symmetric to (unchanged by) the reflection.* Function $\sqrt{\frac{1}{2}}\alpha(\alpha\beta-\beta\alpha)$ is, however, changed to $\sqrt{\frac{1}{2}}\alpha(\beta\alpha-\alpha\beta) = -\sqrt{\frac{1}{2}}\alpha(\alpha\beta-\beta\alpha)$, i.e. it is reversed in *sign only* on reflection, and is said to be *antisymmetric*. The $F_z = -\frac{1}{2}$ block can be treated similarly. We have thus arrived at a new set of functions constructed from those of AMX so as to be symmetric (s) or antisymmetric (a) to the reflection. These are listed below. The s notation of the first column is a convenient short-hand description of the function set out in the second column. The F_z value is attached to s as a subscript in each case (similarly for the "a" functions).

F_z	Symmetric		Antisymmetric
$s_{3/2}$	$\alpha\alpha\alpha$		
$1\ s_{1/2}$	$\sqrt{\frac{1}{2}}\alpha(\alpha\beta+\beta\alpha)$		
$2\ s_{1/2}$	$\beta\alpha\alpha$	$a_{1/2}$	$\sqrt{\frac{1}{2}}\alpha(\alpha\beta-\beta\alpha)$
$1\ s_{-1/2}$	$\alpha\beta\beta$	$a_{-1/2}$	$\sqrt{\frac{1}{2}}\beta(\alpha\beta-\beta\alpha)$
$2\ s_{-1/2}$	$\sqrt{\frac{1}{2}}\beta(\alpha\beta+\beta\alpha)$		
$s_{-3/2}$	$\beta\beta\beta$		

These are in fact the correct wave functions for the AX_2 case. The reduced number of transitions relative to AMX can now be explained most simply by involving a *second selection rule*, common to all forms of spectroscopy:

transitions are allowed only between states of the same symmetry.†

Eight transitions are thus allowed, and although some are coincident in the AX_2 case the true number can be seen from the second order splitting

* The factor $\sqrt{\frac{1}{2}}$ is introduced for normalization (cf. the AB Case, p. 68).

† See p. 65 for first selection rule.

in the weakly coupled AB_2 case. In view of the second non-mixing rule (p. 81), we have reduced the mixing problem in the AB_2 case to 2×2 interactions only by using the above "symmetry functions" as a basis rather than the simple product functions of AMX. All transition energies and intensities can then be expressed in the simple form (Table 5.4, in which $J = J_{AB}$).

Transition $2s_{-1/2} \rightarrow 2s_{1/2}$ is forbidden in AX_2 by the selection rule $\Delta F_z = \pm 1$, but becomes weakly allowed in AB_2 as the only combination line. Neither of the antisymmetric states is involved in mixing, so that transition $a_{-1/2} \rightarrow a_{1/2}$ remains stationary at ν_A throughout.

If positive quantities C_+, C_- and angles θ_+, θ_- are defined by the equations

$$
\begin{aligned}
C_+ \cos 2\theta_+ &= \tfrac{1}{2}(\nu_A - \nu_B) + \tfrac{1}{4}J \\
C_+ \sin 2\theta_+ &= J/\sqrt{2} \\
C_- \cos 2\theta_- &= \tfrac{1}{2}(\nu_A - \nu_B) - \tfrac{1}{4}J \\
C_- \sin 2\theta_- &= J/\sqrt{2}
\end{aligned}
$$

TABLE 5.4

Line positions and relative intensities for the AB_2 case

Transition	Assign-ment	Position	Relative intensity
1. $2's_{1/2} \rightarrow s_{3/2}$	A	$\nu_A + \nu_B + \frac{3}{4}J + C_+$	$[2 \sin \theta_+ - \cos \theta_+]^2$
2. $2's_{-1/2} \rightarrow 1's_{1/2}$	A	$\nu_B + C_+ + C_-$	$[2 \sin (\theta_+ - \theta_-) + \cos \theta_+ . \cos \theta_-]^2$
3. $a_{-1/2} \rightarrow a_{+1/2}$	A	ν_A	1
4. $s_{-3/2} \rightarrow 1's_{-1/2}$	A	$\nu_A + \nu_B - \frac{3}{4}J + C_-$	$[2 \sin \theta_- + \cos \theta_-]^2$
5. $1's_{-1/2} \rightarrow 1's_{1/2}$	B	$\nu_B + C_+ - C_-$	$[2 \cos (\theta_+ - \theta_-) + \cos \theta_+ . \sin \theta_-]^2$
6. $1's_{1/2} \rightarrow s_{3/2}$	B	$\nu_A + \nu_B + \frac{3}{4}J - C_+$	$[2 \cos \theta_+ + \sin \theta_+]^2$
7. $2's_{-1/2} \rightarrow 2's_{1/2}$	B	$\nu_B - C_+ + C_-$	$[2 \cos (\theta_+ - \theta_-) - \sin \theta_+ . \cos \theta_-]^2$
8. $s_{-3/2} \rightarrow 2's_{-1/2}$	B	$\nu_A + \nu_B - \frac{3}{4}J - C_-$	$[2 \cos \theta_- - \sin \theta_-]^2$
9. $1's_{-1/2} \rightarrow 2's_{1/2}$	Comb.	$\nu_B - C_+ - C_-$	$[2 \sin (\theta_+ - \theta_-) + \sin \theta_+ . \sin \theta_-]^2$

From this table we can write very simple rules for finding the spectral parameters (see also Fig. 5.6):

$$
\begin{aligned}
\nu_A &= \nu_3 \\
\nu_B &= \tfrac{1}{2}(\nu_5 + \nu_7) \\
J_{AB} &= \tfrac{1}{3}(\nu_4 + \nu_8 - \nu_1 - \nu_6)
\end{aligned}
$$

In the majority of cases these rules will suffice in the analysis of an AB_2 spectrum: where doubt obtains, recourse must be had to the calculation of line intensities as in Table 5.4. Additionally the position of the combination line 9 provides a check.

8. The ABB′ case

If two nuclei have identical chemical shifts (ν_B) whilst being coupled differently to a third nucleus A, the system would be denoted ABB′. This could not arise through molecular symmetry, but only through an improbable chance coincidence of shifts. However, the concept of magnetic equivalence ($\nu_B = \nu_B'$) with spin-coupling non-equivalence ($J_{AB} \neq J_{AB}'$) is an important one, which will be used in Chapter 7. ABC cases with close similarity between two of the shifts have occasionally been analysed as approximations to ABB′. Although this was once thought to be an easier case to treat theoretically, it is now clear from calculated spectra[6] that the case ABB′, with no useful plane of symmetry as in AB_2, may have up to 15 lines with no obvious simplification relative to ABC. Similarly, the spectrum of the high field limiting case AXX′ would differ from the first order spectrum of AX_2 (cf. the case AA′XX′ in Chap. 7).

References

1. J. D. Roberts, "An Introduction to Spin-Spin Splitting in High Resolution N.M.R. Spectra", Benjamin, New York (1961).
2. E. O. Bishop and R. E. Richards *Mol. Phys.* **3**, 114 (1960).
3. (a) C. A. Reilly and J. D. Swalen, *J. Chem. Phys.* **32**, 1378 (1960).
   (b) S. Castellano and J. S. Waugh, *J. Chem. Phys.* **34**, 295, **35**, 1900 (1960).
   (c) X.-T. Liang, *Scientia Sinica* **13**, 589 (1964).
4. E. O. Bishop, Mol. Phys. **10**, 45 (1965).
5. D. E. McGreer and M. M. Mocek, *J. Chem. Educ.* **40**, 358 (1963).
6. K. B. Wiberg and B. J. Nist, "Interpretation of NMR spectra". Benjamin, New York (1962).

CHAPTER 6

THE CALCULATION OF ABX AND RELATED SPECTRA

C. N. BANWELL

Introduction

In this chapter we shall consider in some detail the way in which a certain type of n.m.r. spectrum may be analysed—by which we mean the calculation of the precise coupling constants and chemical shifts of the resonating nuclei in the molecule which gave rise to the spectrum.

It has already been shown that in spectra where all the chemical shifts are large compared to the coupling constants, e.g. AX, AMX etc., the analysis is first order: the n.m.r. parameters can be read directly from the spectrum. Thus in the AMX case the three coupling constants are given by the repeated separations of lines taken in pairs, while the chemical shifts are found as the mid-point of each quartet (cf. Chapter 5, p. 73). When the chemical shifts become comparable in magnitude with the coupling constants, however, the analysis is no longer first order and in the limit, AMX becomes the almost unapproachable ABC. It is nevertheless instructive to consider the intermediate case, that of ABX, and it is this system which is our concern for the greater part of this chapter. We shall see that some information can be read directly from the spectrum, the rest must be calculated from measured data in a straightforward but non-trivial way.

Figure 6.1 shows how the ABX system arises as a case intermediate between AMX and ABC.

System: AMX
Analysis: "first-order"

System: ABX
Analysis: Possible without computer

System: ABC
Analysis: Computer required

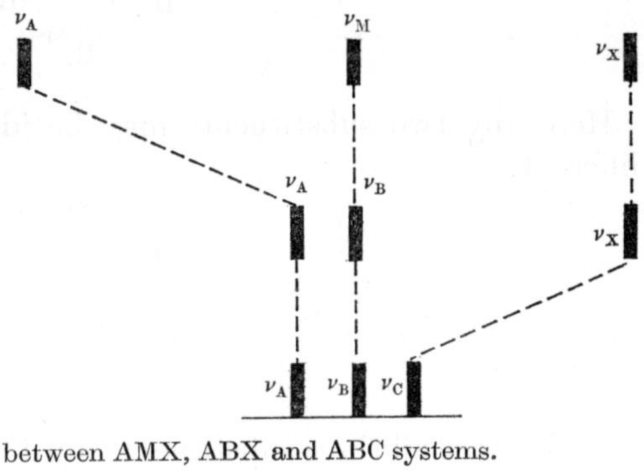

FIG. 6.1. Relationship between AMX, ABX and ABC systems.

ABX-Type Molecules

Before discussing the analysis, it is useful to note the types of chemical system which might be expected to give rise to an ABX spectrum. The following list is by no means exhaustive, and other examples will arise in the problems at the end of the book.

1. Vinyl compounds

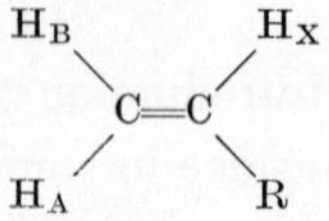

Here the substituent, R, affects the geminal hydrogen (H_X) most strongly and so places its chemical shift well removed from those of the *cis* and *trans* hydrogens. H_A and H_B are non-equivalent, but in general their resonances fall close together.

2. 1,2,3-, 1,2,4-, and 1,3,5-Trisubstituted benzenes

(a)

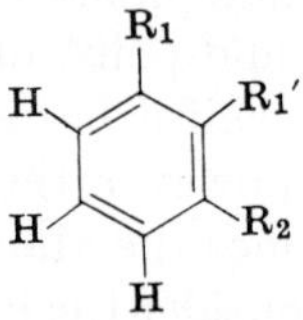

Here R_1 and R_1' may be the same or different, but R_2 must differ from both.

(b)

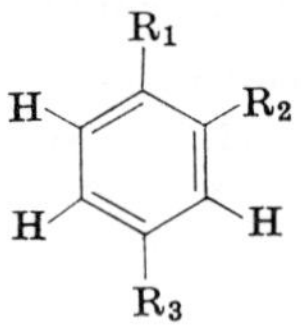

Here any two substituents may be identical, or all three may be different.

(c)

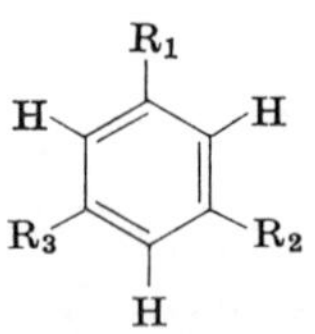

In this case all three substituents must differ.

3. 1,1,2-Trisubstituted ethanes, CH_2P—CHQR

In these molecules the methylene protons are never identical even if free rotation about the C—C bond is assumed, and thus the system may be ABX.

Non-equivalence of this sort can be seen most easily from the various conformations of the molecule shown in Fig. 6.2. Thus in (a) H_1 is placed *trans* to Q and *gauche* to R, and so is in a different chemical environment from H_2 which is *trans* to R and *gauche* to Q. Rotation can never simply interchange their positions (cf. (b) and (c)), and so these protons are never equivalent.

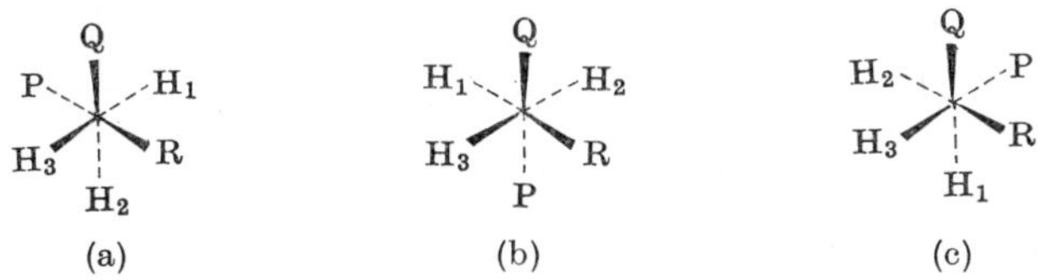

FIG. 6.2.

4. Substituted furans, pyrroles, etc.

It should be pointed out that the above molecules do not necessarily give rise to ABX spectra. An ABC case might well arise unless one of the substituents differs considerably from the others in electron withdrawing or releasing power or in its magnetic effects. Equally as will be seen from example No. 13 of the problems, an AMX case is also possible.

First Order Approach to ABX Spectra

An idea of the general appearance of an ABX spectrum may be gained by considering Fig. 6.3. Taking the sections in order we have the following:

(a) The positions of resonance of the three nuclei A, B and X in the absence of all coupling.

(b) Coupling of the nuclei A and B only. If, for convenience, we assume that J_{AB} is small, then we expect the AB portion of the spectrum to show an AB-type quartet while the X resonance will be unaltered. The coupling constant J_{AB} will, as usual, be reflected in the line spacing as shown.

(c) Coupling between A and X, when the X resonance is split to a doublet with a separation S_{AX}. The symbol S_{AX} is used since we do not know *a priori* whether the line separation will be *equal* to J_{AX} or merely dependent upon it (generally, $S_{AX} \neq J_{AX}$). Similarly the A resonance, already a close doublet due to coupling with B, is split further to a pair of doublets separated by S_{AX}. Note carefully the intensity distribution in the A quartet resulting from this: the "weak, strong" resonance pair of A in (b) becomes "weak, strong, weak, strong" due to splitting by J_{AX}.

(d) Coupling between B and X, resulting in a 1:1:1:1 quartet for the X resonance which contains the two separations S_{AX} and S_{BX} each twice. The B resonance also becomes a "strong, weak, strong, weak" quartet containing, in addition to J_{AB}, the separation S_{BX} twice repeated.

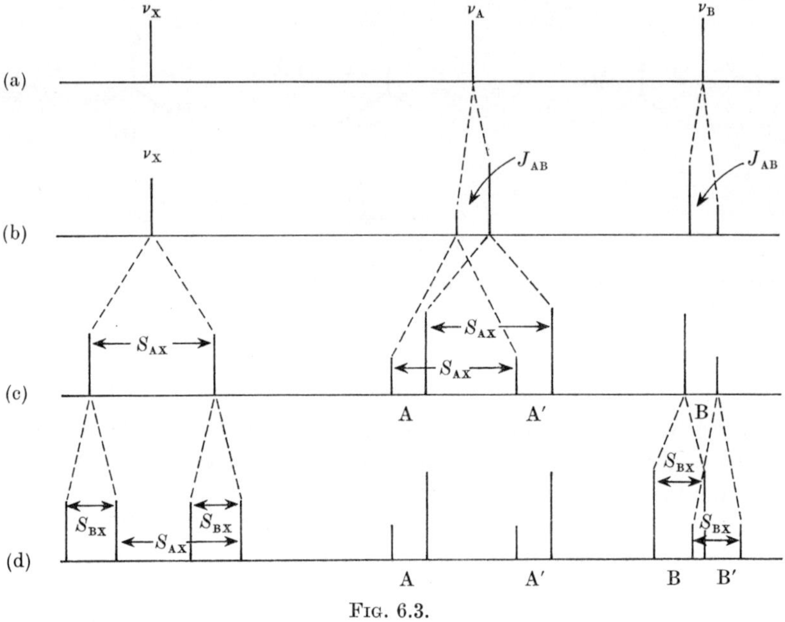

FIG. 6.3.

It should be particularly noted in Fig. 6.3 that the ABX spectrum, approached in this simple way, shows two AB-type quartets in the AB region; in Fig. 6.3 (d) these have been labelled AB and A′B′.

In Fig. 6.4 examples of three actual ABX spectra are shown. Figure 6.4 (a) illustrates an ABX spectrum which follows closely the pattern derived above, i.e. four lines of approximately equal intensity in the X region, and two AB-type quartets. In Fig. 6.4 (b) the X region seems to show six lines, whilst the AB region shows one very clear AB-type quartet as well as a broad, intense resonance which may be recognized as an AB-quartet with a very small chemical shift between the A and B resonances. Finally in Fig. 6.4 (c) the X region clearly shows six lines

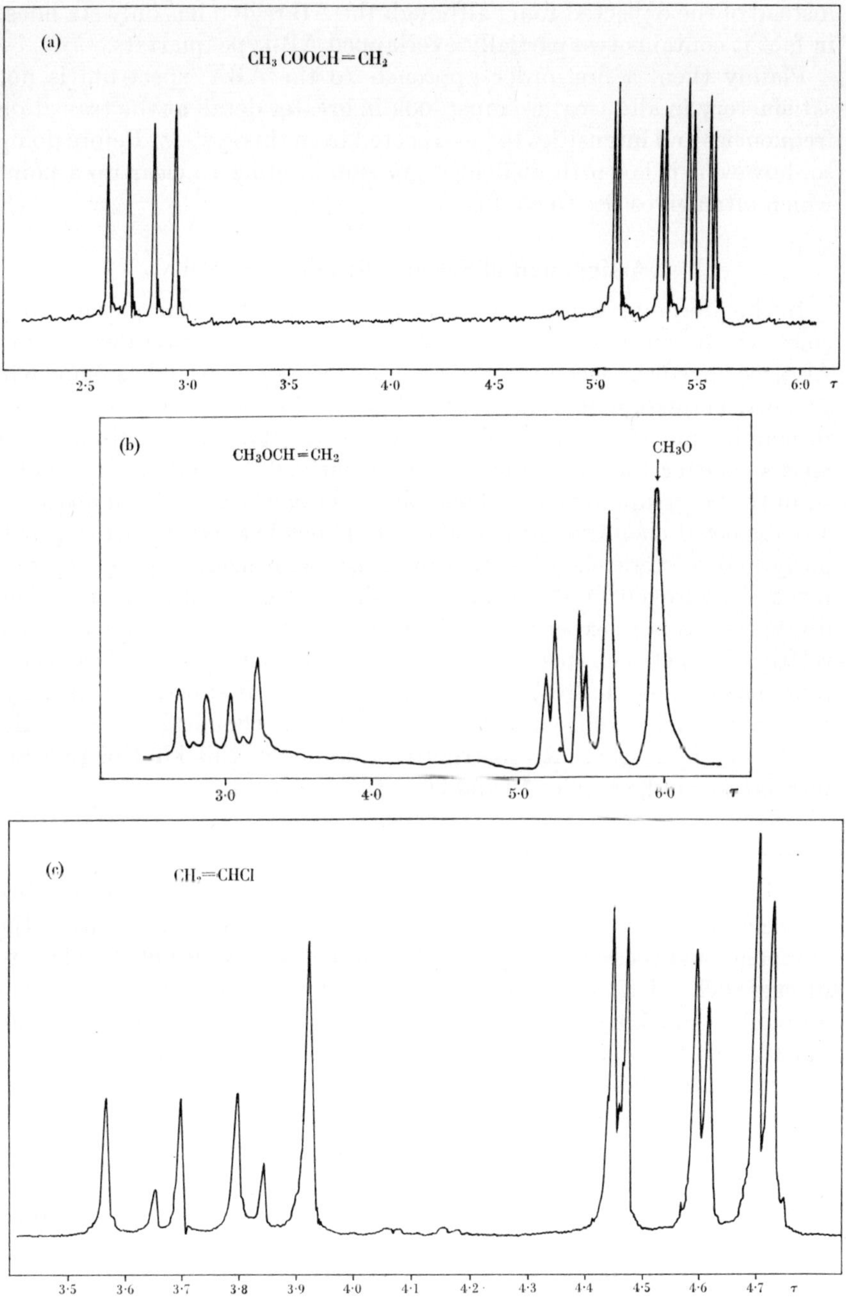

FIG. 6.4. (a) part of the 60 Mc/s spectrum of pure vinyl acetate, $CH_3CO.O.CH{=}CH_2$ (the resonance of the methyl nuclei occurs at 7·98τ); (b) the 40 Mc/s spectrum of vinyl methyl ether, $CH_3.O.CH{=}CH_2$, (10% v/v in carbon tetrachloride) and (c) the 60 Mc/s spectrum of pure vinyl chloride, $CH_2{=}CH.Cl$.

instead of the expected four; although the AB region has only six lines, in fact it contains two partially overlapped AB-type quartets.

Plainly then, a first-order approach to the ABX spectrum is not satisfactory in all cases; we must look in greater detail at the transition frequencies and intensities to be expected from this system. Before doing so, however, it is worth a slight digression in order to clear up a point which often gives rise to confusion.

Assignment of Spectral Groups to Nuclei

In the preceding sections we have mentioned some examples of molecules which can give rise to an ABX spectrum e.g. vinyl derivatives. Also, we have briefly commented on ABX spectra such as those shown in Fig. 6.4. The assignment of AB and X groups in the spectrum to the corresponding nuclei in the molecule however, is not always an obvious nor an easy matter. Such assignments particularly, do *not* follow from analysis of the spectrum, but must be made either on the grounds of observed and expected chemical shifts and on couplings between nuclei in previously known stereochemical relationships, or, if necessary, by replacement of hydrogen by deuterium in order to see which resonance then disappears. It is an experimental matter to deduce the proper one-to-one relationship between spectral groups and molecular nuclei, and each case must be considered afresh; thus a change in the R substituent in vinyl molecules might well cause a reversal of the chemical shifts of the methylene hydrogens. An instructive example of this kind of reversal may be seen in Examples 19 and 20 of the problems.

Complete ABX Spectra

It is a comparatively straightforward matter to derive the line frequencies and intensities expected from an ABX system in terms of the symbolic quantities ν_A, ν_B, ν_X, J_{AX}, J_{BX} and J_{AB}. This is not the place to discuss such a derivation, but the interested reader will find an excellent account in pp. 130–138 of "High Resolution Nuclear Magnetic Resonance".[1] We reproduce here the table of frequencies and intensities which these authors obtained (Table 6.1). For convenience certain abbreviations are used; these are defined by the equations

$$\nu_{AB} = \tfrac{1}{2}(\nu_A + \nu_B) \qquad (6.1)$$

$$\left.\begin{aligned} D_+ \sin 2\varphi_+ &= \tfrac{1}{2}J_{AB} \\ D_+ \cos 2\phi_+ &= \tfrac{1}{2}(\nu_A - \nu_B) + (J_{AX} - J_{BX})/4 \end{aligned}\right\} \qquad (6.2)$$

$$\left.\begin{aligned} D_- \sin 2\varphi_- &= \tfrac{1}{2}J_{AB} \\ D_- \cos 2\phi_- &= \tfrac{1}{2}(\nu_A - \nu_B) - (J_{AX} - J_{BX})/4 \end{aligned}\right\} \qquad (6.3)$$

By squaring and adding the pairs of equations (6.2) and (6.3), explicit expressions for D_+ and D_- can be obtained:

$$D_{\pm} = \tfrac{1}{2}\{[(\nu_A - \nu_B) \pm \tfrac{1}{2}(J_{AX} - J_{BX})]^2 + J_{AB}^2\}^{1/2} \quad (6.4)$$

We must note particularly the following

(i) D_+ and D_- are always positive, and $D_+ \geqslant D_- \geqslant 0$

(ii) $180^9 \geqslant \varphi_+, \varphi_- \geqslant 0^0$.

TABLE 6.1

Frequencies and intensities of ABX spectra

Line No.	Assignment	Energy (i.e. frequency)	Rel. intensity
1	B	$\nu_{AB} + \frac{1}{4}(-2J_{AB} - J_{AX} - J_{BX}) - D_-$	$1 - \sin 2\varphi_-$
2	B	$\nu_{AB} + \frac{1}{4}(-2J_{AB} + J_{AX} + J_{BX}) - D_+$	$1 - \sin 2\varphi_+$
3	B	$\nu_{AB} + \frac{1}{4}(+2J_{AB} - J_{AX} - J_{BX}) - D_-$	$1 + \sin 2\varphi_-$
4	B	$\nu_{AB} + \frac{1}{4}(+2J_{AB} + J_{AX} + J_{BX}) - D_+$	$1 + \sin 2\varphi_+$
5	A	$\nu_{AB} + \frac{1}{4}(-2J_{AB} - J_{AX} - J_{BX}) + D_-$	$1 + \sin 2\varphi_-$
6	A	$\nu_{AB} + \frac{1}{4}(-2J_{AB} + J_{AX} + J_{BX}) + D_+$	$1 + \sin 2\varphi_+$
7	A	$\nu_{AB} + \frac{1}{4}(+2J_{AB} - J_{AX} - J_{BX}) + D_-$	$1 - \sin 2\varphi_-$
8	A	$\nu_{AB} + \frac{1}{4}(+2J_{AB} + J_{AX} + J_{BX}) + D_+$	$1 - \sin 2\varphi_+$
9	X	$\nu_X - \frac{1}{2}(J_{AX} + J_{BX})$	1
10	X	$\nu_X + D_+ - D_-$	$\cos^2(\phi_+ - \phi_-)$
11	X	$\nu_X - D_+ + D_-$	$\cos^2(\phi_+ - \phi_-)$
12	X	$\nu_X + \frac{1}{2}(J_{AX} + J_{BX})$	1
14	Comb. (X)	$\nu_X - (D_+ + D_-)$	$\sin^2(\phi_+ - \phi_-)$
15	Comb. (X)	$\nu_X + (D_+ + D_-)$	$\sin^2(\phi_+ - \phi_-)$

Note: the line numbering follows Pople, Schneider and Bernstein[1], p. 134. Line 13 is here omitted since it always has zero intensity.

Characteristics of ABX Spectra

Several conclusions relevant to the discussion may be drawn directly from the Table.

1. The X region

(1) This may consist of up to six lines; two of which (Nos. 14 and 15) are combination lines (cf. Chapter 5, p. 74) but since, if they are intense enough to be observed, they always appear in the X region, they may be classified as X lines. Also we must note that

$$\begin{aligned}(\text{Intensity of line } 10) + (\text{Intensity of line } 14) &= \cos^2(\phi_+ - \phi_-) + \sin^2(\phi_+ - \phi_-)\\ &= 1\\ &= (\text{Intensity of line } 11)\\ &\quad + (\text{Intensity of line } 15).\end{aligned}$$

Thus if lines 14 and 15 are observable (i.e. if $\sin^2(\phi_+ - \phi_-)$ is much greater than zero), then their intensity is taken from lines 10 and 11. Lines 9 and 12 always show the same intensity, however.

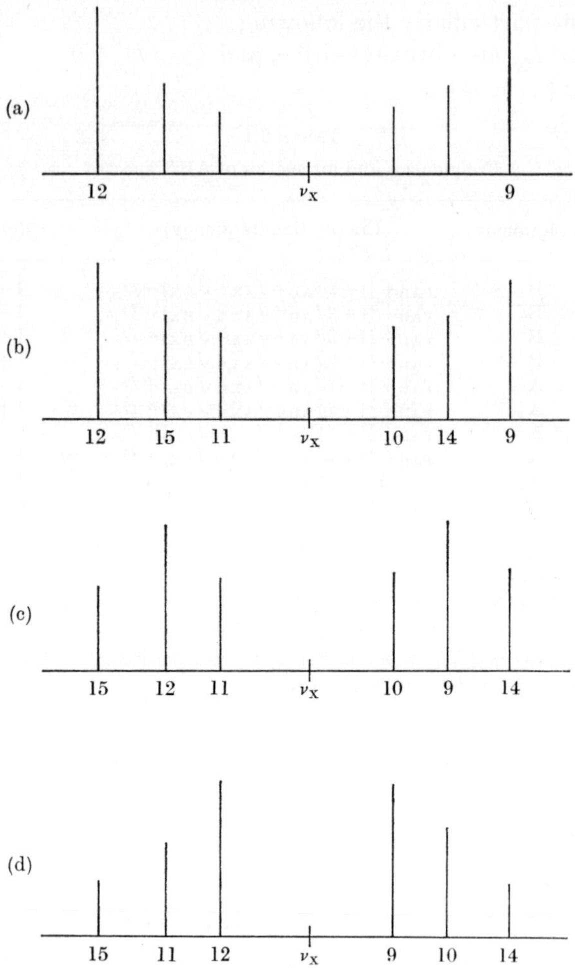

FIG. 6.5.

(2) The X region is symmetrical about ν_X and the value of this therefore can be read directly as the midpoint of the X part of the spectrum. This is a sextet if lines 14 and 15 are visible and a quartet if they are not. Consider the two cases in more detail separately.

(i) Six line X spectrum. In this case the region may have the appearance of Fig. 6.5 (a), and we can immediately assign lines 9 and 12 since these are the two strongest. From Table 6.1 we see that their separation is

$$\mathrm{Sep}^{\mathrm{n}}_{9-12} = \text{Line } 12 - \text{Line } 9 = \nu_X + \tfrac{1}{2}(J_{AX}+J_{BX}) - [\nu_X - \tfrac{1}{2}(J_{AX}+J_{BX})]$$
$$= J_{AX}+J_{BX}$$

Strictly speaking, however, there is no way of telling which is line 9 and which line 12, so we must write

$$\mathrm{Sep}^{\mathrm{n}}_{9-12} = |J_{AX}+J_{BX}| \tag{6.5}$$

where the modulus bars, $|\ldots.|$, indicate that we measure the *magnitude but not the sign* of the quantity enclosed.
Similarly from Table 6.1,

$$\mathrm{Sep}^{\mathrm{n}}_{10-11} = 2(D_+ - D_-) \tag{6.6}$$

$$\mathrm{Sep}^{\mathrm{n}}_{14-15} = 2(D_+ + D_-) \tag{6.7}$$

and, since D_+ and D_- are always positive, $\mathrm{Sep}^{\mathrm{n}}_{14-15} > \mathrm{Sep}^{\mathrm{n}}_{10-11}$, and lines 14 and 15 always lie outside lines 10 and 11. Thus we can complete the assignment of all the X lines as in Fig. 6.5 (b).

The six-line X spectrum may also have the appearance of sections (c) and (d) in Fig. 6.5. In these cases the line assignments must be made as shown, following the rules given above.

(ii) Four line X spectrum. All lines must have the same intensity, and we cannot know *a priori* which pair to assign to 9 and 12 and which to 10 and 11 (or to 14 and 15). We shall return to this problem with a worked example after dealing with the AB region. The chemical shift ν_X can still be read directly, however, since the four-line X spectrum is symmetrical.

2. The AB region

(3) From the Table we see that four of the A and B lines involve D_+ only, viz. 2, 4, 6 and 8, whilst four involve D_- only, viz. 1, 3, 5 and 7. These make up the two symmetrical AB-type quartets which we may expect to find in this region, and in general they will differ in their intensity distribution—the intensity of the 2, 4, 6, 8 quartet depends upon the value of ϕ_+, while that of the 1, 3, 5, 7 depends on ϕ_-. It may be shown easily from the table moreover that the mid-points of the two quartets are separated by $\tfrac{1}{2}|J_{AX}+J_{BX}|$; thus we may immediately check that the two quartets have been chosen correctly by comparing their separation with that between lines 9 and 12 in the X region (cf. Fig. 6.6).

Note that considerable overlapping and interleaving of lines 2, 4, 5 and

7 can occur for some values of $|J_{AX}+J_{BX}|$; this does not normally interfere with the analysis, however.

(4) $$\text{Sep}^{n}_{1-3} = \text{Sep}^{n}_{2-4} = \text{Sep}^{n}_{5-7} = \text{Sep}^{n}_{6-8} = |J_{AB}| \qquad (6.8)$$

and we can, therefore, read $|J_{AB}|$ directly from the spectrum.

(5) Further

$$\left.\begin{aligned}\text{Sep}^{n}_{1-5} &= \text{Sep}^{n}_{3-7} = 2D_{-}\\ \text{Sep}^{n}_{2-6} &= \text{Sep}^{n}_{4-8} = 2D_{+}\end{aligned}\right\} \qquad (6.9)$$

these values provide another cross-check between AB and X regions.

(6) Finally the value of ν_{AB} is given by the mean of all eight lines in the AB region, or alternatively as the mean of the midpoints of the two AB quartets.

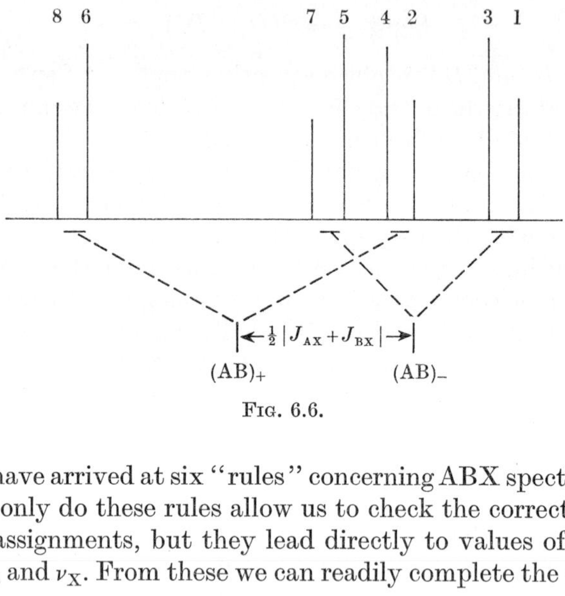

FIG. 6.6.

Thus we have arrived at six "rules" concerning ABX spectra, (1) to (6) above. Not only do these rules allow us to check the correctness of our initial line assignments, but they lead directly to values of $J_{AX}+J_{BX}$, D_{+}, D_{-}, J_{AB} and ν_X. From these we can readily complete the analysis, as shown below.

Worked Example I: X Region Contains Six Lines

For this we choose the spectrum shown at (b) in Fig. 6.4; the observed spectral frequencies and relative intensities are given in Table 6.2, and the spectrum is redrawn as a line spectrum in Fig. 6.7. To simplify the numbers, the frequencies are given with respect to the line at τ 5·65 taken as 0.00.

We begin the analysis by making the unique assignment of the X region—i.e. lines 9 and 12 are the two most intense, lines 14 and 15 lie outside lines 10 and 11. Rule (2) then allows the following measurements to be made:

$$|J_{AX}+J_{BX}| = \mathrm{Sep}^{n}_{9-12} = 21{\cdot}06 \text{ c/s}$$
$$2(D_{+}-D_{-}) = \mathrm{Sep}^{n}_{10-11} = 6{\cdot}01 \text{ c/s}$$
$$2(D_{+}+D_{-}) = \mathrm{Sep}^{n}_{14-15} = 12{\cdot}10 \text{ c/s}$$

from which it follows that

$$D_{+} = 4{\cdot}53 \text{ c/s};\ D_{-} = 1{\cdot}52 \text{ c/s}$$

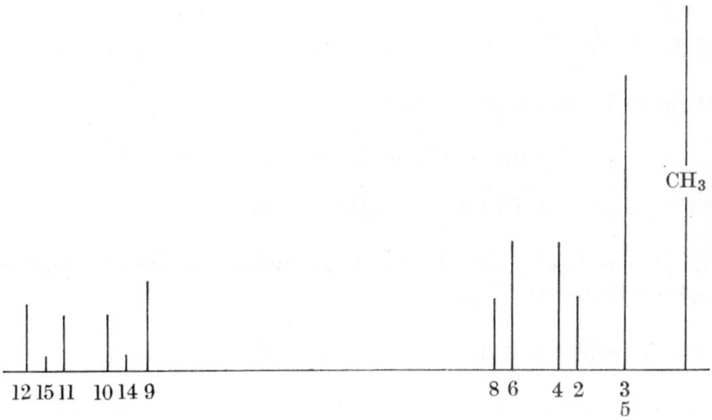

FIG. 6.7.

TABLE 6.2

Line No.	Assignment	Frequency c/s	Rel. int.	Designation
1	B	—	—	(AB)$_-$
3	B	0·00	3·32	
5	A	0·00		
7	A	—	—	
2	B	4·83	0·84	(AB)$_+$
4	B	7·10	1·32	
6	A	14·10	1·32	
8	A	16·25	1·00	
9	X	94·00	1·37	
14	X	98·55	0·20	
10	X	101·57	0·92	
11	X	107·58	0·83	
15	X	110·65	0·10	
12	X	115·06	0·89	

For the $(AB)_+$ and $(AB)_-$ quartets, the four lines to the low field end of the region make an obvious quartet which we may assign to lines 2, 4, 6 and 8. The intensity of the remaining line implies that it contains the whole of the other quartet which cannot be resolved because of overlap. Thus lines 1 and 7 are too weak to be observed, while lines 3 and 5 are superimposed. We may deduce the positions of lines 1 and 7, however, since we know (rule (4)) that the 1–3, 5–7, 2–4 and 6–8 separations are all equal to $|J_{AB}|$. The 2–4 and 6–8 separations are, respectively, 2·27 and 2·15 c/s, so averaging these we have

$$J_{AB} = 2{\cdot}21 \text{ c/s}$$

and line 1 is at $-2{\cdot}21$ c/s, line 7 at $+2{\cdot}21$ c/s.

On this basis, checks for the correctness of the assignment may be made.

(a) (Rule (3)) Separation between midpoints of quartets =

$$\tfrac{1}{2}[\text{Line } 4+\text{Line } 6)-(\text{Line } 3+\text{Line } 5)]$$
$$= 10{\cdot}60 \text{ c/s} = \tfrac{1}{2}|J_{AX}+J_{BX}|$$

From this $|J_{AX}+J_{BX}| = 21{\cdot}20$ c/s in reasonably good agreement with the value derived from the X region.

(b) (Rule (5)) $\text{Sep}^{n}_{2-6} = 9{\cdot}27$ c/s; $\text{Sep}^{n}_{4-8} = 9{\cdot}15$ c/s

These should be identical and equal to $2D_+$ (= 9·06 c/s from X region). The agreement is again satisfactory.

Also $$\text{Sep}^{n}_{1-5} = \text{Sep}^{n}_{3-7} = 2{\cdot}21 \text{ c/s}$$

This should be identical with $2D_-$ (= 3·04 c/s from X); here the agreement is poor, but we are handicapped by not knowing the exact positions of lines 3 and 5 under the rather broad band contour.

The line assignment of Fig. 6.7 is a reasonable one and we may proceed with the calculation using the following averaged values:

$$D_+ = 4{\cdot}6 \text{ c/s} \qquad D_- = 1{\cdot}5 \text{ c/s}$$
$$|J_{AB}| = 2{\cdot}2 \text{ c/s} \quad |J_{AX}+J_{BX}| = 21{\cdot}1 \text{ c/s}$$

From equations (6.2)

$$D_+ \sin 2\phi_+ = \tfrac{1}{2}J_{AB}, \text{ and therefore } \sin 2\phi_+ = 1{\cdot}1/4{\cdot}6 = 0{\cdot}249$$

and $$2\phi_+ = 14°24' \quad (\text{hence } \cos 2\phi_+ = 0{\cdot}969)$$

or $$2\phi_+ = 165°36' \; (\text{hence } \cos 2\phi_+ = -0{\cdot}969)$$

Note carefully that two angles ($2\phi_+$) are consistent with $\sin 2\phi_- = 0{\cdot}249$. Similarly, $\sin 2\phi_- = 1{\cdot}1/1{\cdot}5 = 0{\cdot}734$

and $$2\phi_- = 47°12' \quad (\cos 2\phi_- = 0{\cdot}679)$$

or $$2\phi_- = 132°48' \quad (\cos 2\phi_- = -0{\cdot}679)$$

It is usually possible at this point to determine which combination of these angles is the correct one on the basis of the intensity distribution in the X region. Table 6.3 shows the reasoning.

TABLE 6.3

	$2\phi_+$	$2\phi_-$	$(\phi_+ - \phi_-)$	$\sin^2(\phi_+ - \phi_-)$ (i.e. intensity of lines 14, 15)	$\cos^2(\phi_+ - \phi_-)$ (i.e. intensity of lines 10, 11)
(a)	14° 24′	47° 12′	−(16° 24′)	0·08	0·92
(b)	14° 24′	132° 48′	−(59° 12′)	0·74	0·26
(c)	165° 36′	47° 12′	59° 12′	0·74	0·26
(d)	165° 36′	132° 48′	16° 24′	0·08	0·92

This indicates that (a) and (d) are consistent with the observed intensities of lines 10, 11, 14 and 15 (cf. Table 6.2). while (b) and (c) are not. In fact sets (a) and (d) are identical except for a trivial reversal of sign, so we may take as our angles:

$$2\phi_+ = 14°24'; \cos 2\phi_+ = 0{\cdot}969$$

$$2\phi_- = 47°12'; \cos 2\phi_- = 0{\cdot}679$$

From equations (6.2) and (6.3) it then follows that

$$D_+ \cos 2\phi_+ = 4{\cdot}6 \times 0{\cdot}969 = \tfrac{1}{2}(\nu_A - \nu_B) + (J_{AX} - J_{BX})/4$$

and $$D_- \cos 2\phi_- = 1{\cdot}5 \times 0{\cdot}679 = \tfrac{1}{2}(\nu_A - \nu_B) - (J_{AX} - J_{BX})/4$$

and hence: $$\nu_A - \nu_B = 5{\cdot}48 \text{ c/s}; J_{AX} - J_{BX} = 6{\cdot}88 \text{ c/s}$$

Since $|J_{AX} + J_{BX}| = 21{\cdot}1$ c/s, therefore $J_{AX} = 14{\cdot}0$, $J_{BX} = 7{\cdot}1$ c/s. Further the mean of all the AB lines is at 5·29 c/s, and thus $\nu_A + \nu_B = 2\nu_{AB} = 10{\cdot}58$ c/s, from which we derive

$$\left.\begin{aligned}\nu_A &= 8{\cdot}03\\ \nu_B &= 2{\cdot}55\end{aligned}\right\} \text{c/s from line 3,5 at } 0{\cdot}00$$

As lines 3 and 5 fall at τ 5·65, we have

$$\tau_A = 5{\cdot}65 - 8{\cdot}03/40 = 5{\cdot}45; \qquad J_{AX} = 14{\cdot}0 \text{ c/s}$$
$$\tau_B = 5{\cdot}65 - 2{\cdot}55/40 = 5{\cdot}59; \qquad J_{BX} = 7{\cdot}1 \text{ c/s}$$
$$\tau_X \text{ (directly from spectrum)} = 3{\cdot}04; \; J_{AB} = 2{\cdot}2 \text{ c/s}$$

The Relative Signs of Coupling Constants

Thus far we have not considered the possibility that coupling constants may be either positive or negative. In the case of ABX spectra, Table 6.1 shows that a change in the sign of J_{AB} will not alter the appearance of the spectrum (indeed, it is for just this reason that we wrote $|J_{AB}|$ in equation (6.8)). However, J_{AX} and J_{BX} occur in the Table always in combination as $(J_{AX}+J_{BX})$, and alteration of the *sign* of one of these couplings will result in a change in the *magnitude* of their sum; we would, therefore, expect the spectrum to change. Thus we may hope to determine the *relative* signs of J_{AX} and J_{BX}, but not, of course, their absolute signs nor, indeed, the sign of their sum.

In the above worked example we *measured* the value of $\Sigma J = J_{AX}+J_{BX}$ from the spectrum as Sep^{n}_{9-12}, while we were able to *calculate*

$$\Delta J = J_{AX} - J_{BX}$$

from our D_+ and D_- values. If, as above, we assume J_{AX} and J_{BX} to be of the same sign, then we have

$$\Sigma J \text{ (obs.)} = 21{\cdot}1 = J_{AX}+J_{BX}$$
$$\Delta J \text{ (calc.)} = 6{\cdot}9 = J_{AX}-J_{BX}$$

from which we find $J_{AX} = 14{\cdot}0$, $J_{BX} = 7{\cdot}1$ c/s.
If, on the other hand, we had started with the assumption that J_{AX} is positive and J_{BX} negative, then we would write

$$\Sigma J \text{ (obs.)} = 21{\cdot}1 = J_{AX}-J_{BX}$$
$$\Delta J \text{ (calc.)} = 6{\cdot}9 = J_{AX}+J_{BX}$$

and therefore $J_{AX} = 14{\cdot}0$, $J_{BX} = -7{\cdot}1$ c/s.

Thus both sets of coupling constants (*signs attached*) are consistent with the frequency data. The only way we may resolve the ambiguity is to use the *intensity* data also—as was done with the similar ambiguity over the sets of angles $2\phi_+$ and $2\phi_-$. The interested student is recommended to calculate the complete ABX spectrum to be expected from the data given above with J_{AX} and J_{BX} (a) the same sign, and (b) of opposite sign. The calculation is rapidly accomplished if values for D_+ and D_- are first obtained by replacing $(\nu_A - \nu_B)$, $(J_{AX} - J_{BX})$ and J_{AB}

in equation (6.4), when $\sin 2\phi_+$ and $\sin 2\phi_-$ can be calculated from equations (6.2) and (6.3), and the frequencies and intensities of the spectral lines written down from Table 6.1. It will be found that only the case with J_{AX} and J_{BX} of the same sign gives intensity data consistent with that observed (Table 6.2).

Worked Example II: X Region Contains only Four Lines

An ABX spectrum of this type (Example 17 of the problem) is shown in Fig. 6.8. In this case we cannot say immediately whether it is the inner or the outer pair of lines in the X region which must be numbered 9 and

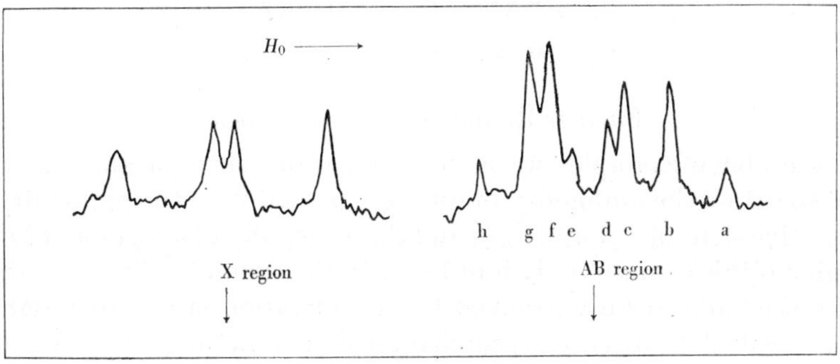

FIG. 6.8.

12, but the assignments in the AB region are more obvious after a few moments' consideration. If we label each line a, b, c, ... in sequence from high field downwards, as in the figure, we see that possible AB-type quartets are shown by lines a, b, c and e and by lines d, f, g and h. An alternative assignment, feasible on the grounds of line separations, would be a, b, g, h and d, e, f, g. This assignment can, however, be rejected immediately from consideration of the line intensities, since the quartet d, e, f, g does not show the characteristic "weak, strong, strong, weak" sequence of an AB system. Thus we have the following assignment.

Label	a	b	c	e	d	f	g	h	
Line No.	2	4	6	8	1	3	5	7	
Frequency	0	5·0	8·5	13·5	10	15	16·5	21·5	c/s

From which we can calculate immediately

Rule (4)	$\|J_{AB}\|$	=	5·0 c/s
Rule (3)	$\|J_{AX}+J_{BX}\|$	=	18·0 c/s
Rule (5)	D_+	=	4·3 c/s
Rule (5)	D_-	=	3·3 c/s

From these we may proceed to the assignment of the X region since we know (Rule (2)) that $\text{sep}^{\text{n}}_{9-12} = |J_{\text{AX}}| + J_{\text{BX}} = 18$ c/s, while $\text{sep}^{\text{n}}_{10-11} = 2(D_+ - D_-) = 2$ c/s. The outer pair of lines in the X region are separated by 18·2 c/s and must, therefore, be assigned to 9 and 12, while the inner pair are separated by 1·7 c/s, which is in reasonable agreement with their assignment to lines 10 and 11. Lines 14 and 15, which should be separated by $2(D_+ + D_-) = 15{\cdot}2$ c/s are plainly too weak to be observed.

Once an assignment has been made and checked in this way, the calculation can proceed exactly along the lines of Example 1. It is left to the student to show that the parameters derived are (see pp. 262–268)

$$\tau_{\text{A}} = 7{\cdot}58; \quad J_{\text{AX}} = 7{\cdot}65 \text{ c/s}$$
$$\tau_{\text{B}} = 7{\cdot}67; \quad J_{\text{BX}} = 10{\cdot}35 \text{ c/s}$$
$$\tau_{\text{X}} = 6{\cdot}72; \quad J_{\text{AB}} = 5{\cdot}0 \text{ c/s}$$

Further Remarks on Ambiguities

In the above examples we have met various sorts of ambiguity in ABX spectra—the ambiguity of angles, $2\phi_+$ and $2\phi_-$; the ambiguity of the relative signs of J_{AX} and J_{BX}; and the ambiguity of assignment in the X region if this contains only four lines. In the examples discussed these ambiguities were readily resolved by consideration of intensity data—either by calculation of the expected intensities as in the case of the angles or relative signs, or merely by inspection in the case of the alternative assignment in Example II. It may be, however, that one or more of these ambiguities cannot be resolved in this way for a particular spectrum. We may not be able to decide, because of experimental error or because of poor resolution of overlapping lines, which of two calculated spectra best fits the observed data. When this happens several alternative steps may be taken. Briefly these are as follows.

(a) To redetermine the spectrum in solution or, where a solvent has already been used, to change the solvent for one of markedly different polarity or dielectric constant. The effect is usually to alter the relative chemical shifts of the nuclei but not the J values. Analysis of the new spectrum might then lead to an unambiguous result or, if alternative results are still possible, only one of them may have J values sufficiently close to the original set to be acceptable.

(b) To use a spectrometer operating at a different frequency. Here each chemical shift is changed by the same proportionate amount (a result not necessarily true of (a)), while again the coupling constants are not affected.

(c) To use double resonance techniques to remove each coupling in turn and thus simplify the spectrum. Alternatively one hydrogen atom

A, B or X may be replaced by deuterium when the spectrum should again simplify to AB and a coupling constant can be uniquely determined.

4. Guess at the correct values by analogy with similar molecules whose spectra are unambiguous.

Deceptively Simple Spectra[2]

It can sometimes happen that an ABX-type molecule gives rise to a spectrum containing fewer than the expected number of lines. These spectra look simple but are, in fact, no simpler to analyse, and it may even prove more difficult to extract the required information from them.

Two cases arise.

I. The X region shows a 1:2:1 triplet whilst the AB region shows the normal eight lines (Fig. 6.9)

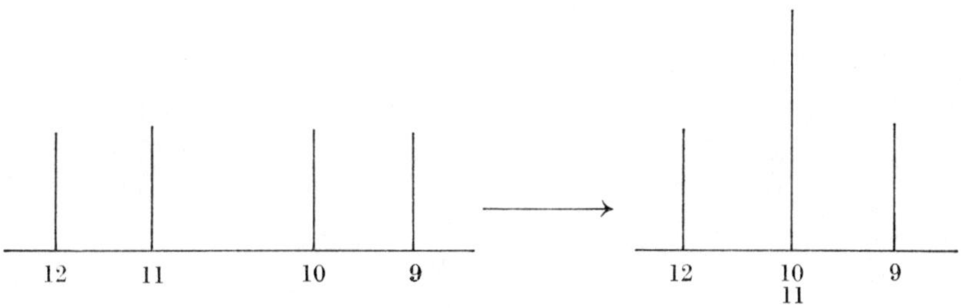

FIG. 6.9.

Collapse of the four line X spectrum (lines 10 and 11 coincident) will arise only if $D_+ = D_-$ (Table 6.1). This may happen in one of two ways

(i) If $J_{AX} = J_{BX}$ $\therefore D_+ = \frac{1}{2}\{(\nu_A - \nu_B)^2 + J_{AB}^2\}^{\frac{1}{2}} = D_-$

or (ii) if $\nu_A = \nu_B$ $\therefore D_+ = \frac{1}{2}\{(J_{AX} - J_{BX})^2 + J_{AB}^2\}^{\frac{1}{2}} = D_-$

Therefore an X triplet arises *either* because the AX and BX couplings are equal *or* because the A and B chemical shifts are equal. In both cases the AB region shows the usual eight lines and a full analysis is possible. The ambiguity as to whether (i) or (ii) is true in a given case can, in principle, be resolved by a calculation of intensities and comparison with those observed.

Note that the system obtained in (ii) is described as AA′X.

II. The X region shows a 1:2:1 triplet, the AB region a 1:1 doublet (Fig. 6.10).

If, simultaneously, $J_{AX} = J_{BX}$ and $\nu_A = \nu_B$, then the spectrum takes on a very simple appearance. We can see immediately that:

$$D_+ = D_- = \frac{1}{2}\{J_{AB}^2\}^{\frac{1}{2}} = \frac{1}{2}J_{AB}$$

and hence $\sin 2\phi_+ = \sin 2\phi_- = 1$.

Not only does the X region collapse to a triplet, but also lines 1, 2, 7 and 8 in the AB region have zero intensity (since $1-\sin 2\phi_+=0$) *and* lines 3 and 5, and lines 4 and 6 coincide. The amount of information which can be obtained from this is very limited; the chemical shifts ν_X and $\nu_A(=\nu_B)$ are given by the centres of their respective multiplets, whilst the separation between lines gives $\frac{1}{2}|J_{AX}+J_{BX}|$. J_{AB} is not obtainable.

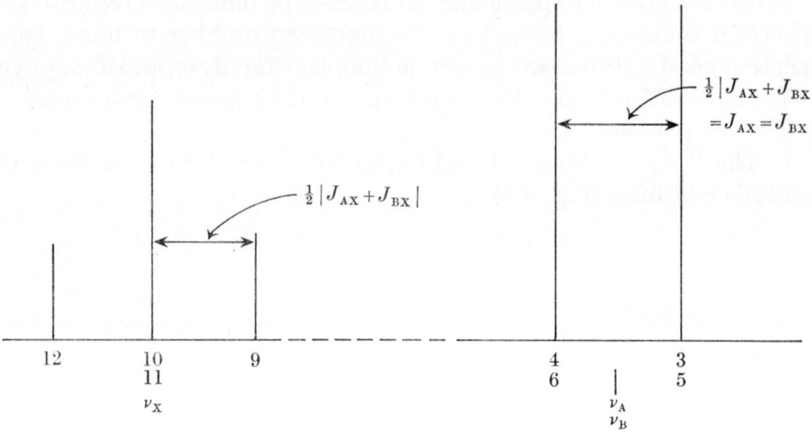

FIG. 6.10.

The conditions $\nu_A=\nu_B$ and $J_{AX}=J_{BX}$ are, of course, those for an A_2X system. It should be carefully noted, however, that a five-line spectrum similar to that of Fig. 6.10 will be obtained even if these conditions are not exactly satisfied, provided that $J_{AB}\gg\frac{1}{2}(J_{AX}-J_{BX})$ and $J_{AB}\gg\nu_A-\nu_B$, since then $D_+\rightarrow\frac{1}{2}J_{AB}$.

References

1. J. A. Pople, W. G. Schneider and H. J. Bernstein, "High Resolution Nuclear Magnetic Resonance". McGraw-Hill, New York (1959).
2. R. J. Abraham and H. J. Bernstein, *Canad. J. Chem.* **39**, 216 (1961).

CHAPTER 7

ANALYSIS OF MORE COMPLEX SPECTRA: FOUR AND MORE SPIN-COUPLED NUCLEI

E. O. Bishop

Four Spin Systems

1. The general ABCD case

The simple rules for predicting the appearance of first order spectra show that in the high-field limit there would be 32 fundamental lines, all of equal intensity, comprising four octets each associated with a spin-flip of one nucleus. The difficulty of exact analysis in a general ABCD case, however, is demonstrated by the mixing diagram of Fig. 7.1 from which

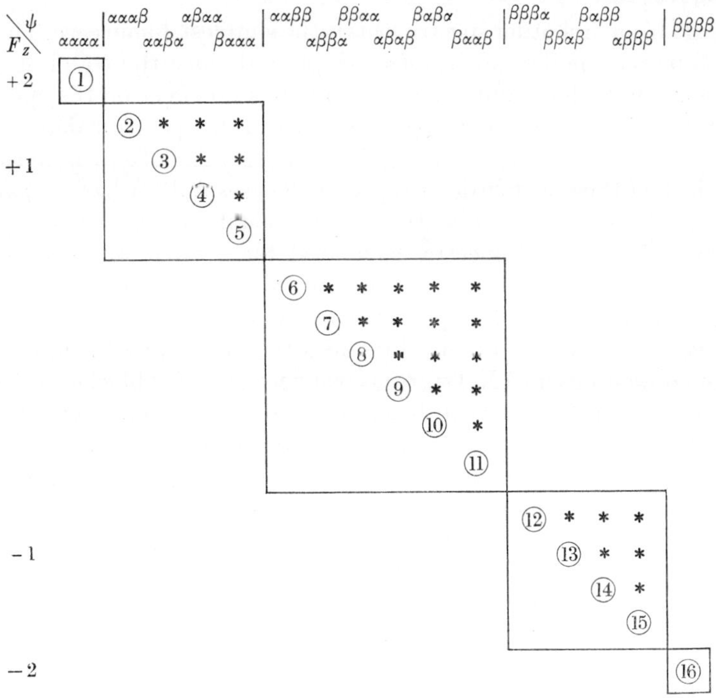

FIG. 7.1. Block mixing diagram for the general ABCD case.

it is seen that $2(4+24)=56$ transitions are allowed by the selection rule $\Delta F_z = \pm 1$. Of these, $56-32=24$ must be combination lines, and it is unlikely that many of these will be observed in a given spectrum unless the system is quite strongly coupled. The mixing problem is a formidable one—involving six of the first order states at a time for $F_z=0$, and two "4×4 blocks" for $F_z=+1, -1$ (cf. Chapter 5).

The remarks of Chapter 5 concerning weakly coupled ABC cases can be extended to ABCD and indeed to most nuclear spin systems. On the one hand, if the spectrum does not depart too drastically from the first order condition, then a reasonable approximation to the various values for ν and J can be obtained by a first order treatment. On the other hand an ABCD spectrum in which all coupling interactions are strong, (i.e. $|J_{ij}| \gtrsim |\nu_i - \nu_j|$) will be much too difficult to analyse without a computer, and will not be considered here.

There are, however, a number of four spin systems in which the spectrum and its analysis are simplified by the presence of *some* first order interactions (cf. ABX) and/or by symmetry of the spin system (cf. AB_2). These are now considered.

2. The AB_3 case

Apart from first order spectra, this is the simplest 4-spin case. There are only four parameters (ν_A, ν_B, J_{AB}, J_{BB}), and since the coupling J_{BB} is between completely equivalent nuclei it has no effect on the spectrum. Apart from a scaling factor, the appearance of the spectrum depends only on the ratio $|J_{AB}/\nu_A - \nu_B|$; it is unaffected by the sign of J_{AB}, as in the AB_2 case. In fact these conclusions apply to *all* cases A_pB_q, where p, q are any integers. A suspected AB_3 spectrum may usually be picked out by comparison with a set of spectra calculated for a range of values of this ratio.[20a, 21] The spectrum contains only fourteen fundamental lines (6 assigned to A and 8 to B_3), plus two combination lines which are always very weak. By virtue of the three-fold rotation axis of symmetry, all mixing of first order (AX_3) states is restricted to 2×2 blocks, and all line positions and intensities can be given in simple algebraic form.[1] Figure 7.2 is taken from this reference, which discusses the proton resonance spectrum of methyl mercaptan as an example. Analysis is straightforward; for example the line separation $\nu_3 - \nu_5$ gives $|J_{AB}|$ directly.* A 1,1-disubstituted ethane $PQ\cdot CH—CH_3$ ($P=Q$ or $P\neq Q$) would conform to the case AB_3 or possibly AX_3, given fairly rapid rotation about the C—C bond (otherwise it would be ABCD).

The rules for first order spectra, given in Chapter 4, show that the limiting AX_3 spectrum would contain six lines only, comprising a 1:3:3:1

* See also p. 125.

quartet from the A proton, and a 1:1 doublet, of three times the total intensity, from B_3. It is readily seen that the top spectrum (weakly coupled AB_3) of Fig. 7.2 is approaching this limit. A familiar example which is even closer to AX_3 is the proton resonance spectrum of acetaldehyde, in which "second order" effects cannot usually be detected at all.

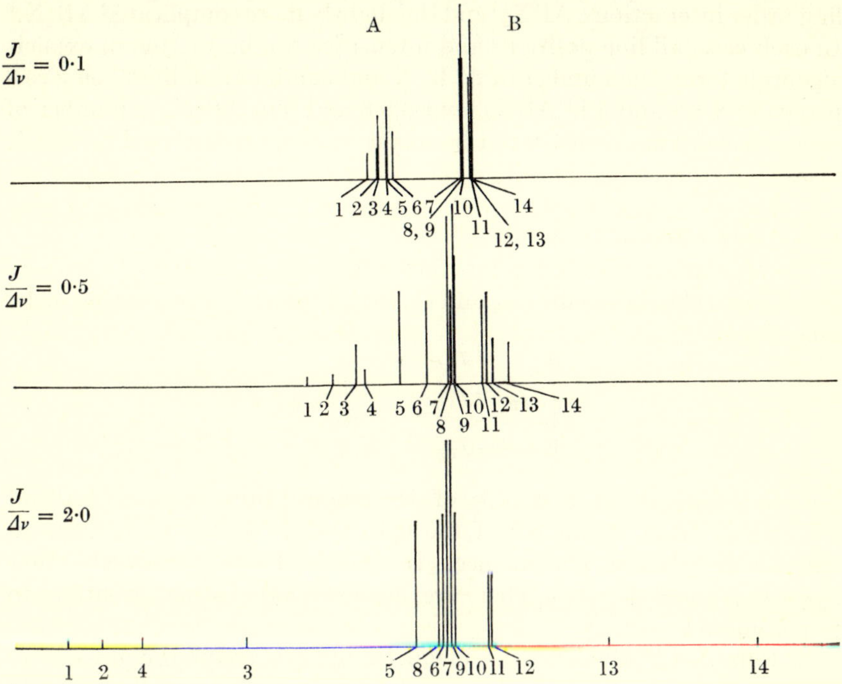

FIG. 7.2. Calculated AB_3 spectra for several values of $J/\Delta\nu = |J_{AB}/(\nu_A - \nu_B)|$ (from Ref. 1).

3. Some other four-spin cases

If all four chemical shifts are different and only one pair has a non-first order interaction, we have the case ABMX. Simple algebraic expressions can again be given for the energies and intensities of all 32 fundamental and 8 combination lines.[2a] This situation applies to the combined proton and fluorine resonance spectra of 1-fluoro-2,4-dinitrobenzene.[2b] If the chemical shift of M is moved towards X then, relabelling, we have the slightly more complicated case ABXY. Most of the transitions can be given in simple algebraic form, but there is now one 4×4 block.[3a] In special cases, all lines can be given explicitly.[3b] If J_{AB} and J_{XY} are large compared to $(\nu_A - \nu_B)$ and $(\nu_X - \nu_Y)$ respectively, then there will be a deceptively simple spectrum of two triplets resembling that of A_2X_2.

Where proton M of ABMX is displaced towards AB, we have the case ABCX, which will clearly be more complicated than ABC, and there are in fact four 3×3 blocks.[4a] 2-substituted pyridines generally fall into this category, since the remaining 2-proton (X) is displaced to low field relative to the others by its proximity to the ring nitrogen.[4b]

Two 4-spin systems are simplified both by symmetry and by virtue of first order interactions, ABX_2[5] and the slightly more complicated AB_2X.[6] In each case, all line positions and intensities can be written in explicit algebraic form, the number of A, B, X and combination lines being respectively 8,8,8 and 4 in ABX_2, and 8,8,8 and 6 in AB_2X. A number of symmetrically *m*-disubstituted benzenes have been analysed as AB_2X cases.[7]

4. Magnetic equivalence in pairs

The remaining possibility for four spin-coupled nuclei is that there may be two chemical shifts, ν_A and ν_B, each common to two of the nuclei thus:—

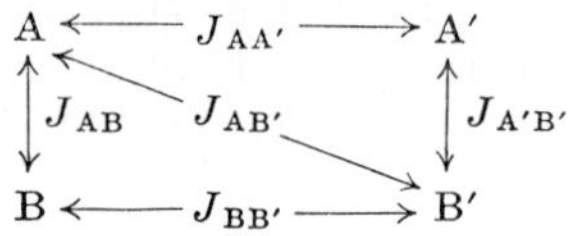

We will assume that the equality of the chemical shifts in pairs (AA′ and BB′) arises from symmetry of the spin system rather than from an improbable chance coincidence. It then follows necessarily that $J_{AB} = J_{A'B'}$ and $J_{AB'} = J_{A'B}$. There are, however, two distinct situations to be considered.

(i) Complete equivalence (i.e. magnetic and spin coupling equivalence) of the nuclei in pairs. Each of the nuclei at ν_A is then coupled to exactly the same extent to each of those at ν_B, i.e. $J_{AB} = J_{AB'}$. The system is denoted A_2B_2, or A_2X_2 in the first order limit. It is an example of the general case A_pB_q mentioned under AB_3, and the spectrum is independent of $J_{AA'}$ and $J_{BB'}$, being determined, apart from a scaling factor, entirely by the ratio $|J_{AB}/\nu_A - \nu_B|$.

(ii) Magnetic but not spin-coupling equivalence in pairs. Here, $J_{AB} \neq J_{AB'}$, and the system will be denoted AA′BB′, or AA′XX′ in the high field limit (vide ABB′, Chapter 5). This convention is not always adopted, and in many papers one will find (i) and (ii) both described as variations of A_2B_2. Case (ii) is clearly more complicated than (i) since, not only are J_{AB} and $J_{AB'}$ to be considered as separate parameters, but also the spectrum may be affected by $J_{AA'}$ and $J_{BB'}$.

The spectra of cases A_2X_2, A_2B_2, AA′XX′ and AA′BB′ have a common characteristic by which they are easily recognized viz. they are all *completely symmetrical* about the mean chemical shift. This property is shared only with the two spin AB case amongst commonly occurring proton spin systems, and the latter is easily distinguished by having only four lines. More confusion has arisen in the literature, however, from misinterpretation of these than of any other simple spin system, as situation (ii) may often give a deceptively simple spectrum resembling that of (i). Since (ii) is of fairly common occurrence, each case is considered in detail.

5. The A_2X_2 case

The simple splitting rule predicts that the A_2X_2 spectrum consists simply of two triplets, identical in appearance and with the components of each multiplet present in the intensity ratio 1:2:1, centred upon ν_A and ν_X (Fig. 7.3 (a)). Analysis to yield ν_A, ν_X and J_{AX}, the component spacing, is first order. 1,2-Disubstituted ethanes PCH_2CH_2Q and flexible ring compounds containing the entity $-CH_2—CH_2-$ often approximate to the A_2X_2 case, with some distortion of the intensity pattern, provided the substituents P and Q are sufficiently different in character, and provided there is reasonable freedom of rotation about the central C—C bond. An example occurs in the problems at the end of this book (Problem No. 7). Rigorous A_2X_2 cases are difficult to find; the combined ^1H and ^{19}F spectra of 1,1-difluoroallene would satisfy the necessary conditions [Ref. 1, Chapter 5].

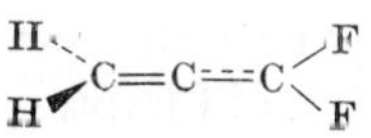

6. The A_2B_2 case

All lines in the A_2X_2 spectrum are composed of two or more coincident transitions. On decreasing the chemical shift difference to the A_2B_2 condition, the spectrum splits up into 14 lines as in Fig. 7.6 (i) (f). All non-first order effects upon the A transitions must, by symmetry, have an equivalent effect upon the B transitions, so that the whole spectrum remains symmetrical. Only one half is shown in Fig. 7.6 (i) (f), in which the relation between the three groups of lines and the first order triplet is evident. There remains one case of two coincident transitions per half spectrum, forming the strongest observed line unless the system is very strongly coupled. This occurs at 15 c/s in the figure, and gives immediately the chemical shift of the corresponding pair of nuclei. There are no combination lines. Since the appearance depends only on the ratio $|J_{AB}/(\nu_A - \nu_B)|$, the conformity or otherwise of a given experimental spectrum to that of A_2B_2 may be checked easily by comparing against a series of spectra calculated for a range of values of this ratio[8b, 20b].

Analysis to obtain ν_A, ν_B, and $|J_{AB}|$ is fairly straightforward; using the high symmetry of the system, most transitions can be expressed in simple algebraic form with the exception of those associated with one instance of 3×3 mixing. It is very difficult to find valid examples of the A_2B_2 case, *sensu stricto*. The spectra of some 1,2-disubstituted ethanes P—CH_2—CH_2—Q closely resemble this, but the resemblance may be misleading (see below).

7. The AA′XX′ case

If $J_{AX} \neq J_{AX'}$, then there arises an interesting situation. The rules of Chapter 4 governing the appearance of a first order spectrum fail to predict unambiguously what the AA′XX′ spectrum will look like. We might suppose (a) that the spectrum would consist of two triplets, as A_2X_2, but with the components this time spaced by $\frac{1}{2}(J_{AX}+J_{AX'})$. However, this is incorrect. Again, one might envisage (b) a doublet spaced by J_{AX} at each chemical shift, with each component further split into a doublet by $J_{AX'}$. This is correct, however, only if $J_{AA'} = J_{XX'} = 0$. In practice, since nuclei A,A′ (and likewise X,X′) are *not* equivalent in a spin-coupling sense, the coupling constants $J_{AA'}$ and $J_{XX'}$ serve to complicate the spectrum. This is the first time we have encountered a high field limiting case which does *not* give a "first order" spectrum as the term is usually understood (see beginning of Chapter 5). Thus, some mixing of basic functions is involved even when these are chosen to make best use of the plane of symmetry of the spin system (c.f. AB_2 in Chapter 5, and see under A_2X_2 in Ref. 1. of Chapter 5 for a full discussion). Consequently, the parameters must be calculated indirectly (and with some ambiguities, see below). This situation always arises if there is magnetic equivalence without spin-coupling equivalence (cf. AXX′ in Chapter 5).

The AA′XX′ and AA′BB′ cases can be described most easily in terms of four composite parameters in place of the straightforward J values, thus:

$$N = J_{AX}+J_{AX'} \qquad K = J_{AA'}+J_{XX'}$$

$$L = J_{AX}-J_{AX'} \qquad M = J_{AA'}-J_{XX'}$$

The effect of successive introduction of these quantities is shown in Fig. 7.3.[8] Only one half of the spectrum is shown. When $L = 0$ we have the case A_2X_2, with a triplet half-spectrum in which the values of K and M are immaterial (Fig. 7.3 (a)). As mentioned above, if L is finite but K and M are zero, we have a quartet half-spectrum (Fig. 7.3 (b)), but as soon as K is finite the central doublet is split into six lines (Fig. 7.3 (c)), with further splitting if $M \neq 0$, (Fig. 7.3 (d)). Each half-spectrum is symmetrical about the corresponding chemical shift, and usually has ten lines.

The strongest doublet (each component of which still consists of two coincident transitions) gives N directly, and the weaker lines are grouped as two symmetrical quartets giving K and M directly. The weaker components of these quartets may lie outside the strong doublet.

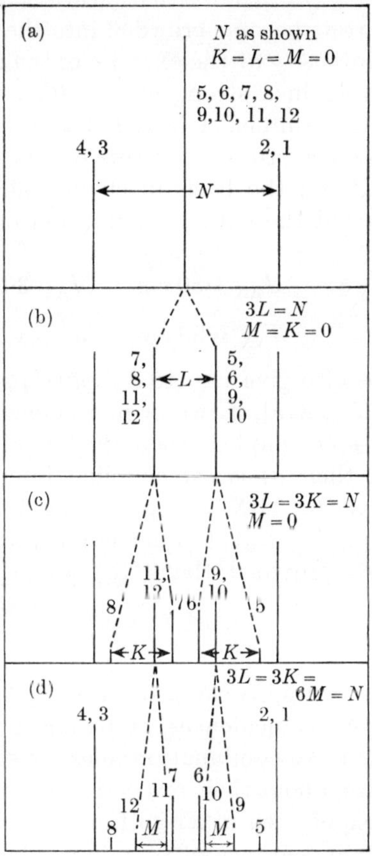

FIG. 7.3. Half-spectra for the cases A_2X_2 and $AA'XX'$ transitions 5 to 12 are split under the influence of K, L, and M. Transitions 1 to 4 are not affected by this but are split by N. (From Ref. 8.)

Figure 7.7 (c) illustrates such a case in which the outer lines are in fact too weak to be observed. The lack of resemblance to A_2X_2 is striking. Only the spectrum of the CH_2 groups is shown. All line positions and relative intensities can be given in explicit algebraic form, and the value of L may be found indirectly. There are, however, certain ambiguities;

it is not possible to tell which is the "*K*-quartet" and which the "*M*-quartet", so the relative signs of $J_{AA'}$ and $J_{XX'}$ cannot be found. Furthermore, in view of the symmetry of the complete spectrum, one cannot tell $J_{AA'}$ from $J_{XX'}$, or J_{AX} from $J_{AX'}$.

If one progressively increases the values of K and M relative to L beyond the stages of Fig. 7.3, a new situation arises. The inner transitions of the K and M quartets become crowded into the centre of each half-spectrum, and gain intensity at the expense of the outer members. The effect is to build up again a strong central line, so that the pattern becomes a "deceptively simple" triplet liable to be misinterpreted as A_2X_2 with $J_{AX}=J_{AX'}$. The proton spectrum of furan is of this type, and re-appraisal in 1961,[3a] on the basis of slight splitting of the central "triplet" line, corrected the earlier assumption of equal 2,3 and 2,4 coupling.

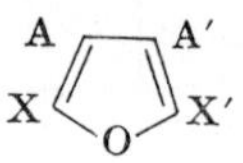

J_{AX} 1·80 c/s $J_{AA'}$ 3·53 c/s

$J_{AX'}$ 0·80 c/s $J_{XX'}$ 1·55 c/s

Other spin systems also give rise to *deceptively simple spectra* under similar conditions. In general, if two or more nonequivalent nuclei are themselves strongly spin-coupled, then they tend to function as an equivalent group in their *apparent* coupling interactions with other nuclei. This was met in the ABX case in Chapter 6., where the X region approximates to a triplet instead of a quartet (i.e. X *appears* to be equally coupled to A and B) provided that $J_{AB} \gg |\nu_A - \nu_B|$ and also $J_{AB} \gg \frac{1}{2} |J_{AX} - J_{BX}|$.

8. The AA′BB′ case

On moving the chemical shifts closer together to give the AA′BB′ case, the remaining transition coincidences are removed, giving a total of 12 lines per half-spectrum. No combination lines appear. The half-spectra cease to be symmetrical about their respective chemical shifts, but they are still mirror images of each other so that the spectrum as a whole retains its distinctive appearance of symmetry about the mean of the chemical shifts. The energies and intensities of half the lines can be expressed in simple algebraic form, but the remainder involve a 4×4 block of mixing of first order states. All parameters, including the relative signs of the coupling constants, can be found—the greatest difficulty being the initial one of numbering the lines correctly. For this purpose, it helps to have a series of calculated spectra for comparison, but the number of variables (six) is too great to permit easy coverage of the total range of spectral patterns which may correspond to an AA′BB′ system. The effects of varying N, L, K and M one at a time are shown in some calculated series of spectra in Ref. 8. In practice, many examples fall into some special category, in which the lines form a distinctive pattern and can be assigned in a predictable manner. Three such categories are now considered.

1. One pair of nuclei (A,A′) is much more strongly coupled than the other, $|J_{AA'}| \gg |J_{BB'}|$, and at the same time $|J_{AB}| \gg |J_{AB'}|$. This situation is encountered, for example, in *ortho*-disubstituted benzenes and 1,4-disubstituted butadienes:

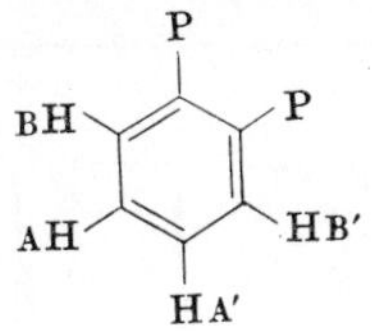

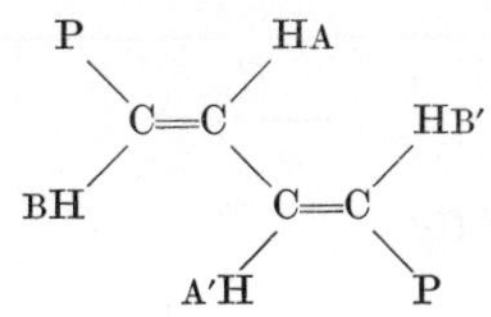

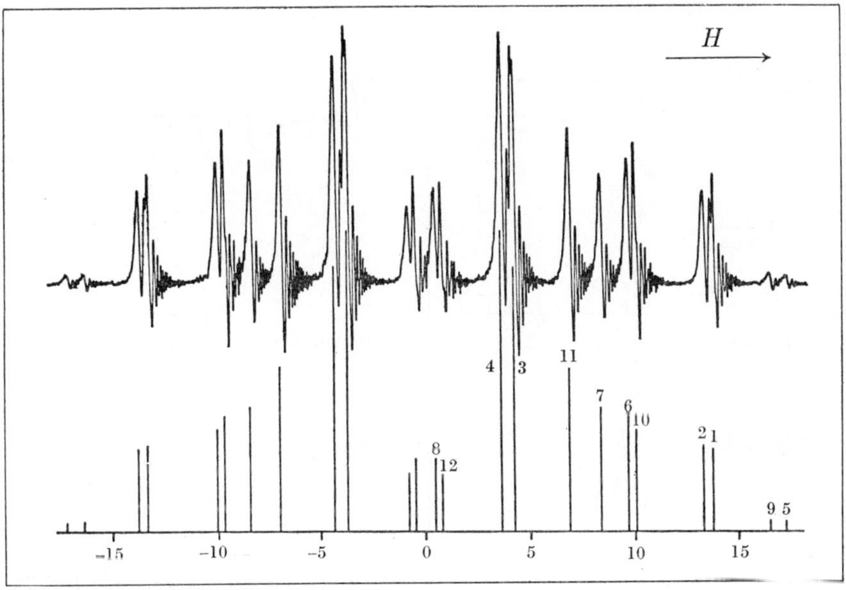

FIG. 7.4. The proton resonance spectrum of *o*-dichlorobenzene at 60 Mc/S. $\Delta\nu = 15{\cdot}8$, $K = 7{\cdot}8$, $L = 6{\cdot}6$, $M = 7{\cdot}2$, $N = 9{\cdot}6$ c/s. (From Ref. 8.)

In both cases, substituents P must be the same. The situation is discussed in detail by Pople, Schneider and Bernstein,[22] and analyses of ortho disubstituted benzenes are described in Refs. 7 and 8. There is a marked tendency for the spectral lines to group in pairs. This is well illustrated by the case of *o*-dichlorobenzene shown in Fig. 7.4, although it must be remembered that in a slightly less well resolved spectrum some of the pairs of lines would merge.

2. $|J_{AB}| \gg$ all other coupling constants. The system approximates to two nearly independent two-spin systems, AB and A′B′. This occurs notably in the case of *para*-disubstituted benzenes,

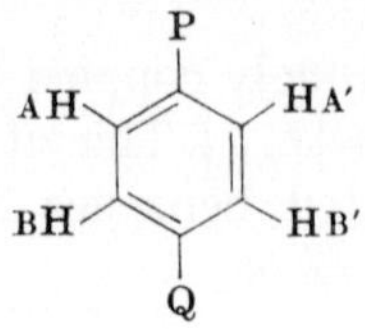

$$J_{AB} \approx 7\text{–}10 \text{ c/s}$$
$$J_{AA'} \approx J_{BB'} \approx 2\text{–}3 \text{ c/s}$$
$$J_{AB'} \approx 0\text{–}1 \text{ c/s}$$

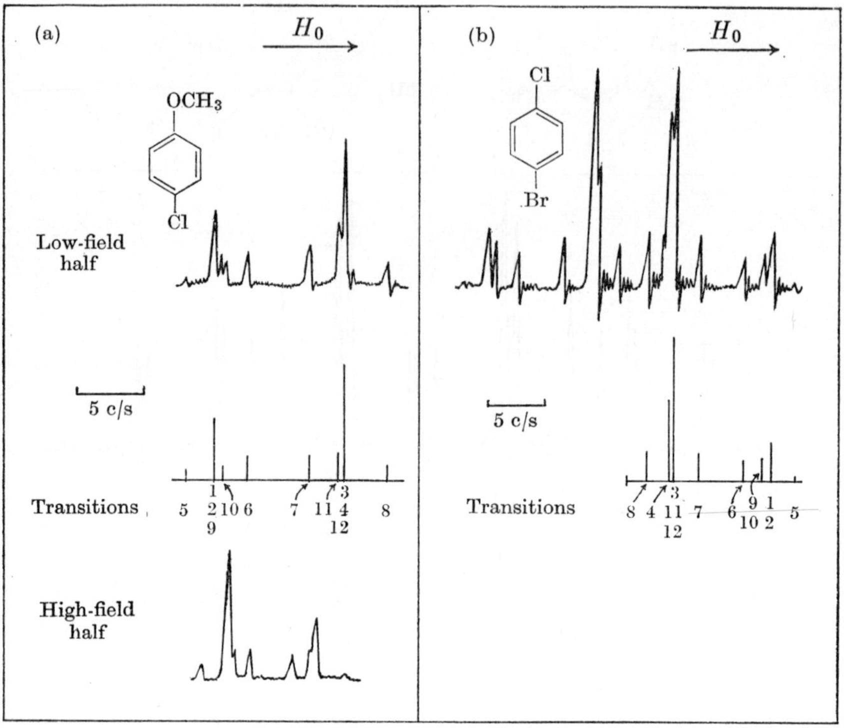

FIG. 7.5. Examples of AA′BB′ spectra of *para*-disubstituted benzenes (from Ref. 7). (a) *p*-chloroanisole, $\Delta\nu = 25{\cdot}0$, $K = 5{\cdot}6$, $L = 8{\cdot}5$, $M = 0{\cdot}6$, $N = 9{\cdot}1$ c/s; (b) *p*-bromochlorobenzene, $\Delta\nu = 10{\cdot}1$, $K = 5{\cdot}0$, $L = 8{\cdot}0$, $M = 0{\cdot}0$, $N = 8{\cdot}6$ c/s.

Here, $P \neq Q$ of necessity, since there would otherwise be no chemical shift differences. Many spectra of this type have been analysed[7, 8, 9] and two examples from Ref. 7 are shown in Fig. 7.5. These show the recognizable resemblance to the AB quartet which would be expected in absence of any cross-ring coupling. The latter, however, causes the four strongest lines to show some evidence of splitting, and also to be flanked on both sides by satellites at distances corresponding roughly to the cross-ring coupling. The general effect is a grouping of lines into *threes* rather than pairs as before.

Figure 7.5 (b) illustrates an important limitation of spectral analysis already discussed in Chapter 6. If we label the low field protons AA′ and the high-field ones BB′ as is customary, we can deduce the spectral parameters by analysis. However, since the spectrum is symmetrical, it is *not* possible to tell whether the AA′ labels should be applied to the protons *ortho* to Cl or *ortho* to Br—this can be settled only by comparison with data on related compounds. In the case of *p*-chloroanisole the slight broadening of the high field lines gives a clue to the assignment, since we might expect very slight spin coupling between the methoxy group and protons *ortho* to it.

3. In the case of 1,2-disubstituted ethanes, it is clear that the *trans* conformation of an ethane derivative P—CH_2—CH_2—Q will be an example of the case AA′BB′ (or sometimes AA′XX′)

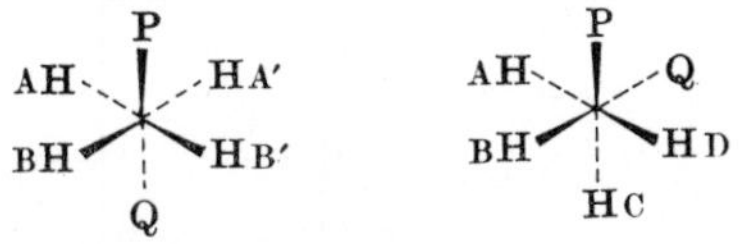

(P ≠ Q)

whereas a less stable *gauche* conformation would conform only to the general case ABCD. What is not so obvious, however, is that in the usual case of rapid rotation about the central C—C bond we do *not* get complete averaging to give a genuine A_2B_2 case with $J_{AB}=J_{AB'}$. The chemical shifts of the pair A,A′ above are identical as in the *trans* conformation (similarly for the pair B,B′), since the two *gauche* forms must make equal contributions to the time average. However, the time average value of J_{AB} is not necessarily the same as $J_{AB'}$, and the system conforms to the case AA′BB′. The problem of rotational averaging is discussed in Chapter 6 (see also Ref. 10). Note that the substituents P and Q must be different.

The proton resonance spectra of 1,2 disubstituted ethanes are considered very thoroughly by Hirst and Grant.[11] The series of calculated spectra in Fig. 7.6 are qouted from this reference, and show the effect of variation in $L=J_{AB}-J_{AB'}$ or $M=J_{AA'}-J_{BB'}$, taking reasonable values for the other parameters. The mean of $J_{AA'}$ and $J_{BB'}$ is taken as -12 c/s, the value of J_{HH} in methane, giving $K=-24$ c/s; whilst the mean of J_{AB} and $J_{AB'}$ is taken as $+6$ c/s, the typical value for vicinal coupling in a freely rotating ethyl group (q.v.), giving $N=+12$ c/s. Often, the spectra are liable to misinterpretation. Any of the cases shown in the lower left-hand or upper right-hand half of Fig. 7.6 could easily be mistaken for the strict A_2B_2 case (Fig. 7.6; $L=0$), since the closest lines in the spectrum

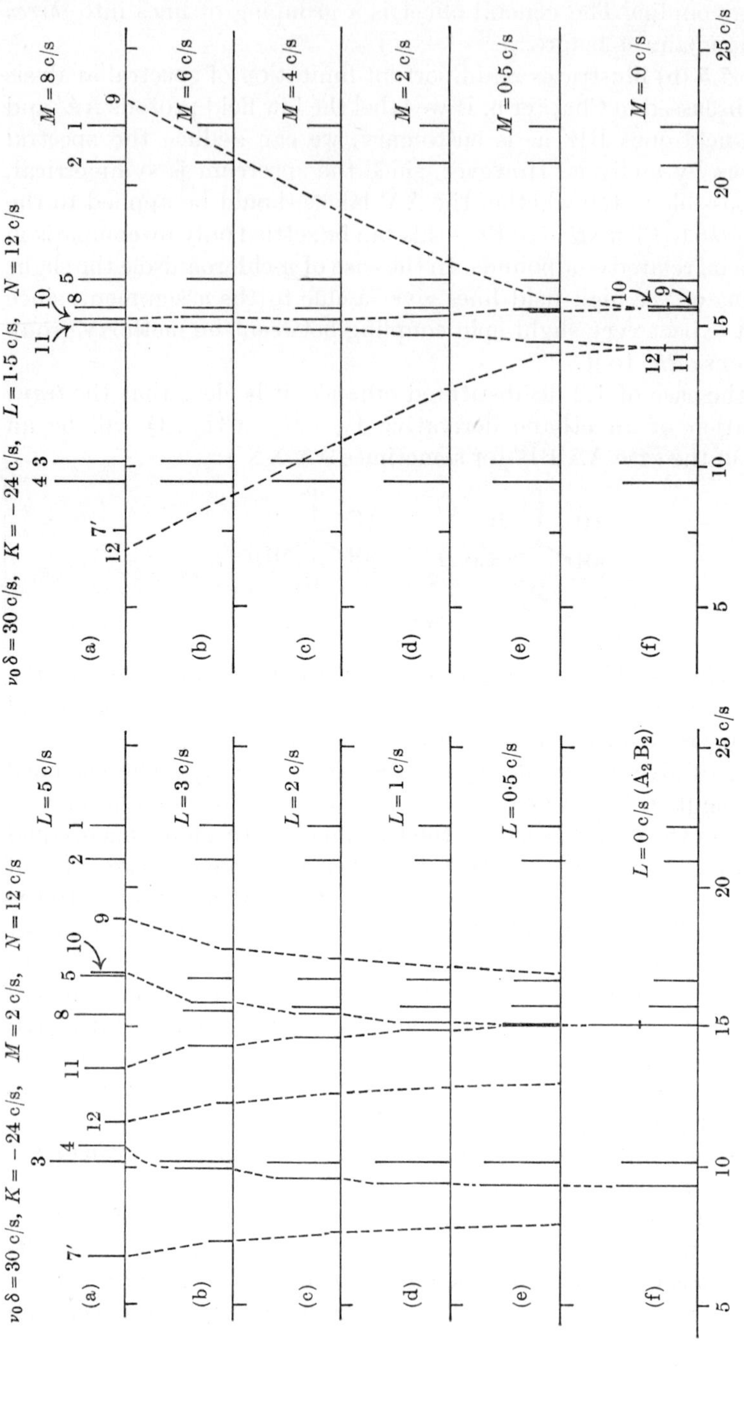

FIG. 7.6. Upper field half of some calculated AA′BB′ Spectra typical of 1,2-disubstituted ethanes. (From Ref. 11.)

(i) Variation of $L = J_{AB} - J_{AB'}$. (ii) Variation of $M = J_{AA'} - J_{BB'}$.

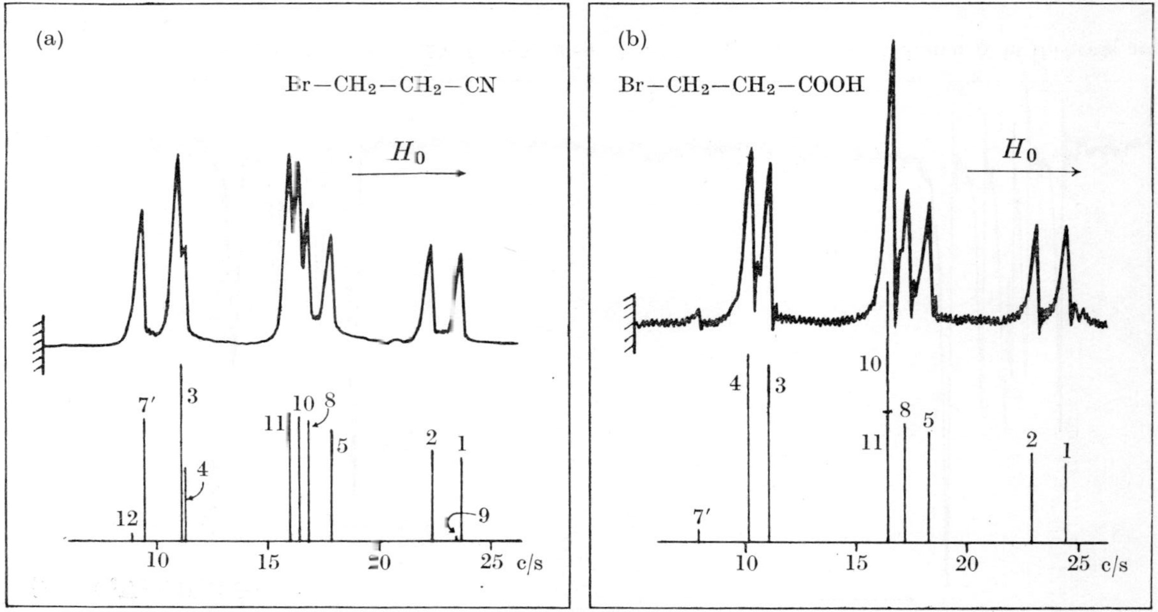

FIG. 7.7. 1,2-Disubstituted ethanes: (a) and (b) high field AA′BB′ half spectra from Ref. 11. Scale is shown in c/s from centre of spectrum taken as zero: (c) an example approximating to AA′XX′. For (a) $\Delta\nu = 32{\cdot}4$, $K = -27{\cdot}9$, $L = 2{\cdot}4$, $M = 7{\cdot}1$, $N = 12{\cdot}7$ c/s. For (b) $\Delta\nu = 34{\cdot}9$, $K = -27{\cdot}6$, $L = 1{\cdot}3$, $N = 13{\cdot}4$ c/s, $M = ?$.

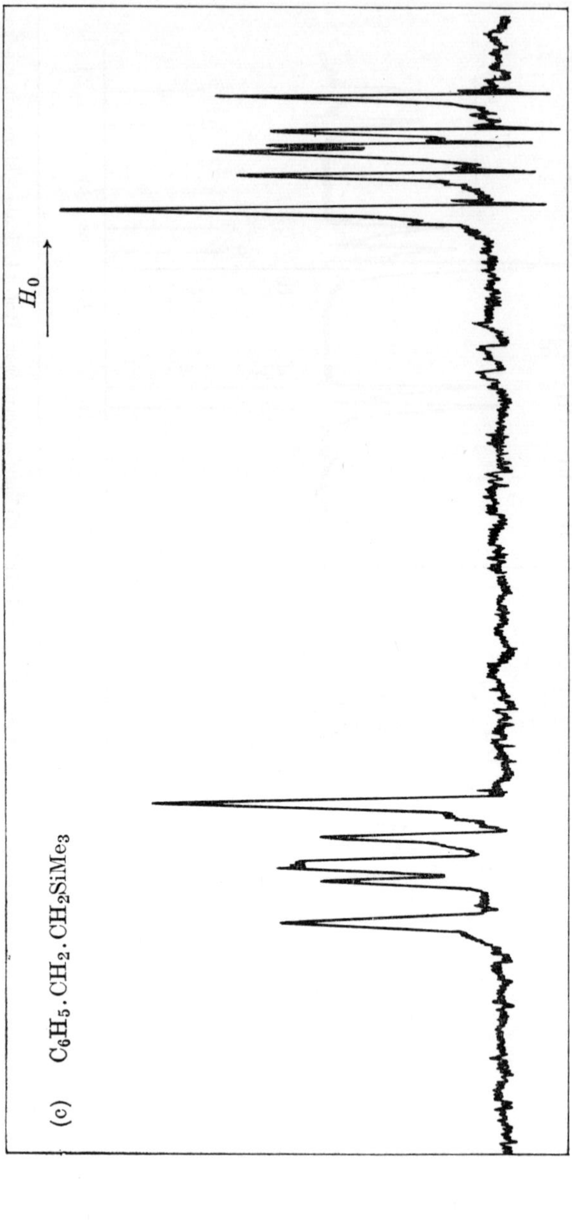

FIG. 7.7 *continued.* (c) $\Delta\nu \approx 105\frac{1}{2}$, $K = -25$, $L = 6{\cdot}5$, $M = 0$, $N = 17{\cdot}2$ c/s (values of both L and N in this case are unusually large[12]).

may not in practice be resolved. Appreciable non-equivalence ($J_{AB} \neq J_{AB'}$) might thus be overlooked, particularly when $|K|, |M| > |L|$, i.e. the same conditions under which AA′XX′ is liable to be confused with A_2X_2. Several experimental cases are shown in Fig. 7.7.[11] Non-equivalence is evident in Fig. 7.7 (a), but in Fig. 7.7 (b) is recognized only by the presence of the low intensity line labelled 7′. Figure 7.7 (c) is an unusual case in which one CH_2 group is displaced considerably to high field by proximity to Si (cf. $SiMe_4$).[12]

Systems of More than Four Spins

Complete spectral analysis of these is very difficult unless (a) some chemical shifts and coupling constants are equal (symmetry of the spin system); (b) some coupling interactions are close to first order; (c) some of the coupling constants are undetectably small. In fact, a *completely* general proton case ABCDE (F....), with all coupling constants finite and no first order interactions, would be rare, since J_{ij} values tend to be smaller than $\nu_i - \nu_j$ and usually attenuate rapidly with an increasing number of bonds between the nuclei. One or other of the above simplifying factors will usually operate. Either the protons will be spaced out through the molecule, when coupling between those widely separated can be ignored, or they will be clustered in localized groups which will tend to have high symmetry. One may cite the case of vinyl methyl ether (Chapter 6), where coupling across the ether linkage cannot normally be detected. This fact enabled the proton spectrum of the vinyl group to be analysed as an ABX case, whilst the protons of the methyl group made up an independent spin system of high symmetry (A_3 case) giving an additional single line spectrum. The more first order interactions there are in a spectrum, the easier it is to interpret—hence the chief advantage in using a high magnetic field. For many complex compounds, a *complete* analysis of the proton spectrum is quite impracticable, yet a great deal of structural information can be obtained by analysis of parts of the spectrum which arise from simple proton groups isolated in a spin coupling sense. The skeletal CH_2 protons of steroids, for example, are very similar in chemical shift, and spin coupling between them causes a broad unresolved band at $\tau \sim 7$–9. Against this background absorption, the resonance of small substituent groups stands out clearly, often as single sharp lines (OMe, OAc, methylene ketal etc.), and can be interpreted separately. Methods of simplifying complex spectra are summarized later.

Two commonly encountered five-spin systems are now discussed. Each constitutes a compact grouping of five protons, all of which are mutually coupled.

1. Monosubstituted benzenes: isotope substitution

The protons of a benzene ring form an almost isolated spin system, since coupling to a side chain is very small. A proton on carbon directly attached to a benzene ring is coupled to the ortho ring hydrogens only to the extent of $J \lessdot 0{\cdot}3$ c/s, a value sufficient to cause slight unresolved broadening of the resonance lines of each. Longer range coupling between side chain and ring is undetectable.

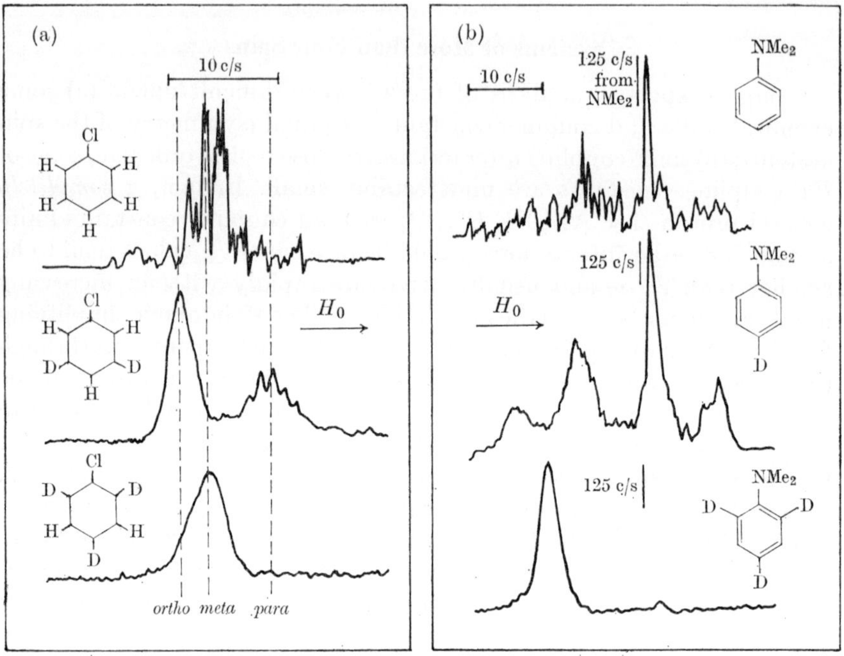

FIG. 7.8. Proton resonance spectra of (a) chlorobenzene at 60 Mc/s, (b) *NN*-dimethylaniline at 29·92 Mc/s and deuterated analogues.

Unfortunately there is no unique appearance to the proton resonance spectrum of a monosubstituted benzene. It conforms to the case AA′BB′C, which can give rise to 110 spectral lines, and is often quite strongly coupled at available magnetic fields. Usually, therefore, it causes one or two groups of closely packed lines, rising sharply in intensity towards the centre of the pattern, within the range 1·7 to 3·7 τ. Two examples are given in Fig. 7·8.

In view of the theoretical interest of substituent effects in aromatic systems, indirect means have been sought to obtain these spectral parameters. Use of *isotope substitution* to simplify the spectra has been the most successful. Magnetic resonance is a property of an isotope rather than an element, so that specific replacement of hydrogen by deuterium at certain sites in the molecule removes the corresponding lines from the proton resonance spectrum. Deuterium can cause its own n.m.r. absorption, having a spin quantum number $I = 1$, but since the magnetogyric ratio is much smaller than that for protons it will occur in a totally different part of the spectrum. The proton spectrum *is* affected by spin coupling to deuterium, but each J_{HD} is a factor of only 1/6·5 that of the original J_{HH}. The case of chlorobenzene and deuterated analogues[13] is shown in Fig. 7.8 (a). Deuteration at both *meta* positions removes the largest coupling interactions (between protons on adjacent carbon atoms), and leaves two distinct absorption regions of intensity ratio 2:1, clearly identified with the *ortho* and *para* protons respectively. The broadening and residual fine structure is due to coupling between the *ortho* and *para* protons, and of each to deuterium. 2,4,6-d_3-chlorobenzene exhibits only a single broadened resonance peak due to the equivalent meta protons. In Fig. 7.8 (b), replacement of the single *para* proton of *NN*-dimethylaniline by deuterium gives reasonable values for three of the parameters.[15] The remaining spectrum is essentially an AB quartet, broadened by cross-ring coupling (cf. *para*-disubstituted benzenes) and by coupling to deuterium. There are also some unresolved "impurity" lines due to the undeuterated compound. Analysis as an AB case gives ν_A, ν_B and J_{AB} but does not indicate which chemical shift belongs to the ortho and which to the *meta* protons. This is immediately obvious from the spectrum of 2,4,6-d_3-dimethylaniline. An alternative technique is catalytic deuteration at *random* sites.[14] In favourable cases H is replaced largely by D, so that in any particular molecule there is unlikely to be more than one remaining H atom. The proton resonance spectrum then collapses to a series of narrow bands at each chemical shift.

2. The ethyl group

Five protons are here grouped in a symmetrical manner, and rotation about the C—C bond is sufficiently fast to cause the system to conform to the case A_2B_3. This is another instance of the general case A_pB_q, in which the spectrum is independent of J_{AA} and J_{BB} and, apart from a scaling factor, depends only on the ratio $|J_{AB}/\nu_A - \nu_B|$. The value of J_{AB} is about $+6$ c/s, although the sign is irrelevant to the appearance of an A_2B_3 spectrum. The chemical shift difference $\nu_A - \nu_B$ depends largely on the electronegativity of the atom linking the ethyl group to the rest of the molecule (Ref. 16, p. 65). If the electronegativity is high (O, S, N, tert-C etc.), then the chemical shift ν_A of the methylene protons is displaced considerably to lower field than ν_B, and the system approximates to the

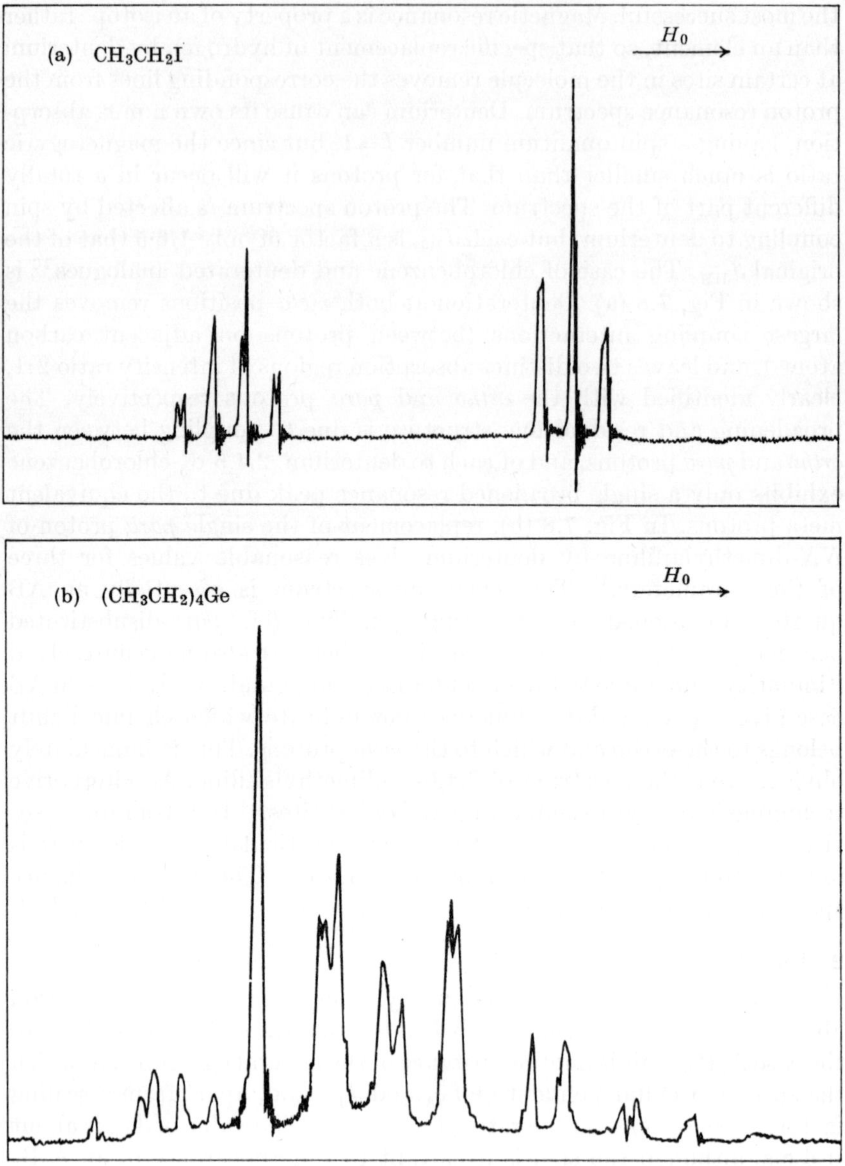

FIG. 7.9.

A_2X_3 condition. The ethyl group then gives one of the most easily recognized spectral patterns in proton resonance spectroscopy, a quartet of intensity ratios 1:3:3:1 for the methylene protons, and a 1:2:1

triplet, 2–3 ppm to higher field, for the methyl protons. The spacing between the multiplet components is $|J_{AX}|$, and the ratio of the *total* intensities of the methylene and methyl resonances is 2 : 3. Departure from the A_2X_3 condition is usually sufficient, however, to cause a distinctive "second-order" hyperfine splitting of each component as in Fig. 7.9 (a). In an ethyl group bonded to a metalloid or metal, the chemical shift of the methylene protons is displaced abnormally high field, close to and sometimes even beyond that of the methyl protons, as in the case of Et_4Ge (Fig. 7.9 (b)). The spectrum is then strongly coupled, and the usual distinctive appearance is lost and may be further complicated by spin coupling to the metal. Where this last is absent, it should be possible to recognize a strongly coupled ethyl group by comparison with a series of spectra calculated for known values of the ratio $|J_{AB}/\nu_A - \nu_B|$.[20c, 21] In Fig. 7.9 (b) this ratio is −0·46 at 60 Mc/s (minus sign indicates the methylene signals are at higher field than the methyl resonances).[12] The pattern is thus reversed relative to those of Refs. 20 (c), 21 with the "centre of gravity" now towards the low field end of the spectrum.

An interesting modification of the spectrum of an ethyl group arises if it is close to an asymmetric centre in an optically active molecule. The $-CH_2$—methylene protons are rendered slightly nonequivalent, in both a magnetic and a spin-coupling sense, and their part of the n.m.r. spectrum then approximates to *two* quartets, slightly displaced relative to each other (for collected references, see under Ref. 16). In theory, this effect would exist even with completely free rotation about the bond separating the CH_2 group from the asymmetric centre, but in most cases conformational preference is probably the dominant factor. Thus, the observed effect is very solvent dependent.[16(a)] It may be observed also in compounds RXR, e.g. diethyl sulphite, if X is not in itself an asymmetric centre but would become so if R were replaced by a different group R′.[16(b)] A similar effect occurs in the proton spectrum of an isopropyl group near an asymmetric centre.

3. Higher alkyl groups

Normal-propyl and isopropyl groups are easily distinguished by their proton resonance spectra (Fig. 7.10). The n-propyl group $-CH_2$—CH_2—CH_3 corresponds to the spin coupling case $A_2B_2C_3$, usually approaching the first order condition $A_2M_2X_3$ but exhibiting second order splitting resembling that of the ethyl group. Parameters for the n-propyl group in different environments are available.[17]

The isopropyl group–$CH(CH_3)_2$ corresponds to the simpler spin system AB_6, generally approaching the first order limit AX_6. The methyl peak is split into a strong doublet by coupling ($J_{AX} \approx +6$ c/s) to the

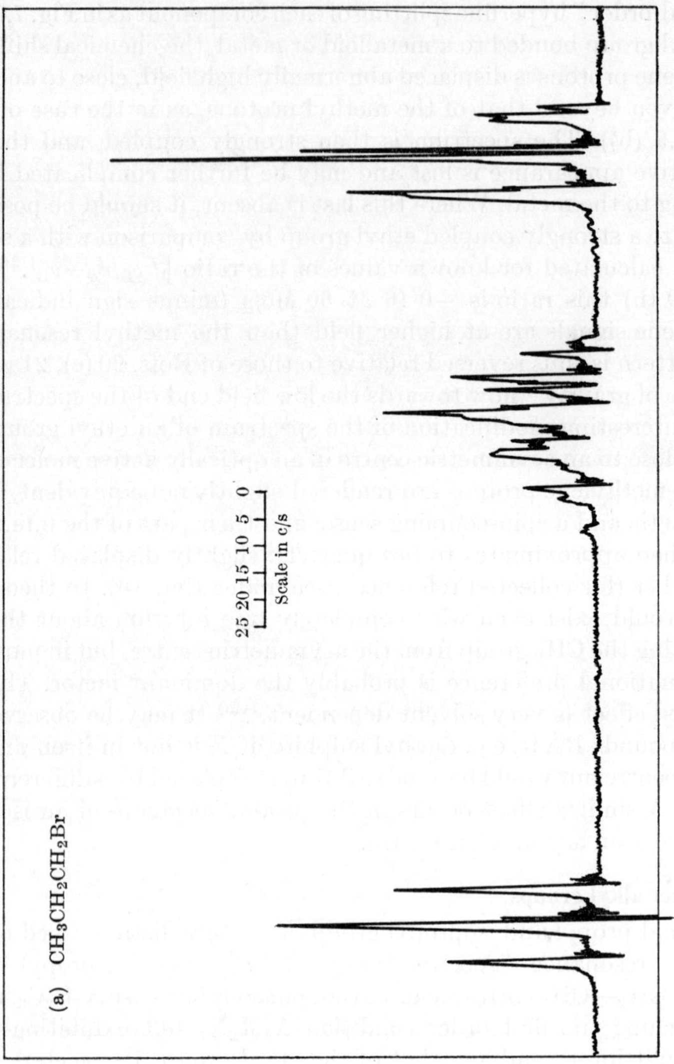

FIG. 7.10 (a).

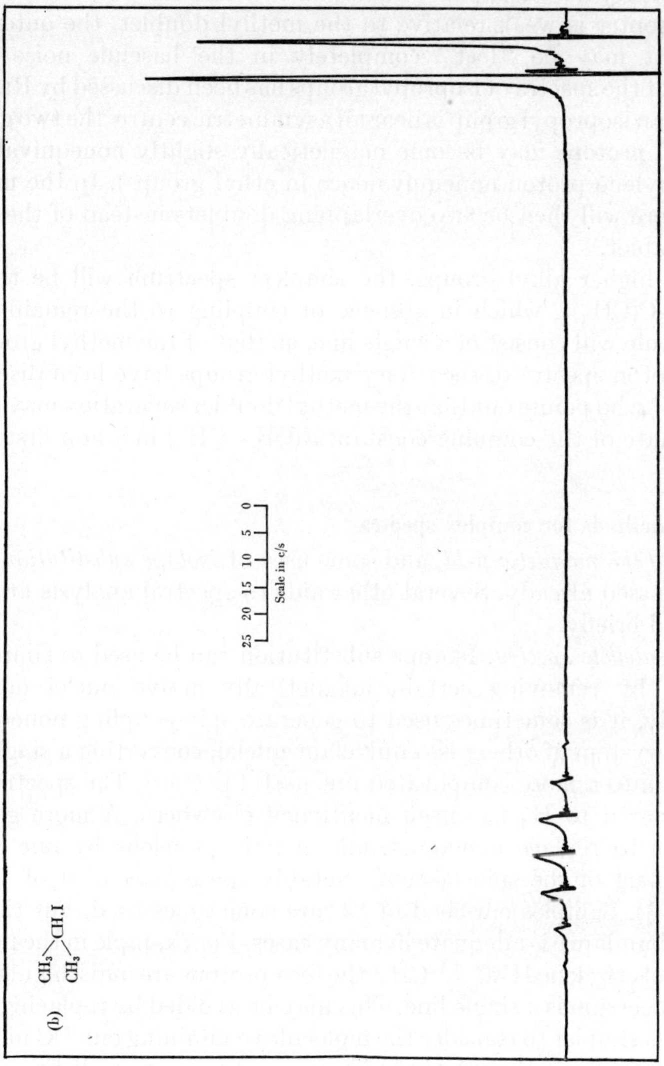

FIG. 7.10 (b).

methine proton, whilst the expected first order spectrum of the latter would be a *septet*. In practice, the components of the latter exhibit second order splitting and in both regions of the spectrum the inner lines gain some intensity at the expense of the outer ones. Since in any case the methine septet is weak relative to the methyl doublet, the outermost component may be "lost" completely in the baseline noise level. Analysis of the spectra of isopropyl groups has been discussed by Ranft.[18]

Where an isopropyl group is near an asymmetric centre, the two groups of methyl protons may become magnetically slightly nonequivalent[16] (c.f. methylene proton nonequivalence in ethyl groups). In the methyl region there will then be two overlapping doublets instead of the usual single doublet.

Of the higher alkyl groups, the simplest spectrum will be that of t-butyl, $-C(CH_3)_3$, which in absence of coupling to the remainder of the molecule will consist of a single line, as that of the methyl group.

The proton spectra of secondary methyl groups have been discussed by Anet,[19] who points out that the methyl doublet separation may give a low estimate of the coupling constant $J(CH—CH_3)$ in a non-first order spectrum.

4. Some methods for complex spectra

Varying the magnetic field, and some uses of *isotope substitution* have been discussed already. Several other aids to spectral analysis are now considered briefly.

^{13}C H satellite spectra. Isotope substitution can be used to simplify a spectrum by removing certain magnetically active nuclei (above). Conversely, it is sometimes used to generate spin-coupling nonequivalence in a system of otherwise equivelant nuclei, converting a single line spectrum into a more complicated but useful pattern. The spectrum of HD compared to H_2 has been mentioned elsewhere. A more general method is to replace a magnetically inactive nucleus by one which becomes part of the spin system—notably the replacement of ^{12}C by ^{13}C (spin $\frac{1}{2}$). Samples enriched in ^{13}C are sometimes used, but the 1% natural abundance is adequate in many cases. For example in the normal molecule of ethylene $H_2^{12}C={}^{12}CH_2$, the four protons are indistinguishable and the spectrum is a single line. This may be avoided by replacing H by D, but it is simpler to consider the molecules containing one ^{13}C nucleus. The spin system is then AA′A″A‴X, since we can now distinguish in spin-coupling between those protons (A,A′ say) adjacent to ^{12}C, and those (A″,A‴) adjacent to ^{13}C (=X). We write AA′ rather than A_2, since these nuclei are coupled differently to A″. This is another example (cf. AA′XX′) of a high field limiting case which does not give a first order

spectrum but a complicated one in which all J_{HH}'s play a part. However, there is one greatly simplifying feature: the geminal $J(^{13}C—H)$ is much larger ($\sim$150–200 c/s) than any other coupling. This causes all lines due to the $^{13}CH_2$ protons to be displaced considerably ($\sim$75–100 c/s) to either side of the spectral centre; to low field when the spin of ^{13}C is α, and the corresponding lines to high field when the ^{31}C spin is β. Lines due to the $^{12}CH_2$ group are much nearer the centre, and usually obscured by the very strong line due to $H_2{}^2C—^{12}CH_2$ molecules. In natural abundance, the low intensity $^{13}CH_2$ proton lines are described as "^{13}CH satellites" to the central peak, and must be carefully differentiated from spinning side bands by varying the spinning speed. The relatively large displacement between $^{13}CH_2$ and $^{12}CH_2$ proton lines has exactly the same effect as a chemical shift difference between them—i.e. either $^{12}CH_2$ resonance looks like one half of a weakly coupled AA′BB′ (nearly AA′XX′) spectrum (the $^{12}CH_2$ half is generally obscured) and can be analysed as such (cf. Ref. 22, p. 163).

Sub-spectral analysis. This is discussed by Diehl, Jones and Bernstein[23] (See Ref. 23 for earlier work). If a nuclear spin system possesses local symmetry, this can be used to classify the spin functions into different symmetry types (see AB_2, Chapter 5). The n.m.r. spectrum can then be regarded as made up of superimposed but non-interacting "sub-spectra", each composed of transitions between states of a particular symmetry (since transitions between states of different symmetry are forbidden). *If* one or more sub-spectra can be *reliably picked out* (a sub-spectrum may have the recognizable pattern of a simpler spin system) then the difficulty of analysis will be greatly simplified, since each sub-spectrum can be treated separately. It may then be possible to obtain all the required parameters without proceeding to a complete analysis. If we take the AB_3 case as an example, four of the sixteen lines have the pattern of an AB quartet. These are numbers 3, 5, 8, and 13, in Fig. 7.2) and can be analysed as such to give all the parameters ν_A, ν_B and J_{AB}. We could also regard the "antisymmetric" transition No. 3 at position ν_A in the AB_2 case (Chapter 5, Fig. 5.6) as a "sub-spectrum" with the pattern of the one nucleus case "A", whilst the ^{13}CH satellites discussed above are also sub spectra of a rather special kind.

Double resonance. This is far too extensive a subject to attempt to cover adequately here, and the reader is advised to study Ref. 24. Essentially, it is an n.m.r. experiment using simultaneously two different radio frequencies, that of the "irradiating" field H_2 being held constant at a pre-selected point in the spectrum whilst the other (H_1 field) is used to observe the rest of the spectrum in the usual way. The experimental equipment is already widely available to chemists, and the technique is making important contributions to many aspects of n.m.r., including spectral analysis. The main applications to the latter are summarized briefly.

The result of a double resonance experiment is markedly dependent on the strength of the irradiating (H_2) field. Fairly *strong* irradiation near a chemical shift position causes the spin coupling between that nucleus (or nuclei) and the rest of the spin system to become ineffective, i.e. it is "*spin decoupled*". This can be done without disturbing couplings within the remaining spin system provided that the chosen chemical shift is well away from those of other nuclei. The main uses in analysis are: (a) to simplify a complicated spectrum by "removing" certain nuclei from the system. It can also be used to sharpen spectral lines broadened by unwanted coupling to nuclei with an electric quadrupole moment—notably ^{14}N (when the H_2 field must be at the ^{14}N resonance frequency); (b) to find which nuclei are coupled to which by observing where multiplets are collapsed in other parts of the spectrum; (c) location and measurement of chemical shifts in hidden parts of a spectrum, e.g. obscured by a solvent peak.

Use of a *weak* irradiating field has another important application (d). The presence of a weak H_2 field between two components of a first order spin multiplet causes a collapse (i.e. merging) of some but not all of the lines in other multiplet splittings associated with the same coupling. The lines collapsed are those sharing an energy level in common with the pair being irradiated. Re-siting the H_2 field at different places enables a detailed line assignment to be given, from which the *relative signs* of the coupling constants can be found. The effects in non-first order spectra are more complicated. This application is in some degree complementary to conventional analysis, which depends on non first order effects to establish relative signs.

References

1. R. J. Abraham, J. A. Pople and H. J. Bernstein, *Canad. J. Chem.* **36,** 1302 (1958).
2. (a) J. Lee and L. H. Sutcliffe, *Trans. Faraday Soc.* **54,** 308 (1958). (b) T. P. Schaefer, *Canad. J. Chem.* **40,** 431 (1962).
3. (a) R. J. Abraham and H. J. Bernstein, *Canad. J. Chem.* **39,** 216 (1961). (b) V. J. Kowalewski and D. G. de Kowalewski, *J. Chem. Phys.* **36,** 266 (1962); N. V. Riggs, *Austral. J. Chem.* **16,** 521, (1963).
4. (a) C. N. Banwell and N. Sheppard, *Proc. Roy. Soc.* **A263,** 136 (1961). (b) W. Brugel, *Z. Elektrochem.* **66,** 159 (1962).
5. F. S. Mortimer, *J. Mol. Spectroscopy* **3,** 335 (1959).
6. R. J. Abraham, E. O. Bishop and R. E. Richards, *Mol. Phys.* **3,** 485 (1960); P. Diehl and I. Gränacher, *J. Chem. Phys.* **34,** 1846 (1961).

7. J. Martin and B. P. Dailey, *J. Chem. Phys.* **37**, 2594 (1962); **38**, 2582 (1963).
8. D. M. Grant, R. C. Hirst and H. S. Gutowsky, *J. Chem. Phys.* **38**, 470 (1963);
9. G. W. Smith, *J. Mol. Spectroscopy* **12**, 146 (1964).
10. J. A. Pople, *Mol. Phys.* **1**, 1 (1958).
11. R. C. Hirst and D. M. Grant, *J. Chem. Phys.* **40**, 1909 (1964).
12. E. O. Bishop, M. A. Jensen, D. Skelton and Miss J. Smithers, unpublished work.
13. H. Spiesecke and W. G. Schneider, *J. Chem. Phys.* **35**, 731 (1961).
14. J. L. Garnett, L. J. Henderson, W. A. Sollich and G. V. D. Tiers, *Tetrahedron Letters* 516 (1961).
15. E. O. Bishop, P. Farrell, S. F. Mason and R. F. M. White, unpublished results.
16. (a) E. I. Snyder, *J. Amer. Chem. Soc.* **85**, 2624 (1963); G. M. Whitesides, D. Holtz and J. D. Roberts, **86**, 2628 (1964). (b) E. O. Bishop, *Ann. Reports* **58**, 67 (1961); E. O. Bishop and R. A. Craig, unpublished results.
17. J. R. Cavanaugh and B. P. Dailey, *J. Chem. Phys.* **34**, 1094 (1961).
18. J. Ranft, *Ann Physik.* **10**, 1 (1962).
19. F. A. L. Anet, *Canad. J. Chem.* **39**, 2262 (1961).
20. K. B. Wiberg and B. J. Nist. "Interpretation of N.M.R. Spectra", Benjamin, New York (1962). (a) p. 301; (b) p. 318; (c) p. 579.
21. P. L. Corio, *Chem. Rev.* **60**, 363 (1960).
22. J. A. Pople, W. G. Schneider and H. J. Bernstein, "High Resolution Nuclear Magnetic Resonance". McGraw-Hill, New York (1959).
23. P. Diehl, R. G. Jones, and H. J. Bernstein, *Canad. J. Chem.* **43**, 81 (1965).
24. J. D. Baldeschwieler and E. W. Randall, *Chem. Rev.* **63**, 81, (1963).

CHAPTER 8

PROTON–PROTON COUPLING AND STEREOCHEMISTRY

R. J. ABRAHAM

We have seen previously how the measurements of proton chemical shifts can provide a wealth of information of use to the chemist. In this chapter we wish to consider the other molecular parameter which is obtained from an n.m.r. spectrum, namely the coupling constant, and to show how the value of this coupling can provide useful information. We will restrict this discussion to proton–proton coupling constants, as these are of most use to the chemist. Due partly to the wealth of data available, the theory of these couplings is more developed and less speculative than that for other nuclei: thus it provides a reasonable basis for our discussions. Our approach here will be to give a qualitative description of the theories of proton–proton couplings, to show how these theories are supported by experiment, and thus how the values of proton–proton couplings can be used to provide valuable information.

A coupling constant is a measure of the energy of interaction of two nuclei in a molecule and all nuclei with non-zero nuclear spins (I) can in theory couple to each other. In practice such couplings are often not observed for nuclei with $I \geqslant 1$, as these nuclei possess quadrupole moments which, by interacting with the electric field gradient at the nucleus, can decouple the nuclear spin. Thus coupling constants have most often been measured between nuclei with $I=\frac{1}{2}$, e.g. ^{1}H, ^{13}C, ^{19}F, ^{31}P. The coupling between two nuclei A and B is given the symbol J_{AB}, measured in c/s, though for some theoretical purpose a reduced coupling constant $K_{AB}=(2\pi/\hbar\gamma_A\gamma_B).\ J_{AB}$ with units of cm^{-3} is more convenient.[1] This removes the nuclear moments γ_A and γ_B and thus is only dependent on the electronic environment of the molecule. The coupling constant can be of either sign and the definition of sign is conventional. If we consider the two hydrogen nuclei H_A and H_B in a molecule, there are two possible orientations of the nuclei, shown in Fig. 8.1 (a), (b).

$H_A\uparrow$---------$\downarrow H_B$ (a) $H_A\uparrow$---------$\uparrow H_B$ (b)

FIG. 8.1.

(Note that there is no external magnetic field, thus the orientation represented as $\uparrow\uparrow$ is identical to $\downarrow\downarrow$). By convention, if 8.1 (a) is of lower energy than (b), we say that the coupling J_{AB} is positive; if (b) is of lower energy than (a), then J_{AB} is negative.

In simple first order spectra in which the values of the observed splittings equal the couplings, only the numerical value of J can be obtained. In more complex spectra which require detailed analysis, both the magnitudes and *relative* signs of the couplings can be found. However, the *absolute* signs of the couplings can only be obtained indirectly, either by orienting the molecules in the liquid,[2] or less certainly, by comparing the couplings with others predicted as positive by theory.[3] Fortunately both these methods give the same answer, thus the signs of all proton–proton couplings can now be related to these known absolute signs. This has been performed in the following discussion.

Before discussing the values of the couplings between various protons, it is convenient to consider the mechanisms by which these couplings occur. How do the nuclear spins interact? There are four possible types of interaction.

In a stationary molecule the two protons concerned will interact via a direct dipole–dipole mechanism in which the magnetic field produced by one nucleus will affect the other. This interaction is very large, for example for two hydrogen nuclei 2Å apart, the interaction energy may be as much as 30 kc/s. In addition the interaction depends on the relative orientation of the two nuclei and on the external field. Thus the width of the peaks in the proton resonance spectrum of a solid will be of this same order (ca 30 kc/s), which will completely obscure any chemical shift changes. In liquids and gases, however, the rapid molecular motion of the molecules averages these interactions to zero, and thus we obtain the narrow bands characteristic of high resolution spectra.

The proton–proton couplings which are found in these high resolution spectra will be due to a mechanism involving the molecular electrons, i.e. these are electron coupled interactions. There are three possible mechanisms in this case.[4] (1) The nuclear magnetic moments interact with the orbital electronic currents. (2) There is a dipole interaction between the nuclear magnetic moments and the electron spin magnetic moment. (3) There is an additional interaction between the nuclear magnetic moments and the electron spins in S orbitals due to the non-zero value of the electron wave function at the nucleus. This term is known as the contact term.

For proton–proton couplings, interactions 1 and 2 are negligible compared with 3. Thus for the HD molecule, the observed coupling of 43·5 c/s is made up of ca. 40 c/s from (3), with ca. 3 c/s from (2) and 0·5 c/s from (1).[4] However, this is not the case for other nuclei, and the evaluation of interactions (1) and (2) represents a considerable theoretical problem.

The coupling mechanism for proton–proton couplings can be visualized in terms of the two hydrogen nuclei in the hydrogen molecule.

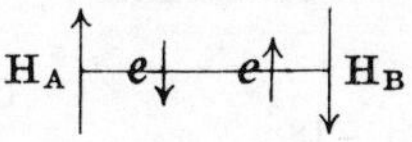

H_A will interact with the 1*S* electron via the contact term and this means that the antiparallel orientation of the nuclear and electron spins is favoured. By the Pauli exclusion principle the electrons in the molecule will tend to line up antiparallel, as shown, and the other electron will interact with the nucleus H_B resulting in the favoured orientation of nuclei shown to give a positive coupling constant. In an exactly analogous fashion, the ^{13}C—H couplings are predicted positive and this result has been used to obtain the "absolute" signs of proton–proton couplings.[3]

The quantum mechanical description of the above interaction may be presented as follows (for a more detailed description, see Ref. 4, Chapter 8). The ground state wave function does not involve the nuclear spins, thus the effect of the nuclear spins is to perturb this wave function slightly. Perturbation is very small as the energies involved are minute (10 c/s is equivalent to ca 10^{-10} cal/mole) and the new wave function is given by mixing excited state wave functions into the ground state. Such excited states can be represented in valence bond theory as non-classical structures and examples of these will be given later. The amount of mixing which occurs is inversely proportional to the energy difference between the ground and excited states. The value of this is often not known and must be estimated as accurately as possible. In addition, other approximations have to be made in many cases (e.g. that all excited states have the same energy), and thus the precise values of the couplings obtained by theory are often uncertain. Of greater importance usually is the trend or relationship between the couplings and other molecular parameters predicted by theory. In this sense, theoretical predictions have been of immense value in understanding and correlating the wealth of available experimental data on proton–proton couplings, even though there are still gaps in the overall picture as we shall see later.

In the remainder of this chapter we will discuss the interpretation of proton–proton couplings. The values of these couplings are conveniently summarized in Table 8.1 where it should be noted that ranges of values are given in many cases. This does not mean that there are no couplings lying outside these ranges, such cases will be dealt with later, but merely

TABLE 8.1

Characteristic proton–proton coupling constants

Compound or structural type	Coupling range (c/s)	Compound or structural type	Coupling range (c/s)
OPEN CHAIN			
CH_2 (>C<, H, H geminal)	−10 to −18	>C═CH—CHO	7 to 9
—N═C(H)(H)	8 to 16	>CH—SH	6 to 8
		>CH—OH	4 to 8
		>CH—NH—	4 to 8
>CH—CH<	5 to 10	>CH—C—CH<	0 to 1
>CH—CH═C<	5 to 10	>CH—C═CH—	0 to −2
$CH_3.CH_2$—	7 to 8	>CH—C≡CH	−2 to −3
H(H)C═C(H)H	J_{gem} −3 to +7; J_{cis} 3 to 18; J_{trans} 12 to 24	>CH—C═C—CH<	0 to 2
>CH—CHO	1 to 3	>CH—C≡C—CH<	2 to 3

Compound or structural type	Coupling range (c/s)		
CYCLIC			
Aromatic and Olefinic	*ortho*	*meta*	*para*
Benzene derivatives	5–9	2–3	0–1
	J_{23}	J_{34}	J_{24} / J_{25}
Furan	1·8	3·5	0·8 / 1·6[41]
Pyrrole	2·6	3·4	1·4 / 2·1[41]
Thiophen	5·2	3·6	1·3 / 2·7[42]
Cyclopentadiene	5·1	1·9	1·1 / 1·9[11]

H(C═C)H with $(CH_2)_n$ ring							
n=1	2	3	4	5	6	9[43, 44]
ca 1	2·7	5·1	8·8	10·8	10·3	10·7

TABLE 8.1—*continued*

Compound or structural type	Coupling range (c/s)		
Saturated	J_{gem}	J_{cis}	J_{trans}
Cyclopropane	−4·5	9·2*	5·4*[45]
Ethylene oxide	+5·5*	4·5	3·2[7, 24]
Ethylene imine	+2·0*	6·3	3·8[46, 47]
Ethylene sulphide	ca 0*	7·1	5·6[46]
Cyclobutane derivatives	−10 to −17	6 to 11	5 to 9
Cyclopentane derivatives	−10 to −17	7 to 11	2 to 8
Cyclohexane derivatives	−10 to −17	J_{ae} 2 to 6	J_{aa} 8 to 13 J_{ee} 2 to 8

* Value for methyl derivatives.

that these represent convenient limits. In some cases the couplings normally encountered may lie within much narrower ranges than those quoted (for example, the vinyl compounds given range from vinyl fluoride to lithium vinyl). It is convenient to divide the couplings as follows:

geminal couplings H—C—H, over two bonds
vicinal couplings, over three bonds H—C—C—H or H—C=C—H
long range couplings, over four or more bonds.

Geminal Couplings

Selected values of the geminal couplings are given in Table 8.2. In this and subsequent tables, for those cases in which the coupled protons are not immediately obvious, either only the coupled protons will be written down or they will be underlined. It can be seen that these couplings vary considerably in magnitude and can have either sign. They range from −20·3 c/s in malononitrile to 41 c/s in formaldehyde. The first detailed theory of geminal couplings[7] was based on valence bond calculations.[14] This predicted a positive coupling constant in methane and a dependence of the coupling on the $\mathrm{H}\widehat{\mathrm{C}}\mathrm{H}$ angle, with the coupling decreasing as the angle increased from 109° 28′ in methane to 120° in ethylene. This was in reasonable agreement with the available data at that time. However, when the sign of the coupling constant in a saturated CH_2 fragment was shown conclusively to be negative, these ideas were no longer tenable.

TABLE 8.2

Selected examples of geminal coupling constants (c/s)

Molecule	Coupling constant	Molecule	Coupling constant
CH_4	$-12{\cdot}4^{5}$	$CH_2{:}C{:}C<$	-9^{15}
$CH_3.CCl_3$	$-13{\cdot}0^{5}$	$CH_2(CN)_2$	$-20{\cdot}3^{7}$
$CH_3.CO.CH_3$	$-14{\cdot}9^{5}$	$CH_2{=}O$	$+41^{10}$
$CH_3.C_6H_5$	$-14{\cdot}4^{59}$	$CH_2{:}CH_2$	$+2{\cdot}5^{9}$
CH_3Cl	$-10{\cdot}8^{5}$	$CH_2{:}CHF$	$-3{\cdot}2^{8}$
CH_3OH	$-10{\cdot}8^{5}$	$CH_2{:}CHLi$	$+7{\cdot}1^{8}$
CH_3F	$-9{\cdot}6^{5}$	$Br\underline{CH_2}.CH_2OH$	$-10{\cdot}4^{6}$
		$Br.\underline{CH_2}.\underline{CH_2}OH$	$-12{\cdot}2^{6}$
		$Br.CH_2.\underline{CH_2}CN$	$-17{\cdot}5^{6}$
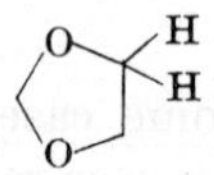	$-8{\cdot}3^{5}$	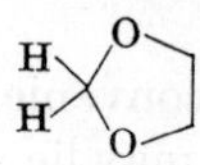	$\sim 0^{12}$
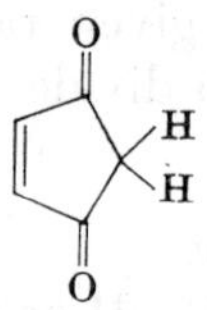	$-21{\cdot}5^{7}$	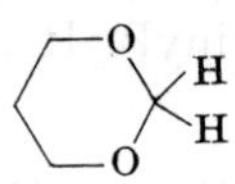	-5–6^{12}
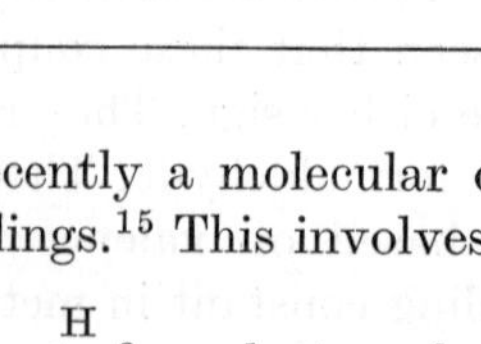 Metacyclophane	$-12{\cdot}0^{17}$	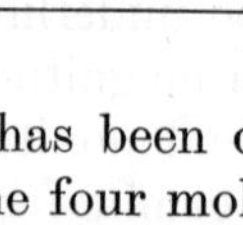	$-13{\cdot}9^{13}$

Recently a molecular orbital theory has been developed for these couplings.[15] This involves considering the four molecular orbitals of a four electron fragment in which there are two occupied levels and two excited levels. It is necessary to consider the mixing of the two ground states with either of the two excited states, and depending on this mixing, it is possible to get either positive or negative contributions to the coupling constant. Such mixing will depend of course on the energy differences between the ground and excited states, and the precise values of these are debatable: the accuracy of such calculations therefore is uncertain. The qualitative predictions however are of great use, and three main ones emerge from the theory.

(a) Increasing the H–C–H angle increases the S character of the orbitals and this gives a positive contribution to the coupling. *Note that this is the reverse of the former theory.*

(b) A substituent which withdraws electrons from the σ bonds (by the inductive effect) will give a positive contribution to the coupling, and vice-versa.

(c) A substituent which withdraws electrons from the π bonds (by the hyperconjugative effect) gives a negative contribution to the coupling.

Furthermore (b) and (c) differ radically in their orientation dependence. The σ-effect has no orientation dependence, whereas in the π-bonds the maximum effect will occur when the $1S$ orbital of the H atoms has maximum overlap with the π-bonds, and there will be no effect when the H atom lies in the nodal plane of the π-orbitals. These principles are easier to comprehend when actual examples are considered, as follows.

Interaction (a) would predict that as the H–C–H angle increases from methane to ethylene, the coupling would become more positive. Furthermore in small ring compounds the H–C–H angles are also larger than the tetrahedral value, and again the coupling should become more positive, this is precisely what is observed; the coupling in cyclopropane ($-4{\cdot}5$ c/s) lies midway between that of methane ($-12{\cdot}4$ c/s) and ethylene ($+2{\cdot}5$ c/s).

The coupling constant in formaldehyde is $+41$ c/s, much more positive than in ethylene ($+2{\cdot}5$ c/s). By the inductive effect the oxygen removes σ-electrons, therefore a positive shift would be expected [(b) above]. Furthermore, the orbitals of the oxygen atom are treated as consisting of one σ- and two π-orbitals and it is in the P_y and P_z orbitals that the two lone pairs of electrons are placed in this discussion. Of these the electrons in the P_z orbital (i.e. perpendicular to the plane of the molecule) will be drawn towards the oxygen atom, and this will have no effect on the coupling since the protons are in the nodal plane of this orbital. To a higher degree of approximation, the electrons in the P_y orbital will be donated to the CH_2 end. This will have a considerable effect on the coupling, as there is now maximum overlap between the P_y orbitals and the hydrogen atoms.

A large positive contribution will result. This explanation can be represented schematically as under

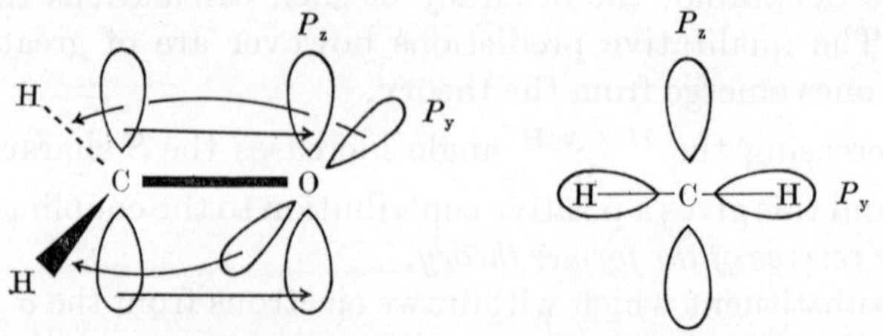

That the positive coupling is formaldehyde was predicted by the theory and verified experimentally shortly after.[15]

By contrast, in allene there is strong hyperconjugation leading to a withdrawal of the π-electrons, and thus to a predicted increase in the negative contribution. The sign of the observed coupling of 9 c/s should therefore be negative. As yet this has not been verified. For substituted vinyl compounds, molecular orbital theory predicts that an electron withdrawing substituent X withdraws electrons from the antisymmetric pair and therefore gives a negative contribution to the geminal coupling: this is experimentally observed. Indeed there is a linear relationship between the geminal coupling in vinyl compounds and the electronegativity (E) of the substituent, thus

$$J_{gem} = 8{\cdot}7 - 2{\cdot}9\,E.$$

The values of J in vinyl compounds range from $-3{\cdot}2$ c/s in vinyl fluoride to $+7{\cdot}1$ c/s in vinyl lithium.[8] Effects (b) and (c) also explain the observed changes in geminal couplings in saturated CH_2 fragments. Thus the coupling in methyl fluoride ($-9{\cdot}6$ c/s) is more positive than that in methane ($-12{\cdot}4$ c/s), a result in accordance with theory since fluorine withdraws electrons by the inductive effect. By contrast, the effect of electron withdrawal by hyperconjugation is shown clearly in Table 8.2. The almost constant change in the geminal coupling of ca. $-1{\cdot}9$ c/s for each adjacent π-bond in a freely rotating methane fragment had previously been noted, and interpreted as due to a σ–π interaction,[16] analogous to the theory of long range couplings (see later).

On both theories, the hyperconjugation contribution will depend on the amount of overlap of the hydrogens $1S$ orbitals and the orbital of the double bond. This overlap *approximately* follows a $\cos^2\phi$ dependence, where ϕ is the dihedral angle between the C—H bond and the π-orbital. Thus the contribution can be represented by an expression of the form.

$$k\cos^2\phi\,.\,\cos^2(120+\phi).$$

The usefulness of the qualitative predictions of an expression of this form may be seen in the following examples.

In a freely rotating fragment, the observed coupling is the average of three couplings, which will have very different contributions.

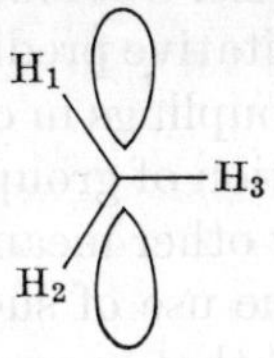

e.g. In toluene, which can be visualized as here, J_{13} and J_{23} will be the same as in methane, as there is no overlap of the π-orbital with H_3. However, J_{12} will be very different, as there is almost maximum overlap. From the value of the coupling in toluene (−14·4 c/s), and taking J_{23} and J_{13} to be the same as in methane, J_{12} would then be −18·4 c/s. In molecules where free rotation cannot occur, a range of couplings between the value for methane and that corresponding to maximum overlap will be found. For example in metacyclophane (Table 2), the conformation of the molecule is such that the CH_2 protons are oriented like H_1 and H_3 in toluene above: thus the hyperconjugative contribution will be small. The observed coupling (−12·0 c/s) is virtually identical to that in methane in full agreement with the theory.

The numerically large couplings in cyclic systems are due to a favourable overlap such as H_1 and H_2 in toluene above. For example, in 3,5-dioxocyclopentene (Table 8.2) the approximately planar ring gives rise to an orientation of this type and thus the coupling of −21·5 c/s is not surprising. In fact a figure larger than this might have been expected as the value of J_{12} in acetone would be ca −20 c/s, and the possibility that the ring angles of 3,5-dioxocyclopentene are considerably distorted cannot be discounted. It would appear that geminal couplings of up to −30 c/s may be possible in favourable orientations. A related orientation dependence can explain the difference in couplings of the methylenedioxy group in the dioxolan and meta dioxan rings (Table 8.2), if the oxygen lone pairs are considered as an electron donating π-orbital. In the six-membered ring, the CH_2 orientation becomes similar to that of H_1 and H_3 in toluene above: one of the protons lies in a node of the π-orbital, thus there is no hyperconjugative contribution. In the five membered, more planar ring, greater overlap occurs giving a positive contribution to the coupling. This difference has been used to differentiate between possible five- and six-membered rings in an unknown compound.[12] It is of interest to note that sulphur is very different from oxygen in its effect on the geminal couplings: in a 1,3-dithian ring the coupling of the methylene dithian protons is slightly more negative than that in methane (Table 8.2).

To summarize, it is clear that the qualitative discussion of the theory given above does interpret most of the observed couplings and therefore may be used in a diagnostic sense. Obviously, many more investigations are necessary before the quantitative predictions can be tested. Even so, the measurement of geminal couplings in cyclic molecules may well give information about the orientation of groups near a CH_2 fragment which would be difficult to obtain by other means.

The major limitations on the use of such couplings is experimental. It has already been mentioned that in many cases the two protons of a methylene group are magnetically equivalent and in such cases deuterium substitution is the only means of obtaining the coupling.

Vicinal Couplings

Vicinal (CH.CH) couplings are those most commonly encountered in proton magnetic resonance spectra, and a large number of these couplings have been measured in a variety of systems.

Inspection of Table 1 shows that the values of these couplings are relatively constant, in contrast to the geminal couplings. They are always positive and range from 0–25 c/s. This range includes couplings through both double and single bonds and also couplings via atoms other than carbon. The range of values arises mainly from the orientation dependence of these couplings, a property first predicted by theory thus providing considerable stimulus to experiment in this field.

The theory was developed by Karplus,[20] simultaneously with the geminal coupling theory, using a valence-bond approach. Both theories predicted positive coupling constants in saturated molecules. When it was discovered experimentally that the vicinal and geminal couplings were of opposite sign, it was not initially clear which coupling was positive and which was negative; one theory was obviously in error. It is now known that geminal couplings in saturated molecules are negative and that the theory thus led to an erroneous result in this case. There is considerable experimental support for the general predictions of the vicinal theory. In this the HCCH fragment is considered as a six-electron six-orbital fragment made up of two hydrogen 1*S* orbitals and four carbon directed (*Sp*) orbitals. These may be represented pictorially as in (I). In the perfect pairing approximations, as in the ground state (II), there will be no coupling because each electron pair in a bond cannot interact with the others. Thus it is necessary to consider "mixing in" some non-perfect pairing structures, of which the two important ones are (III) and (IV).

This leads to a similar result for both saturated and olefinic systems, the differences between them being mainly due to the different dimensions and *Sp* hybridization in the two cases. Of necessity the theory had to make certain approximations such as those mentioned earlier, thus the

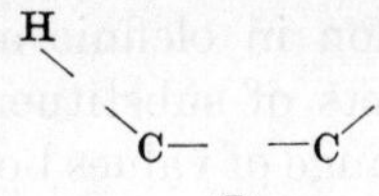

(I)

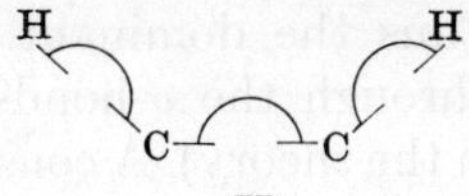

(II)

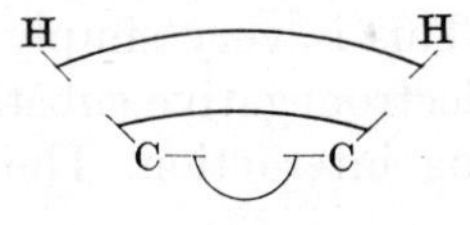

(III)

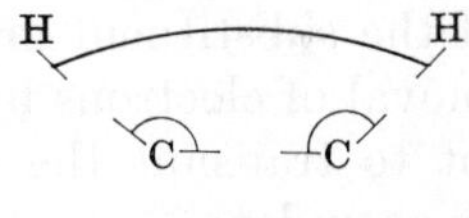

(IV)

actual values of the calculated couplings are approximate. More important however are the relationships between the couplings. The three conclusions which emerge are as follows.

1. Vicinal couplings are always positive;
2. In olefinic systems J_{trans} is larger than J_{cis};
3. In a saturated fragment the coupling is proportional to $\cos^2\phi$ where ϕ is the dihedral angle between the two C—H bonds.

It is convenient to consider first the application of the theory to olefinic systems. The calculated values of the *cis* and *trans* couplings in ethylene

TABLE 8.3

Selected values of vicinal coupling constants (c/s) in olefinic and aromatic systems

Molecule	Coupling constant J_{cis}	J_{trans}	Molecule	Coupling constant
$CH_2{:}CH_2$	11·7	19·0[9]	CH:CH	9·5[9]
$CH_2{:}CH.OR$	6·7	14·2[8]		
cis- and *trans*-CHCl:CHCl	5·3	12·1[18]	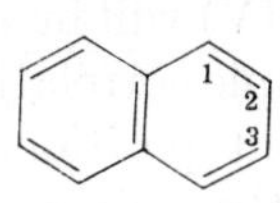	J_{12} 8·6[4] J_{23} 6·0
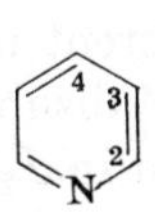	J_{23} J_{34}	5·5[19] 7·5	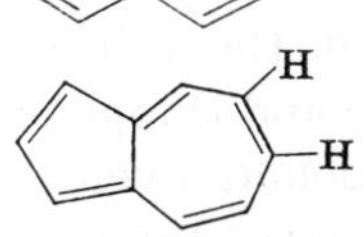	10–12[4]

are 6·1 and 11·9 c/s respectively, and although these are somewhat lower than the observed values (11·7 and 19·0 c/s) (Table 8.3), the *relative* magnitudes of the couplings are in very good agreement. Karplus later calculated the contribution of the π-electrons to this coupling, via a σ–π mechanism (see later) and obtained $+1{\cdot}5$ c/s for both the *cis* and *trans* couplings. Thus the dominant contribution in olefinic and aromatic systems is through the σ-bonds (the effects of substituents were not considered in the theory). A considerable range of values both of J_{cis} and J_{trans} in vinyl compounds is given in Tables 8.1 and 8.3. When sufficient reliable data had been accumulated, it further emerged that the values of these couplings are linearly related to the electronegativity (E) of the first atom of the substituent group.[8] This is very simply rationalized since the removal of electrons by an electronegative substituent leaves fewer present to transmit the coupling interaction. The appropriate equations are as under:

$$J_{trans} = 19 - 3{\cdot}3E$$

$$J_{cis} = 11{\cdot}7 - 4{\cdot}7E$$

In this way reasonable values of J_{cis} and J_{trans} for any vinyl compound can be obtained, and this may be extended to any olefin if it is further assumed that the effects of more than one substituent are additive. Application of this in identifying *cis* or *trans* olefins is obvious.

Another general result not predicted by the theory is the dependence of the *cis* olefin coupling on ring size in cyclic olefins and aromatic compounds. In general, the coupling increases as the ring size increases. This can be seen very clearly in the case of the *cis* olefins (Table 8.1), the values increasing from ca. 1 c/s in the cyclopropene ring to a constant value of ca. 10–11 c/s in seven membered and larger rings. A similar variation occurs in aromatic systems. For example, in the five membered aromatic compounds, the ortho couplings J_{23} and J_{34} lie between 2–5 c/s, in benzene and its congenors the value is 5–9 c/s, whereas in the seven-membered ring of azulenes the coupling is ca 10–12 c/s. Again this can be simply rationalized. As the ring size increases, the distance between the hydrogen atoms decreases, thus the contributions of structures of types (III) and (IV) will increase giving a larger coupling constant. The value of such simple correlations in distinguishing between different ring sizes and different positions of substitution, is considerable. Note in particular that the values of J_{23} and J_{34} are sufficiently different in furan, pyrrole and thiophen to distinguish between a 2,3 and 2,4 disubstituted compound, even though there is considerable variation in J_{23} (but not J_{34}) in the three cases.

Saturated CH.CH Couplings

The application of the theory to these couplings has been of considerable importance in recent years. The result of the theoretical calculations can be written as

$$J = k_1 \cos^2\phi \quad 0° \leqslant \phi \leqslant 90°$$

$$J = k_2 \cos^2\phi \quad 90° \leqslant \phi \leqslant 180°$$

where ϕ is the dihedral angle between the CH bonds and k_1 and k_2 are constants. In the original calculations k_1 had the value 8·5 c/s and k_2 9·5 c/s; a further constant of −0·3 c/s in the original will be neglected in what follows. These equations were in reasonable accord with the available results at that time (1959), which were mainly of the 1H–2H coupling in the pyranose ring (Table 8.4). With the observation of vicinal couplings larger than 9·5 c/s, the maximum calculated value, however, theoretical values had to be modified. Again this is not a serious defect as the theoretical values of k_1 and k_2 are only approximate and the influence of substituents was not considered. It is now recognized that the couplings are affected by the substituents on the CH.CH fragment in a very similar manner to the olefinic couplings. This has been most clearly demonstrated for J_{AV}, the coupling for equal populations of the rotational isomers in a substituted ethane. The coupling can be observed directly in a molecule such as CH_3.CHXY, it has to be deduced for less symmetric ethanes. The coupling is linearly dependent on the sum of the electronegativities of the first atoms of the substituent groups[21] and the appropriate equation is

$$J = 18·0–0·80\Sigma E$$

Some examples illustrating the magnitude of the effect are given in Table 8.4. Although a few studies have been performed on the effect of substituents on the couplings for a given dihedral angle, the orientation dependence of this electronegativity effect is not known in detail. In general the greater the number of electronegative substituents present, the smaller are the observed couplings.

Combination of the $\cos^2\phi$ law and the electronegativity relationship does provide a general explanation of the most of the vicinal couplings shown in Tables 8.1 and 8.4. For example the axial-axial couplings in six-membered rings are of similar magnitude to J_{trans} in substituted ethanes as expected (they both equal $J_{180°}$); they are much greater than the equatorial-axial, and equatorial–equatorial couplings and J_{gauche}, which are all ca $J_{60°}$. In addition there is a general decrease in the coupling with an increasing number of electronegative substituents, the low values of

TABLE 8.4

Selected examples of vicinal couplings across saturated bonds

Molecule	Coupling constant	Molecule	Coupling constant	
			J_{gauche}	J_{trans}
CH_3CH_2Li	8·4[21]			
CH_3CH_3	8·0[29]	$CH_2Br.CH_2Br$	3·4	15·3[24]
$CH_3.CH_2.CH_3$	7·3[21]	$CHCl_2.CH_2Cl$	3	10[24]
CH_3CH_2OH	7·0[21]	>CH.CH:C<	3·5	11·5[25]
$CH_3.CH:CH_2$	6·4[22]	>CH.CHO	0·1	8·3[23]
$CH_3.CH(OH)_2$	5·3[21]			
$CH_3.CHO$	2·9[23]			
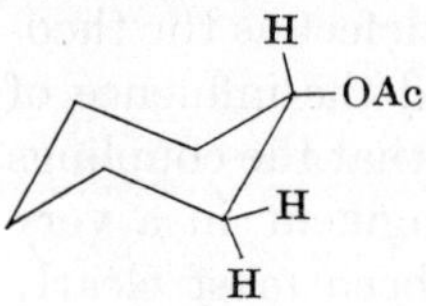	J_{aa} 11·4[26] J_{ae} 4·2	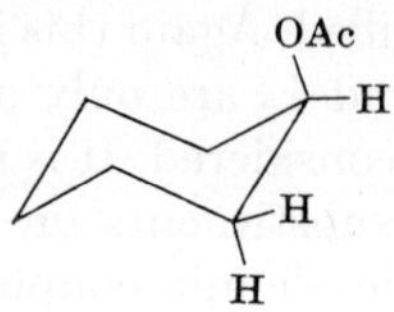	J_{ea} J_{ee}	2·7[26] 2·7
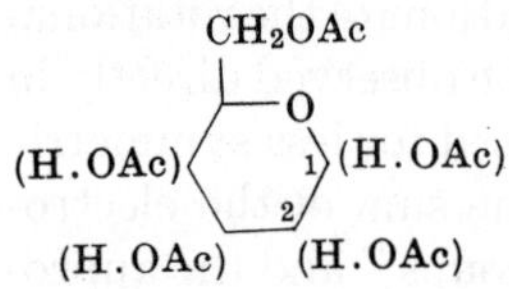	1H, 2H couplings $J_{aa} \approx 8$, J_{ae}, $J_{ee} \approx 2$–3[27]			
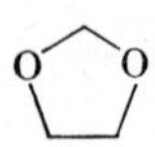	J_{cis} 7·3[24] J_{trans} 6·0	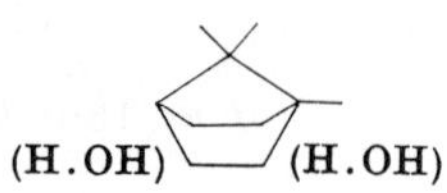	$J_{exo-exo}$ $J_{exo-endo}$ $J_{endo-endo}$	7·7[28] 2·3 8·9

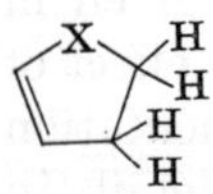	X	CH_2	CO	O	S
	J_{cis}	7·4	7·2	10·7	10·0[29]
	J_{trans}	4·6	2·2	8·3	7·5

the 1H, 2H couplings in the sugar ring being particularly striking. Note that the values of J_{trans} and J_{gauche} in the substituted ethanes are not measured couplings but extrapolated values: they are therefore less accurate than the other values quoted but within these limits, they show the same trends.

The measured couplings in substituted ethanes, which are averages of the couplings in the rotational isomers, lie between these extreme values.

Predictions of the theory can be extended to small ring compounds. In the planar cyclopropane ring J_{cis} ($J_{0°}$) is larger than J_{trans} (Table 8.1), as one would expect, but due to the increase in the 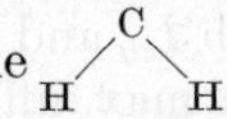 angle in this ring the *trans* dihedral angle is ca 150° which considerably increases the value of J_{trans}, as observed. Also the increasing electronegativity in the series CH_2, S, NH and O is clearly reflected in the values of the coupling constants observed in cyclopropane, ethylene sulphide, imine and oxide.

The same general trend is observed in four-membered rings with $J_{cis} > J_{trans}$, although there is little systematic data on these systems. For five-membered rings, it is generally true that $J_{cis} > J_{trans}$, but since in some cases there is the possibility that the ring will show a considerable degree of distortion, the orders of these couplings could be reversed. Thus it is necessary to consider the possible conformations of the ring before assigning, for example, *cis* or *trans* isomers on the basis of the observed couplings.

The general success of the theory has led to the extension of it to couplings involving a sp^2-hybridized carbon atom, although this case was not covered by the original theory. The observed coupling in propene is close to that in propane suggesting a similar mechanism and also there is undoubtedly a similar orientation dependence, resulting in $J_{trans} \gg J_{gauche}$ (Table 8.4). Even in the case of couplings involving aldehydic protons, for example CH.CHO couplings, which are anomalously low, there is again a similar orientation dependence. This immediately explains the larger couplings observed in unsaturated aldehydes (Table 8.1) as the favourable conformation of the double bonds results in a *trans* orientation of the protons. Even with couplings through an atom other than carbon, some orientation dependence has been observed, e.g. in CH.NH.CO couplings[30] and in CH.OH couplings.[31] It is very likely that J_{trans} and J_{gauche} follow the same trends as before, but the validity of a $\cos^2\phi$ relationship has not been tested.

The general success of the $\cos^2\phi$ law has led to a considerable number of applications of this rule, in particular to conformational problems in five- and six-membered rings. However, doubts have been expressed recently on the quantitative validity of the $\cos^2\phi$ equation. Thus the value of J_{ae} in cyclohexyl acetate is quite different from those of J_{ea} or J_{ee} (Table 8.4) although they should all be $J_{60°}$. (This was first given prominence for steroid alcohols,[32] which have almost identical couplings to those quoted here). Also $J_{endo-endo}$ does not equal $J_{exo-exo}$ in the camphane ring, though both are $J_{0°}$. A reversal of the electronegativity rule is also shown by the couplings in five-membered cyclic olefins (Table 8.4). In cyclopentene and cyclopentenone J_{cis} and J_{trans} are in general agreement with theory: for a planar ring these equal $J_{0°}$ and $J_{120°}$ respectively. Replacement of carbon by oxygen or sulphur, however, considerably *increases* both J_{cis} and in particular J_{trans}. Although the conformations of the molecules may differ slightly, such differences would not account for these values.

Thus the results of the quantitative application of the theory must be treated with caution until more rigorous experimental tests are performed. The chief difficulty in testing the theory being the lack of precise knowledge of the values of the dihedral angles between protons in most molecules. However, the general success of the theory in accounting for the trends in the large number of observed vicinal coupling constants is such that it may be used with confidence in a qualitative sense, e.g. to distinguish between different conformations or different geometric isomers (cf. Example No. 17, Appendix IV).

Long-range Couplings

(a) *σ–π-Couplings*

Although these couplings were formerly considered somewhat rare apart from some exceptional cases, the advent of spectrometers of very high resolution has enabled a large number of these couplings to be observed, and a number of reviews of them are available.[34, 48] These couplings can be conveniently classified into those which proceed through a σ–π-interaction and those which do not. Selected examples of the former are shown in Table 8.5. The two couplings most commonly encountered in this category are the allylic (CH:C.CH) and homo-allylic (CH.C:C.CH) couplings and numerous examples of these have been observed. Following a proposal that these couplings operate through the π-bond,[50] Karplus[35] developed a valence-bond treatment of these couplings which correctly forecasts both their magnitude and sign in many systems.

On this theory, there is a certain amount of overlap of the π-orbital and the hydrogen 1*S*-orbital in HC: and HC.C: systems. The amount of this overlap, or mixing, can be estimated from the observed values of the analogous couplings between an unpaired electron in a *p*-orbital and the corresponding protons in the electron spin resonance spectrum of the free radical. The coupling in a H—Ċ fragment is of opposite sign but approximately equal in magnitude to that in a CH_3—Ċ fragment. Furthermore the coupling in a HC—Ċ fragment is proportional to the amount of overlap of the hydrogen 1*S*-orbital and the *p*-orbital of the electron; i.e. to $\cos^2\phi$, where ϕ is the dihedral angle between the CH bond and the plane of the *p*-orbital. These results have been incorporated into the valence–bond theory of proton–proton couplings, and the calculated

TABLE 8.5

Selected examples of σ–π-couplings

Molecule	Observed coupling J_{cis}	J_{trans}	Calculated coupling[35]
CH_2:CH.CH_3	−1·7	−1·3[22]	−1·7
CHCl:CH.CH_2Cl	−0·5	−0·9[36]	
CH_3CH:C(CH_3).CO_2H	1·2	1·5[37]	+2·0
CH_3CH:C:CHCl	−5·8[38, 39]		−6·7
CH_3CH:C:CHCl	+2·4[38, 39]		
CH:C.CH_2.CH_3	−2·4[40]		−3·7
CH_3.C:C.CH_2.CH_3	+2·6[48]		+2·9
CH_3.C:C.C:CH	1·3[38]		
CH_3.C:C.C:C.CH_3	1·3[38]		
CH_3-Aryl	$J_{CH_3\text{-}H\ ortho}$		−1·10 to −1·5
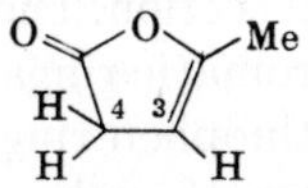	$J_{Me\text{-}3H}$	−1·6[34]	
	$J_{Me\text{-}4H}$	2·7	
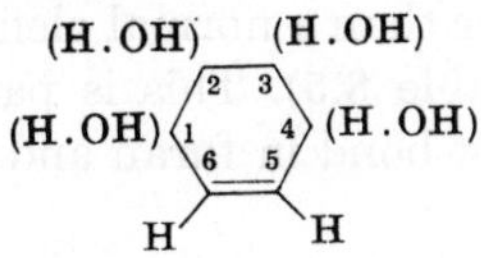	J_{15}	He′ −0·5; Ha′ −2·1[49]	
	J_{14}	e′e′ ca 0; e′a′ 1·3; a′a′ 3·0	

couplings for free rotation about the single bonds are shown with the observed values in Table 5: good general agreement obtains in all cases. One interesting result which emerges is that the coupling is not greatly attenuated over a number of bonds. The theory can be represented schematically for the allylic coupling as follows.

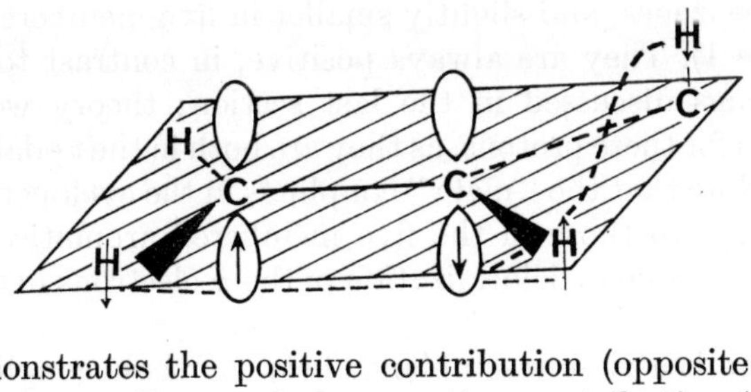

This demonstrates the positive contribution (opposite spin) to the vicinal olefinic couplings and the negative contribution (parallel spin) to the allylic couplings.

The theory also predicts from the e.s.r. couplings, that there will be a $\cos^2\phi$ dependence of the coupling for *each* CH.C: proton which means a $\cos^2\phi$ dependence of the allylic coupling and a $\cos^2\phi_1 . \cos^2\phi_2$ dependence of the homo-allylic (CH.C:C.CH) coupling, due to the involvement of two dihedral angles. Thus the couplings will be very small (ca 0) when the CH protons are in the plane of the double bond (the nodal plane of the π-orbital) and maximum for protons in the dihedral plane of the π-orbital. The large couplings observed in cyclic systems (Table 8.5) are well explained on this basis. The general agreement of the orientation dependence of these couplings with the $\cos^2\phi$ law is similar to that in the vicinal couplings; it probably suffers from the same approximations. However, as these couplings are much smaller, the approximations in the theory are less noticeable. The observed allylic couplings do differ from the predicted ones in those orientations in which the CH.C: proton lies in the plane of the double bond (Table 8.6).[51] This is due to a σ bond coupling over four bonds, which will be considered in the next section. The coupling of a methyl group to the *ortho* protons in many aromatic rings has been observed, particularly in the furan, pyrrole and thiophen rings. This is generally considered as an example of σ-π-coupling, though as the aromatic bond has less double bond character than a normal olefinic bond, the coupling is smaller than normal (Table 8.5). This is particularly noticeable with couplings through the 3–4 bond in furan and thiophen in which the coupling is only ca 0·5 c/s.[33]

(b) *Other long-range couplings*

Many other long range couplings have been observed which do not proceed via the σ–π-mechanism. We shall arbitrarily restrict the following discussion to couplings of a reasonable magnitude, namely, those $\geqslant$ 0·5 c/s.

The earliest and commonest example of such a coupling is the coupling between *meta* protons in aromatic systems. These are of ca 2–3 c/s in substituted benzenes, and slightly smaller in five-membered heteroaromatics (Table 1). They are always positive, in contrast to the normal allylic couplings discussed in the last section; theory would predict zero couplings for these protons, as they are both in the nodal plane of the π-electrons. Note that the "meta" couplings in the cyclopentadiene ring are very similar to those in the five membered aromatics, thus these couplings are not due to delocalization of the π-electrons in the aromatic rings.

Indeed there is a close resemblance between these couplings and the coupling between protons separated by four saturated bonds. In this latter the value is zero for a freely rotating fragment, but for specific orientations of the bonds the coupling may be of either sign. It is largest (> 1 c/s) if the bonds are in a planar zig-zag orientation i.e. 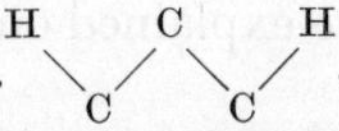 .[5] This coupling has often been observed between diequatorial protons in six-membered rings where the chair conformation allows this planar zig-zag orientation. Other analogous couplings are those between the 2-exo and 3-exo protons in the camphane-2,3-diols (Table 8.6) and the couplings shown in the bicyclo-2,2,1-heptanols (Table 8.6). A similar coupling between equatorial protons in a cyclohexene ring has been shown recently to be positive.[49] If the reasonable assumption is made that all the above couplings over four staggered saturated bonds are positive, then the similarity between these and the *meta* couplings in aromatic systems becomes clear. Thus the major part of the contribution to the aromatic *meta* couplings is undoubtedly through the σ-bonds. The exceptionally large magnitude of the coupling in benzenes could be due to either a small contribution from the π-orbitals or to very favourable angles in the 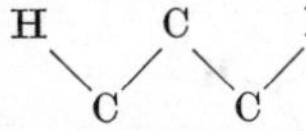 fragment.

Other examples of the planar zig-zag hypothesis over five bonds are the well established couplings of the aldehydic protons to the ring protons in furan, thiophen and benzaldehydes. In contrast to a methyl substituent, in which the largest coupling is with the proton *ortho* to the methyl, the aldehyde proton couples almost entirely with a proton *meta* to the aldehyde group. Some examples are shown in Table 8.6. In all these cases, a planar zig-zag orientation of the coupled protons can be drawn. Of particular interest is the coupling in the 2-substituted benzaldehydes,[5] where the largest coupling is with the 5-proton for all substituents except hydroxyl. In this case the intramolecular hydrogen bond reverses the favoured orientation of the aldehyde group and coupling with the 3 proton now arises. Two examples shown in Table 8.6 make this clear.

An exceptionally large coupling of 7–8 c/s between the α and α' protons in some substituted 2,1,1-bicyclohexanes (Table 8.6) has been observed. Although this coupling does conform to the planar zig-zag hypothesis, the very large J value suggests an additional mechanism. The authors suggest[56] that the coupling occurs through "fairly extensive overlap of the small lobes of orbitals directed 180° away from the direction of the C-5 and C-6 to proton bonds (and consequently pointed towards one another)".

In the above discussion, we have selected those long range couplings which either illustrate a particular theory or hypothesis, or are of interest in themselves. However, it should be remembered that there are many long range couplings, often of small magnitude (< 0·5 c/s), which cannot be explained on either of the above mechanisms.

TABLE 8.6
Selected examples of long range couplings

Molecule	Coupling (c/s)	Molecule	Coupling (c/s)
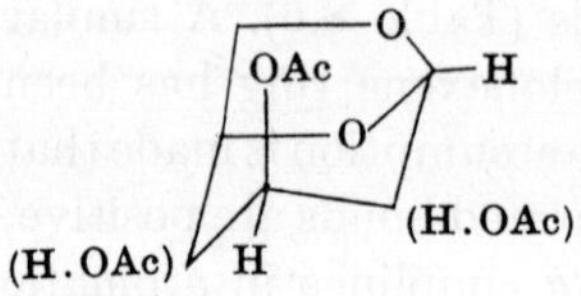	1·3–1·8[52]	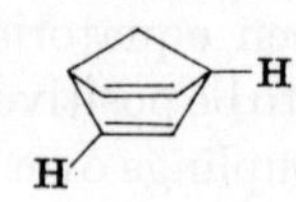	0·8[51]
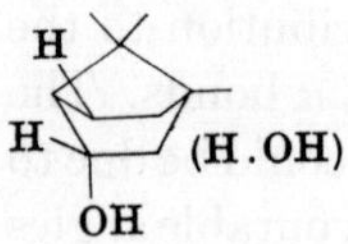	1·0–1·3[28]	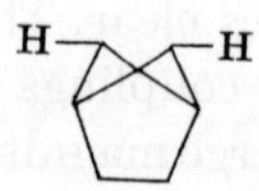	5–7[56, 57]
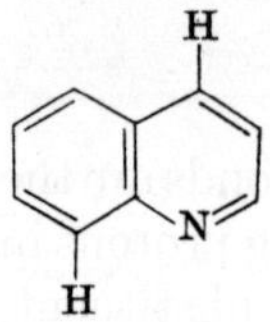	0·8[55]	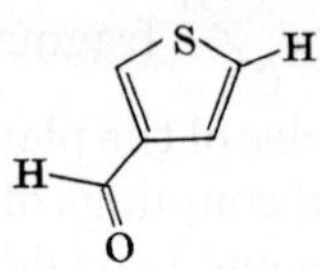	0·4–0·8[33]
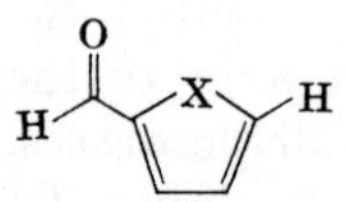	X=O, S $J=0·7–1·4$[33, 54]		
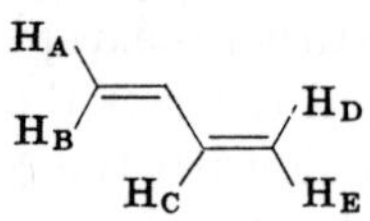	$J_{AE} \approx 1·5$ $J_{BC} \approx 0·7$	$J_{AC} \approx J_{AD} \approx J_{BD} \approx 0·5$[53]	
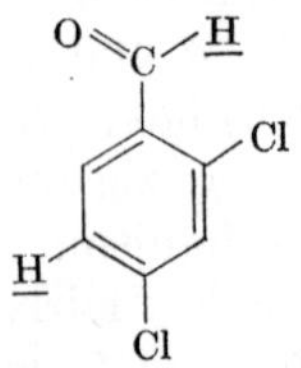	0·7[58]	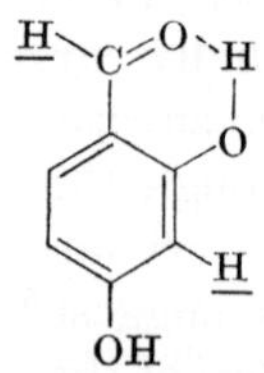	0·6[58]

References

1. J. A. Pople and D. P. Santry, *Mol. Phys.* **8**, 1 (1964).
2. A. D. Buckingham and K. A. McLauchlan, *Proc. Chem. Soc.* **144** (1963).
3. P. C. Lauterbur and R. J. Kurland, *J. Amer. Chem. Soc.* **84**, 3405 (1962).

4. J. A. Pople, W. G. Schneider and H. J. Bernstein, *High Resolution Nuclear Magnetic Resonance*, McGraw-Hill, London (1959).
5. C. A. Banwell and N. Sheppard, *Discuss. Faraday Soc.* **34**, 115 (1962).
6. R. C. Hirst and D. M. Grant, *J. Chem. Phys.* **40**, 1909 (1964).
7. H. S. Gutowsky, M. Karplus and D. M. Grant, *J. Chem. Phys.* **31**, 269 (1958).
8. T. Schaefer, *Canad. J. Chem.* **40**, 1 (1962).
9. R. M. Lynden-Bell and N. Sheppard, *Proc. Roy. Soc.* **A269**, 385 (1962).
10. B. L. Shapiro, R. M. Kopchik and S. J. Ebersole. *J. Chem. Phys.* **39**, 3154 (1963).
11. S. L. Manatt and D. D. Elleman, XLVII Meeting of S.I.P.S., Oct. 1964.
12. T. A. Crabb and R. C. Cookson, *Tetrahedron Letters* 679 (1964).
13. R. J. Abraham and W. A. Thomas, unpublished results.
14. M. Karplus and D. H. Anderson. *J. Chem. Phys.* **30**, 6 (1959).
15. A. A. Bothner-By and J. A. Pople, *J. Chem. Phys.* **42**, 1339 (1965).
16. M. Barfield and D. M. Grant, *J. Amer. Chem. Soc.* **85**, 1899 (1963).
17. H. S. Gutowsky and C. Juan, *Discuss. Faraday Soc.* **34**, 52 (1962).
18. E. B. Whipple, W. E. Stewart, G. S. Reddy and J. H. Goldstein, *J. Chem. Phys.* **34**, 2136 (1961).
19. W. G. Schneider, H. J. Bernstein and J. A. Pople, *Canad. J. Chem.* **35**, 1487 (1957).
20. M. Karplus, *J. Chem. Phys.* **30**, 11 (1959).
21. R. J. Abraham and K. G. R. Pachler, *Mol. Phys.* **1**, 165 (1963).
22. A. A. Bothner-By and C. Naar-Colin, *J. Amer. Chem. Soc.* **83**, 231 (1961).
23. R. J. Abraham and J. A. Pople, *Mol. Phys.* **3**, 609 (1960).
24. N. Sheppard and J. J. Turner, *Proc. Roy. Soc. A***252**, 506 (1959).
25. A. A. Bothner-By, C. Naar-Colin and H. Gunther, *J. Amer. Chem. Soc.* **84**, 2748 (1962).
26. F. A. L. Anet, *J. Amer. Chem. Soc.* **84**, 1053 (1962).
27. R. U. Lemieux, R. K. Kullnig, H. J. Bernstein and W. G. Schneider, *J. Amer. Chem. Soc.* **79**, 1005 (1957).
28. F. A. L. Anet, *Canad. J. Chem.* **39**, 789 (1961).
29. R. J. Abraham, R. C. Sheppard, W. A. Thomas and S. Turner, *Chem. Commun.* **43** (1965).
30. F. A. Bovey and G. V. D. Tiers, *J. Amer. Chem. Soc.* **81**, 2870 (1959).
31. R. J. Abraham, *Compt. rend.* **11**, 589 (1962).
32. D. H. Williams and N. S. Bhacca, *J. Amer. Chem. Soc.* **86**, 2742 (1964).
33. R. A. Hoffman and S. Gronowitz, *Arhiv. Kem* **16**, 563 (1960).
34. S. Sternhell, *Rev. Pure Appl Chem.* (*Australia*) **14**, 15 (1964).
35. M. Karplus, *J. Chem. Phys.* **33**, 1842 (1960).
36. F. S. Mortimer, *J. Mol. Spectroscopy*. **3**, 335 (1959).
37. R. R. Fraser, *Canad. J. Chem.* **38**, 549 (1960).
38. E. I. Snyder and J. D. Roberts, *J. Amer. Chem. Soc.* **84**, 1582 (1962).
39. S. L. Manatt and D. D. Elleman, *J. Amer. Chem. Soc.* **84**, 1579 (1962).
40. P. T. Narasimhan and M. T. Rogers, *J. Chem. Phys.* **33**, 727 (1960).
41. R. J. Abraham and H. J. Bernstein, *Canad. J. Chem.* **39**, 905 (1961).
42. S. Gronowitz and R. A. Hoffman, *Arhiv. Kem* **13**, 279 (1958).
43. G. V. Smith and H. Kriloff, *J. Amer. Chem. Soc.* **85**, 2016 (1963).
44. O. L. Chapman, *J. Amer. Chem. Soc.* **85**, 2014 (1963).
45. D. J. Patel, M. E. H. Howden and J. D. Roberts, *J. Amer. Chem. Soc.* **85**, 3218 (1963).
46. F. S. Mortimer, *J. Mol. Spectroscopy* **5**, 199 (1960).

47. J. I. Musher and R. G. Gordon, *J. Chem. Phys.* **36**, 3097 (1962).
48. R. A. Hoffman and S. Gronowitz, *Arkiv. Kem* **16**, 471 (1960).
49. R. J. Abraham, H. Gottschalck, H. Paulsen and W. A. Thomas, *J. Chem. Soc.* 6268 (1965).
50. R. A. Hoffman, *Mol. Phys.* **1**, 326 (1958).
51. F. S. Mortimer, *J. Mol. Spectroscopy* **3**, 528 (1959).
52. L. D. Hall, and L. Hough, *Proc. Chem. Soc.* **382** (1962).
53. E. O. Bishop and J. I. Musher, *Mol. Phys.* **6**, 621 (1963).
54. R. J. Abraham, and H. J. Bernstein, *Canad. J. Chem.* **37**, 1056 (1959).
55. F. A. L. Anet, *J. Chem. Phys.* **32**, 1274 (1960).
56. J. Meinwald and A. Lewis, *J. Amer. Chem. Soc.* **83**, 2769 (1961).
57. K. B. Wiberg, B. R. Lowry and B. J. Nist, *J. Amer. Chem. Soc.* **84**, 1594 (1962).
58. D. G. Kowalewski and V. J. Kowalewski, *J. Chem. Phys.* **37**, 1009 (1962).
59. M. Barfield and D. M. Grant, *J. Amer. Chem. Soc.* **85**, 1901 (1963).

CHAPTER 9

N.M.R. STUDIES OF NUCLEI OTHER THAN HYDROGEN

J. FEENEY

The basic requirement for an isotope to show a nuclear magnetic resonance effect is that the nucleus of the isotope must possess a magnetic moment, μ. Not all isotopes fulfil this requirement and to date there are about one hundred isotopes known to have magnetic moments. Common examples which fulfil this condition are ^{1}H, ^{7}Li, ^{11}B, ^{13}C, ^{14}N, ^{17}O, ^{19}F, ^{29}Si and ^{31}P while common nuclei with no moments are ^{12}C, ^{16}O and ^{32}S. These latter nuclei have spin numbers $I = 0$ and are known to have even numbers of protons and neutrons in their nuclei.

About 90% of all published n.m.r. studies are concerned with ^{1}H resonance. The next most intensively studied nucleus is ^{19}F whilst not inconsiderable attention has been paid to ^{11}B, ^{13}C, ^{14}N, ^{17}O and ^{31}P. Many other magnetic nuclei have been but briefly studied and it is instructive to consider the various factors which determine whether or not a nucleus can be studied successfully by high resolution n.m.r. These are as follows.

1. The magnetic moment of the nucleus must be large because the natural sensitivity of a nucleus to n.m.r. detection at constant field strength, is proportional to the cube of the magnetic moment (μ^3). Both ^{1}H and ^{19}F nuclei have large moments ($\mu_H = 2{\cdot}793$ and $\mu_F = 2{\cdot}627$ nuclear magnetons) and they are two of the nuclei which are most sensitive to detection. ^{14}N with its low magnetic moment ($\mu_N = 0{\cdot}404$) has a natural sensitivity of only 0·1% of that of the ^{1}H nucleus at constant field strength.

2. The isotope should have a high natural abundance although isotope enrichment of samples may be used to compensate for a low abundance. Factors (1) and (2) are concerned essentially with sensitivity problems and can be overcome to some extent by increasing the size of the sample. Another method is to use a device employing a computer averaging technique (CAT); thus by repeatedly scanning the spectrum of weak signals we can sum electronically any coherent signals whilst disregarding the incoherent (random) noise signals. The magnetic field must be

extremely stable for this technique to work satisfactorily. All methods for increasing sensitivity result in a loss of resolution.

3. The isotope should have spin-number $I = \frac{1}{2}$: higher spin numbers have quadrupole moments associated with them. Such quadrupole moments can interact with fluctuating electric field gradients in the molecule thus providing an efficient relaxation mechanism for the nuclei which leads to excessive line broadening. For example, ^{11}B ($I = 3/2$) and ^{14}N ($I = 1$) both possess quadrupole moments.

4. The nucleus must possess a short relaxation time to avoid saturation when suitable values of H_1 and sweep rate are employed. Relaxation times for ^{19}F are often longer than for ^{1}H although in both cases they are sufficiently short to allow the nuclei to be studied easily. ^{13}C relaxation times are so long that usually it is necessary to examine the nucleus under conditions of rapid passage in the dispersion mode. Although it is more difficult to saturate nuclei under such conditions, the attainable resolution is much less.

For all nuclei other than ^{1}H and ^{19}F there are sensitivity problems of varying degrees and the importance of examining such nuclei at the highest field strength available cannot be overstressed. A considerable amount of work has been expended on the study of nuclei other than hydrogen and the following brief account of n.m.r. studies of the nuclei of interest to organic chemists is in no way intended to be comprehensive. A number of reviews of n.m.r. studies of nuclei other than ^{1}H are available.[1–6]

Fluorine-19

This nucleus has ideal n.m.r. properties: it is the only naturally occurring fluorine isotope, it has a spin number $I = \frac{1}{2}$ and possesses a large magnetic moment. Table 9.1 gives a selection of typical ^{19}F chemical shifts measured from a CF_3COOH external reference[7]; they extend over a range of several hundred ppm and can be used to characterize different molecular environments in fluorine containing compounds.

Fluorine chemical shifts

Fluorine chemical shifts extend over a much wider range (~ 500 ppm) than that encountered for ^{1}H nuclei (~ 15 ppm). Whereas the diamagnetic shielding effect of the electron cloud dominates the shielding mechanism for ^{1}H nuclei, it accounts for only about 1% of the total observed range of ^{19}F chemical shifts. In this case the second order paramagnetic term in Ramsey's expression[8] for nuclear shielding has been shown to be responsible for the large observed chemical shift contributions. The magnitude

TABLE 9·1

Typical ^{19}F chemical shifts measured from CF_3COOH external reference[7]

Group	Range of chemical shifts (ppm)	Group	Range of chemical shifts (ppm)
$—SO_2F$	− 143 to − 117	$—CF_2CFI—$	30
$—C(O)F$	− 107 to − 94	$—CF_2CH_3$	29 to 37
$CF_3S—$	− 37·5 to − 36·5	$—CF_2CF<$	34
$—CF_2I$	− 23 to − 19	$—CF_2N=N—$	33·5
$CF_3C<$	− 22 to − 1	$—CF_2C(O)—$	40 to 54
$—CF_2Cl$	− 18 to − 5	$—CF_2CF_3$	40 to 48
$CF_3CF<$	− 7 to 0	$—CF_2CF_2—$	44 to 52
$—CF_2—O—$	− 1 to 11	$—CF_2CH_2—$	45 to 52
$CF_3CH_2—$	0 to 6	$—CF_2CF_2H$	53 to 55
$CF_3CHF—$	3 to 5	$—CF=CF—$	59 to 82
$CF_3CF_2—$	2·5 to 9	$—CF_2H$	60 to 64
$—CFBr_2$	2·5	$>CFI$	65 to 68·5
$CF_2=CF—$	7 to 50	$>CF=CF_2$	84 to 122
$—O—CF_2CF_3$	15	$>CFCF_3$	85 to 122·5
$—CF_2Br$	15	$—CF(CF_3)_2$	102 to 118
$—CF_2NO_2$	20	$>CFCF_2—$	114 to 120
$—CF=C(CF_3)_2$	21 to 44	$>CFH$	123 to 157

The above data are from a compilation by Brame.[7] Chemical shifts are measured from CF_3COOH external reference: high field shifts are positive.

of this term depends on the magnetic fields set up at the nucleus by the orbital motion of the valency electrons. For hydrogen and its isotopes we need only consider *s*-electrons which are known to exert a zero instantaneous orbital magnetic field at the nucleus and thus the paramagnetic shielding contribution is unimportant. Likewise there will be a zero paramagnetic contribution for the nucleus surrounded by a completely filled *p*-shell as is found in the F^- ion. For a covalently bound fluorine nucleus having a valency shell with unequal populations in its three 2

p-orbitals there is a large paramagnetic shielding contribution and this should be large in the F_2 molecule where the bond is completely covalent. Saika and Slichter[9] have estimated the paramagnetic contribution to the ^{19}F shielding in the F_2 molecule using the expression

$$\Delta\sigma_{para} = -\frac{2e^2\hbar^2}{3m^2C^2}\left(\frac{1}{r^3}\right)_{av}\cdot\frac{1}{\Delta E}$$

where e is the electronic charge, m is the mass of the electron, r is the average distance of the p-electrons from the nucleus and ΔE is an average value for the difference in energy between the ground and excited electronic states. The estimated $\Delta\sigma$ value is in only moderate agreement with the observed shielding difference between the F^- ion and the fluorine nuclei in the F_2 molecule. The paramagnetic shielding contribution arises from mixing of the valency electrons between their ground state and low lying excited states of suitable symmetry, a process induced by the presence of the applied magnetic field. For all magnetic nuclei other than the isotopes of hydrogen there is the possibility of low lying excited states above the valency p- or d-electrons, and thus the paramagnetic shielding contribution is usually the dominant factor controlling the shielding of such nuclei. Most nuclei therefore exhibit chemical shifts ranging over several hundred ppm. Because of the difficulty in detecting experimentally the presence of low lying excited electronic states of suitable symmetry for mixing, it is not surprising that, at present, ^{19}F shielding is not well understood.

In a series of binary fluorides, XF_x, the ^{19}F shielding increases as the electronegativity of X decreases[10]. This is because a decrease in the electronegativity of X increases the ionic character of the bond and thus decreases the paramagnetic contribution to the shielding: the result is that resonance signals are observed at higher field values. Thus in the series, F_2, NF_3, PF_3, AsF_3, SbF_3 and F^- the chemical shifts, measured to high fields of the F_2 fluorine nuclei, are 285, 463·7 469·1, 507·9 and 625 ppm respectively.

However, when we consider the ^{19}F chemical shifts in fluorocarbons the situation is more complex. Examination of Table 9.2 reveals that the ^{19}F shielding in different series of compounds can show opposite trends for the same changes in substituents. Such observations have been explained qualitatively by considering the shielding effects of both the partial ionic character in the CF bonds, and the presence of multiple bonded species (for example $Cl_2C{=}F^+$ species in CCl_3F).

It has been suggested that intramolecular electric fields arising from Van der Waal's dispersion interactions between fluorine atoms and neighbouring polarizable groups, can also make significant ^{19}F shielding

TABLE 9.2

^{19}F chemical shifts of some fluorocarbons measured from $CFCl_3$ internal reference

Compound	Chemical Shifts (ppm) CF_3	CF	Reference
CF_4	69		12
CF_3Cl	33		12
CF_3Br	21		12
CF_3I	5·0		12
$(CF_3)_2CF_2$	82·8	131·2 (CF_2)	13
$(CF_3)_2CFCl$	79·4	141·5	13
$(CF_3)_2CFBr$	77·7	144·0	13
$(CF_3)_2CFI$	75·7	148·6	13

The chemical shifts for the substituted methanes[12] were measured from $CFCl_3$ (external reference). High field shifts are positive. Chemical shifts can be converted to a CF_3COOH reference scale using the relationship

$$\delta^{ext}_{CF_3COOH} = \delta^{int}_{CFCl_3} - 76{\cdot}5 \text{ ppm}$$

contributions.[14] Neighbouring diamagnetic anisotropic shielding effects are only of the same order of magnitude as in ^1H resonance studies (< 2 ppm) and can thus be neglected.

Spin–spin interactions involving fluorine nuclei

In fluorocarbons, the complex mechanism for F—F and H—F coupling involves not only the Fermi contact term (which controls H—H coupling) but also the electron-orbital and dipole electron spin terms. At the moment there is no satisfactory theoretical interpretation of the observed coupling constants. Usually, F—F and H—F coupling constants (Table 9.3) are much larger than the corresponding H—H coupling constants. Vicinal F—F coupling constants vary over a range 0–18 c/s and they have been shown to decrease with increase in the total electronegativity of the substituent groups.[22] An increase in substituent electronegativity decreases the electron density in the C—C bond separating the interacting nuclei. Fluorine–fluorine spin coupling constants in (CF_3CF_2- groups) are very small (0–2 c/s) and are often smaller than the coupling constants between fluorine nuclei on 1,3 or 1,4 carbon atoms. Thus in the ^{19}F spectrum of $(CF_3)_2NCF_2CF_3$, the CF_2 absorption band is a septet ($J = 16$ c/s) clearly indicating that the CF_2 group is coupling (over 4 bonds) with the 6 equivalent fluorine nuclei of the $-N(CF_3)_2$ groups, rather than with the adjacent CF_3 group (over 3 bonds)[15]. Likewise the CF_3 group in perfluoromethylcyclohexane (see Fig. 9.1) couples more strongly with the ring CF_2 groups at the 2 and 6 positions than it does with the adjacent single fluorine.[16] Thus we observe 10 lines grouped as five doublets. The coupling between each pair of the quintet is $J_{(CF_3)(CF_2)_{2,6}} = 13{\cdot}6$ c/s, that between the two components of each doublet is $J_{(CF_3)F_1} = 6{\cdot}1$ c/s.

TABLE 9.3

Some typical F—F and F—H spin coupling constants

Group	H—F Coupling constant (c/s)	Group	F—F Coupling constant (c/s)
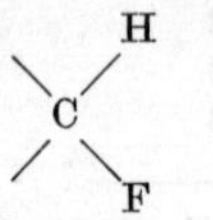	44–81	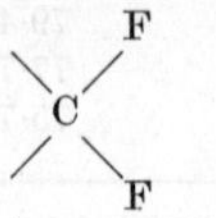	150–300
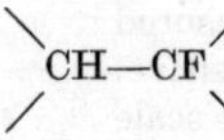	7–13*	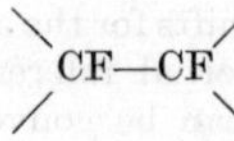	0–18
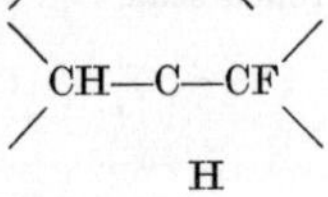	~0*	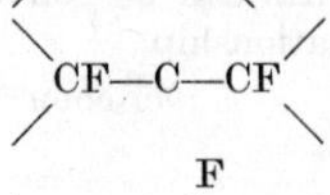	7–10
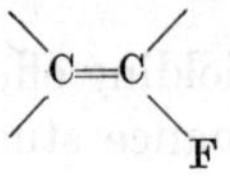	72–81	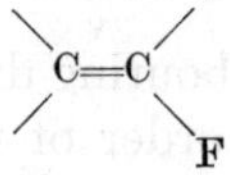	7–87
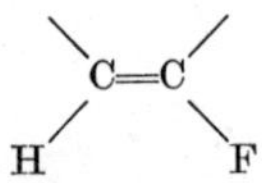	1–8	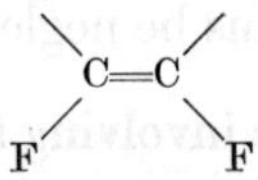	19–58
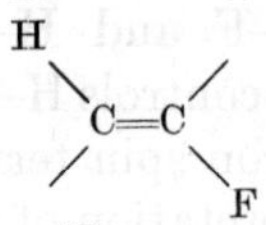	12–40	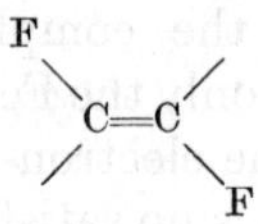	109–131
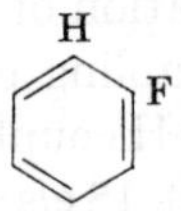	6–10†	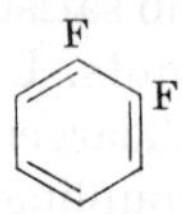	~20
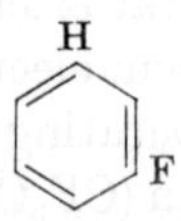	6–8·5†	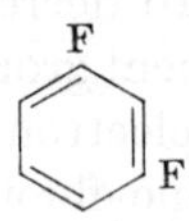	0–7
	~2†		0–15

* Varian Workshop Manual.

† Reference 58.

Further small splittings of 0·94 c/s arise from long-range coupling with other ring fluorine nuclei. Obviously caution is needed in the use of F—F spin coupling constants for assigning the relative positions of the interacting groups of nuclei within a molecule.

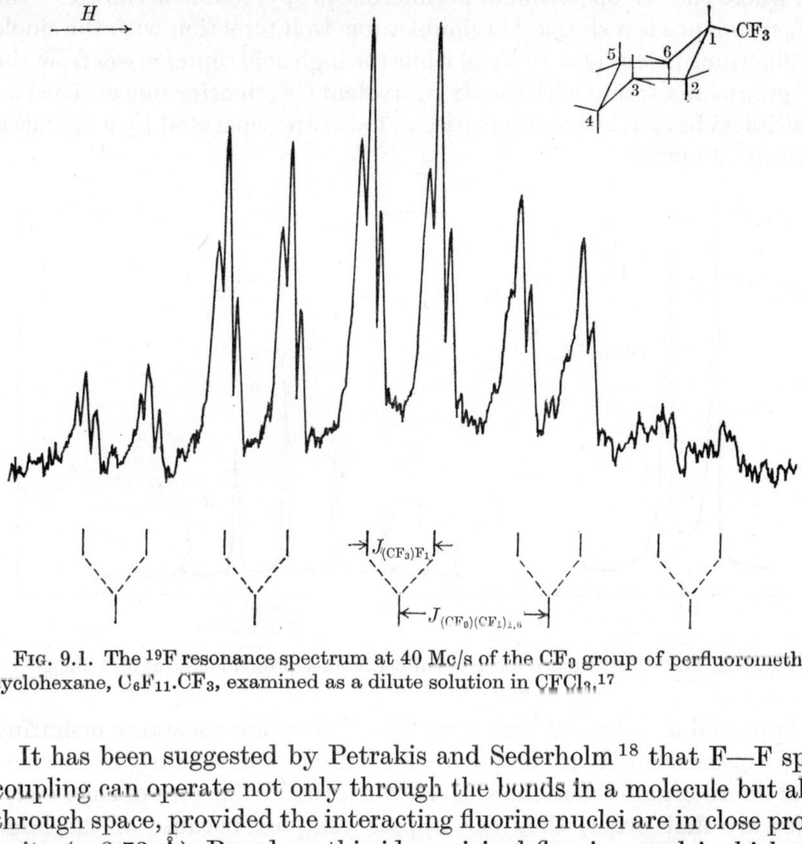

FIG. 9.1. The ^{19}F resonance spectrum at 40 Mc/s of the CF_3 group of perfluoromethylcyclohexane, $C_6F_{11}.CF_3$, examined as a dilute solution in $CFCl_3$.[17]

It has been suggested by Petrakis and Sederholm[18] that F—F spin coupling can operate not only through the bonds in a molecule but also through space, provided the interacting fluorine nuclei are in close proximity ($< 2{\cdot}73$ Å). Based on this idea, vicinal fluorine nuclei which are separated by about 2·73 Å have no through-space contribution to the coupling. Fluorine nuclei on 1, 3 and 1, 4 carbon atoms in fluoroalkanes can come into fairly close proximity (~ 2 Å) in certain conformations, and would be expected to have a large through-space contribution to the coupling: the large observed coupling constants may thus be explained. This correlation between coupling constants and internuclear separation, however, has been criticized in the literature[19, 20] and a greater body of experimental data are obviously required to establish a general relationship of this type.

Although coupling constants involving fluorine nuclei are larger than the corresponding H—H coupling constants they are usually much smaller than the observed ^{19}F chemical shifts. Because of this, ^{19}F resonance spectra are often first order and can be analysed readily using the simple $(2nI+1)$ splitting rule. A typical example is provided in Fig. 9.2 where the ^{19}F spectrum of perfluoroisopropyl iodide is shown[13]: the CF_3 resonance is a simple 1:1 doublet due to interaction with the single CF fluorine nucleus ($J = 12{\cdot}4$ c/s) while the high field septet arises from the CF group interacting with the six equivalent CF_3 fluorine nuclei. The two multiplets have relative intensities 6:1 and are separated by a chemical shift of 73 ppm.

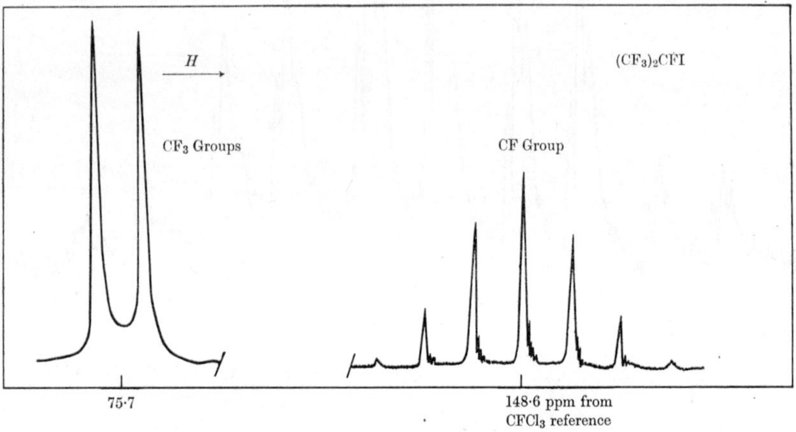

FIG. 9.2. The ^{19}F resonance spectrum of perfluoroisopropyl iodide at 56·4 Mc/s.[13]

Appreciable values of long range F—F coupling constants operating over 5 or more bonds are frequently encountered particularly in unsaturated fluoro-organic molecules. Even when the ^{19}F spectra of such molecules are first order, they can appear complex because of the large number of multiplet splittings. Figure 9.3 shows part of the ^{19}F spectrum of a fluoropentene[23] where the complex pattern can by analysed readily in terms of first order interactions. The F_3 fluorine nucleus couples with all the other magnetic nuclei in the molecule.

The n.m.r. determination of molecular structures of fluoro-organic compounds is usually simpler than the analogous problem for the corresponding hydrogen containing compounds. For example, if the ^1H spectrum of a mixture of trichlorobenzene isomers is examined, a complex pattern extending over less than 1 ppm is obtained. In Fig. 9.4 the spectrum obtained for a mixture of perfluorotrichlorobenzenes is shown.[13]

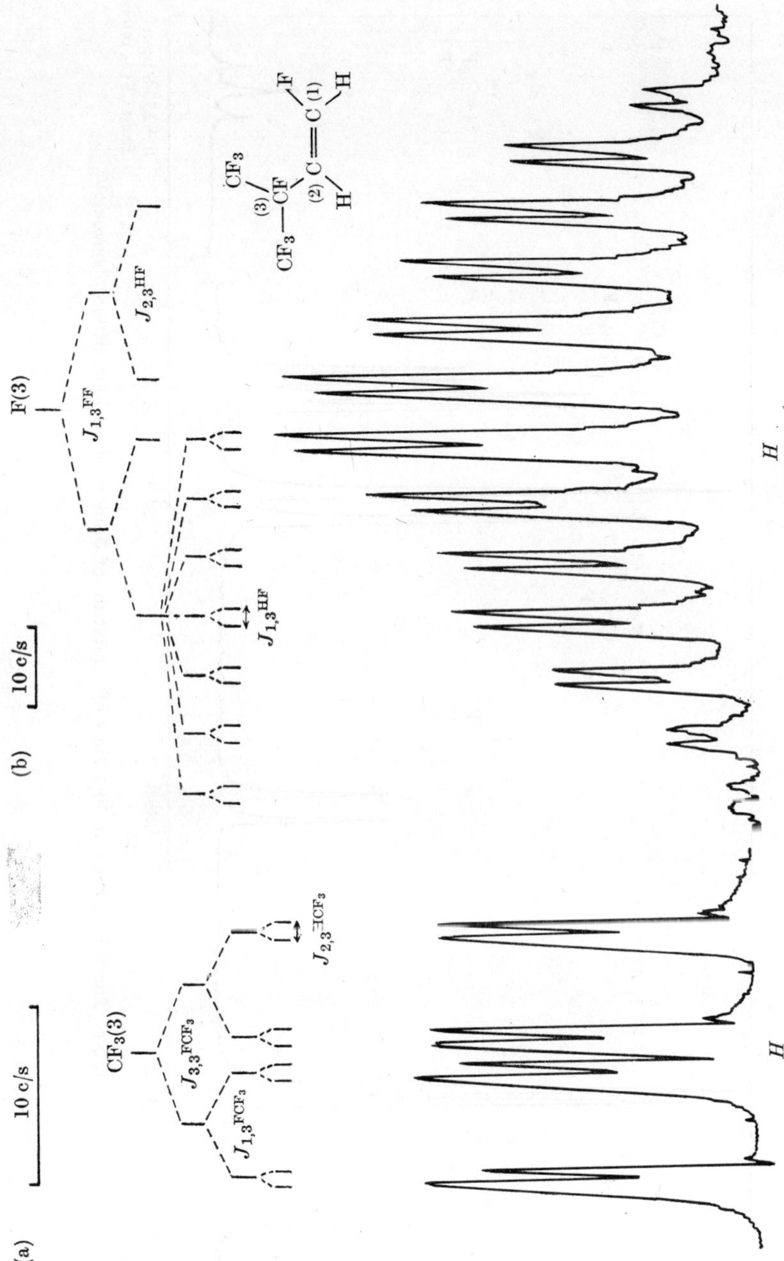

FIG. 9.3. Part of the ^{19}F resonance spectrum at 56·4 Mc/s of *cis*-i-$C_3F_7CH{=}CHF$.[23]

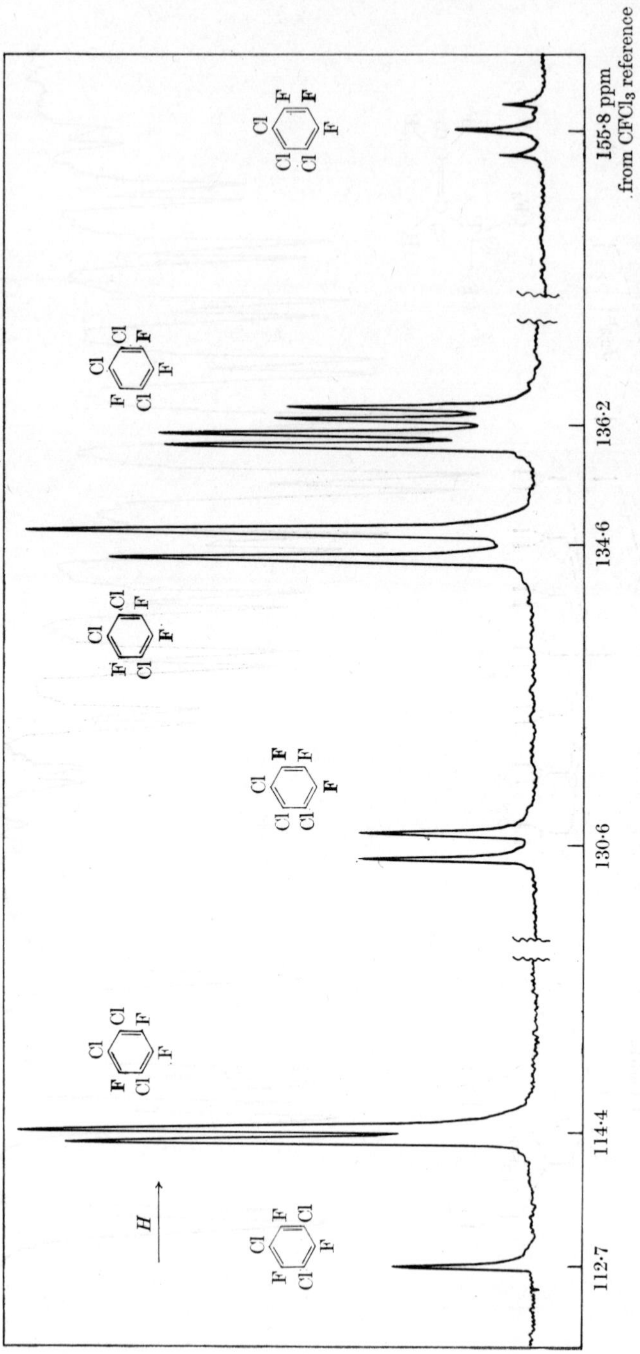

FIG. 9.4. The ^{19}F spectra at 56·4 Mc/s of a mixture of isomers of trichlorotrifluorobenzene.[13]

There is no overlap of bands and the first order spectra can be analysed readily. However, examination of the *ortho-*, *meta-* and *para-*F—F coupling constants (see Table 9.3) reveals a serious overlap in the observed values for J_{FF}^{para} and J_{FF}^{meta}; this lessens their value in determining positions of substitution in fluoroaromatic molecules.

The anisotropic ring currents detected in ^{1}H resonance studies of aromatic molecules make insignificant contributions to ^{19}F shielding in fluoroaromatic molecules. For such molecules an explanation of the ^{19}F chemical shifts has been suggested based on a semi-empirical theory which assumes that the major factors controlling the shielding are the π-electron densities and the C—F π-bond characters.[24, 25] This theory cannot account for the anomalous chemical shifts observed for fluorine nuclei *ortho* to substituents. It has been shown that such anomalous *ortho* shifts can be explained by invoking the presence of an intramolecular electric field effect[14]. This operates via the Van der Waal's dispersion interaction between polarisable substituents such as I, Br and Cl and the ortho fluorine atoms.

Conformational studies

Conformational studies using ^{19}F n.m.r. measurements have two advantages over similar ^{1}H studies:

(i) the various molecular conformations differ only in that the atoms are arranged differently in space and such differences give rise to much larger variations in ^{19}F shielding than in ^{1}H shielding;

(ii) if there are large chemical shift differences between the nuclei in different conformations then it is easier to "freeze-out" the conformers: this is because the minimum conformational lifetimes required for the detection of the spectra of individual conformers (rather than an averaged spectrum) are related to the reciprocals of the chemical shift differences ($1/\delta$) between the conformers. The above considerations also favour the study of slow chemical exchange processes involving fluorine compounds.[26] Many ^{19}F conformational studies of substituted fluoroethanes[27] and fluorocyclohexanes have been reported.[16, 28]

Boron-11

The boron-11 nucleus (natural abundance 81·2%) has a spin number $I = 3/2$ and possesses a considerable quadrupole moment. It has a fairly large magnetic moment and it is possible to obtain strong ^{11}B absorption signals using conventional 5 mm (outside diameter) spinning sample tubes. The other abundant boron isotope, ^{10}B (natural abundance 18·8%) also has a magnetic moment with a spin number $I = 3$. For ^{11}B, the observed chemical shifts extend over ~ 200 ppm and a typical compilation is given in Table 9.4.[29] There have been many applications of ^{11}B resonance measurements to the determination of molecular structures of inorganic compounds. For example it has been shown[30] that the monochloro-derivative of pentaborane-9 is substituted at a basal position. The ^{11}B resonance spectrum of B_5H_8Cl is shown in Fig. 9.5. Had substitution taken place at the apical position, then the spectrum should have shown two resonance signals, one for the substituted apical boron and a doublet for the four equivalent basal fluorine nuclei, each interacting with a terminal hydrogen nucleus. In fact four chemically shifted

TABLE 9.4

^{11}B chemical shifts and coupling constants for a series of boron compounds.[29]

Compound	Chemical shift (ppm)	B—H Coupling constant (c/s)
$B(CH_3)_3$	−68·2	
$B(C_2H_5)_3$	−66·6	
$C_6H_5BCl_2$	−35·9	
BF_3	+6·6	
BCl_3	−29·2	
BBr_3	−22·7	
BI_3	+23·3	
$C_6H_5B(OH)_2$ (in pyridine)	−15·2	
$C_4H_9B(OH)_2$ (in acetone)	−14·3	
$C_6H_5B(OC_2H_5)_2$	−10·4	
$B[N(C_2H_5)_2]_3$	−12·9	
$B(OCH_3)_3$	0	
$B(OC_2H_5)_3$	+0·6	
$B(O\text{-n-}C_3H_7)_3$	+0·5	
$B(O\text{-n-}C_4H_9)_3$	−0·1	
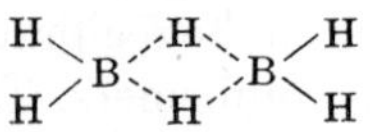	+0·5	137 (terminal) 48 (bridge)
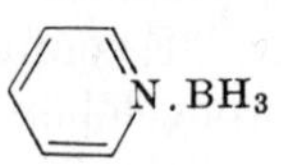	+31·4	90
$(CH_3)_2NH.BH_3$	+32·8	94
$(CH_3)_3N.BH_3$	+24·9	97
$(CH_3)_2PH.BH_3$	+55·6	96
$NaBC_6H_4$ (in H_2O)	+16·1	
$NaB(OCH_3)_4$ (in CH_3OH)	+15·2	
$Al(BH_4)_3$	+55·1	86
$NaBH_4$ (in 0·1 N NaOH)	+61·0	81
$LiB(C{\equiv}CC_6H_5)_4$	+49·4	

Chemical shifts measured from $B(OCH_3)_3$ reference; positive shifts are to high fields of the reference.

multiplets are observed and these are consistent with a basally substituted compound as indicated in the structure shown in Fig. 9.5.

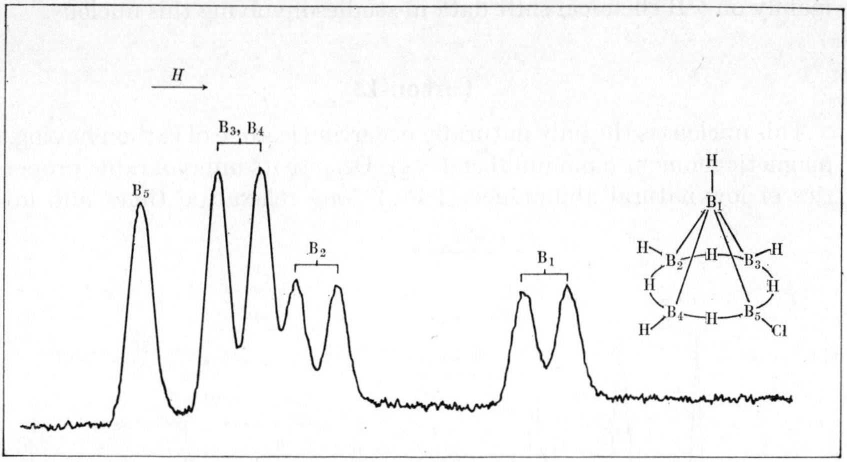

FIG. 9.5. The ^{11}B resonance spectrum of B_5H_8Cl at 23,400 gauss.[30]

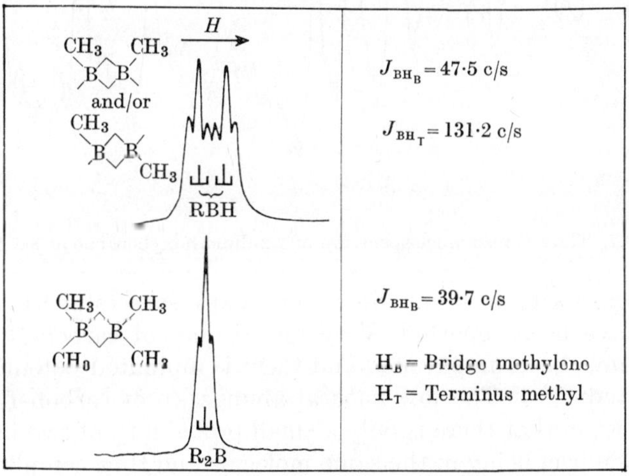

FIG. 9.6. The ^{11}B spectra of two alkyl diboranes.[31]

The bridge hydrogen nuclei are not coupled to the boron nuclei, a feature which is characteristic of the higher boranes. There is considerable quadrupolar broadening of the lines in the spectrum and this prevents small spin–spin splittings from being detected. In the spectra of the two alkyl diboranes (Fig. 9.6), the only fine structure observed is that between directly bonded boron and hydrogen atoms;[31] no B—C—H coupling involving the hydrogen nuclei of the alkyl groups can be detected. Because of the limited nature of ^{11}B coupling constants one must rely mainly on ^{11}B chemical shift data in studies involving this nucleus.

Carbon-13

This nucleus is the only naturally occurring isotope of carbon having a magnetic moment (spin number $I = \frac{1}{2}$). Despite its unfavourable properties of low natural abundance (1·1%), long relaxation times and low

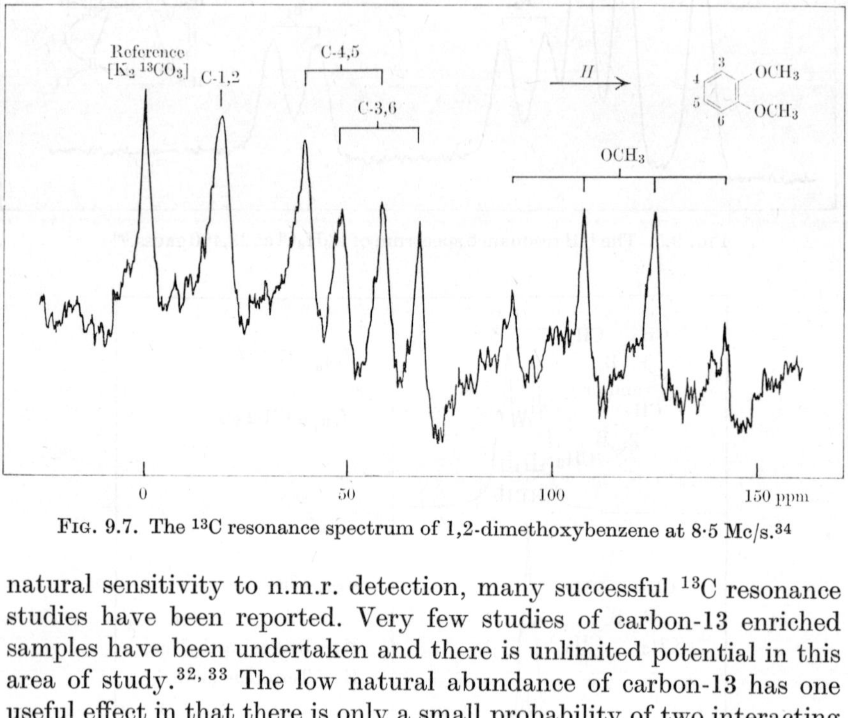

FIG. 9.7. The ^{13}C resonance spectrum of 1,2-dimethoxybenzene at 8·5 Mc/s.[34]

natural sensitivity to n.m.r. detection, many successful ^{13}C resonance studies have been reported. Very few studies of carbon-13 enriched samples have been undertaken and there is unlimited potential in this area of study.[32, 33] The low natural abundance of carbon-13 has one useful effect in that there is only a small probability of two interacting carbon-13 nuclei being in the same molecule and thus complicating the observed spectrum. To overcome the sensitivity problem associated with low natural abundance, large sample tubes (up to 15 mm O.D.) and high values of the radiofrequency field, H_1, are used. Obviously, the nucleus is examined at the highest magnetic field strength available. To avoid saturation, the ^{13}C resonance spectra are normally observed in the dispersion mode using rapid sweep rates (50 to 100 milligauss sec^{-1}) and H_1 values of up to 200 milligauss.[3, 35] Figure 9.7 shows the ^{13}C resonance spectrum of 1,2-dimethoxybenzene and it is typical of ^{13}C spectra recorded under rapid sweep, dispersion mode conditions.[34] Saturation effects cause considerable distortion of the signal and the attainable resolution is quite low ($\sim$10 c/s). Lauterbur[34] has made the major contribution to ^{13}C resonance studies and Table 9.5 includes ^{13}C chemical shift data taken from his work.

Studies of alkanes,[3] alkenes,[38] aromatics,[36] ketones[38, 39] and many other series of compounds have been made. ^{13}C chemical shifts often

TABLE 9.5

^{13}C chemical shifts and coupling constants in a series of compounds

Compound[3]	Chemical shift (ppm)	Compound[38]	Chemical shift (ppm)
CCl_4	81	CS_2	$-15{\cdot}6$
$CHCl_3$	98	$*C(CH_3)_4$	150·3
CH_2Cl_2	124	$C(*CH_3)_4$	146·2
CBr_4	205	Olefins	$\sim$25 to 75
$CHBr_3$	168	Acetylenes	$\sim$90 to 110
CH_2Br_2	156	Carbonyls	-40 to $+30$
CH_3Br	165	Benzene	49·3
CHI_3	316	Alkanes	140 to 180
CH_2I_2	240		
CH_3I	198		
$C_6H_5{*}COOCH_3$	11·1		
$C_6H_5{*}CH_3$	156·9		
$*CH_3CH_2Cl$	159·9		

Compound	C—H Coupling constant (c/s)	Reference	Compound	Coupling constant (c/s)	Reference
CH_3CH_3	126	42	$(CH_3)_3C{*}COOH$	$J_{CCCH} = 4{\cdot}2$	41
CH_3F	149	42	$(CH_3)_2CH{*}COOH$	$J_{CCH} = 5{\cdot}2$	41
$CHCl_3$	209	42	$*C{=}{*}CH_2$	$J_{CC} = 70$	32
$>C{=}{*}CH_2$	160	40	$C_6H_5{*}C{\equiv}{*}CH$	$J_{CC} = 176$	32
$C_6H_5C{\equiv}{*}CH$	251	32	$C_6H_5{*}CH_2{*}CH_3$	$J_{CC} = 34$	32
C_6H_6	160	3	CF_3X	$J_{CF} \approx 275$	21

Chemical shifts are quoted in ppm from $CH_3{*}COOH$ external reference. Values given in the right hand column have been converted from a CS_2 reference scale using the relationship

$$\delta_{CS_2} = \delta_{CH_3{*}COOH} + 15{\cdot}6 \text{ ppm}$$

* Indicates the ^{13}C nucleus involved in the measurement.

depend simply on the electronegativity of the attached substituents, but for easily polarizable groups, dispersion effects are also important.[3, 37]

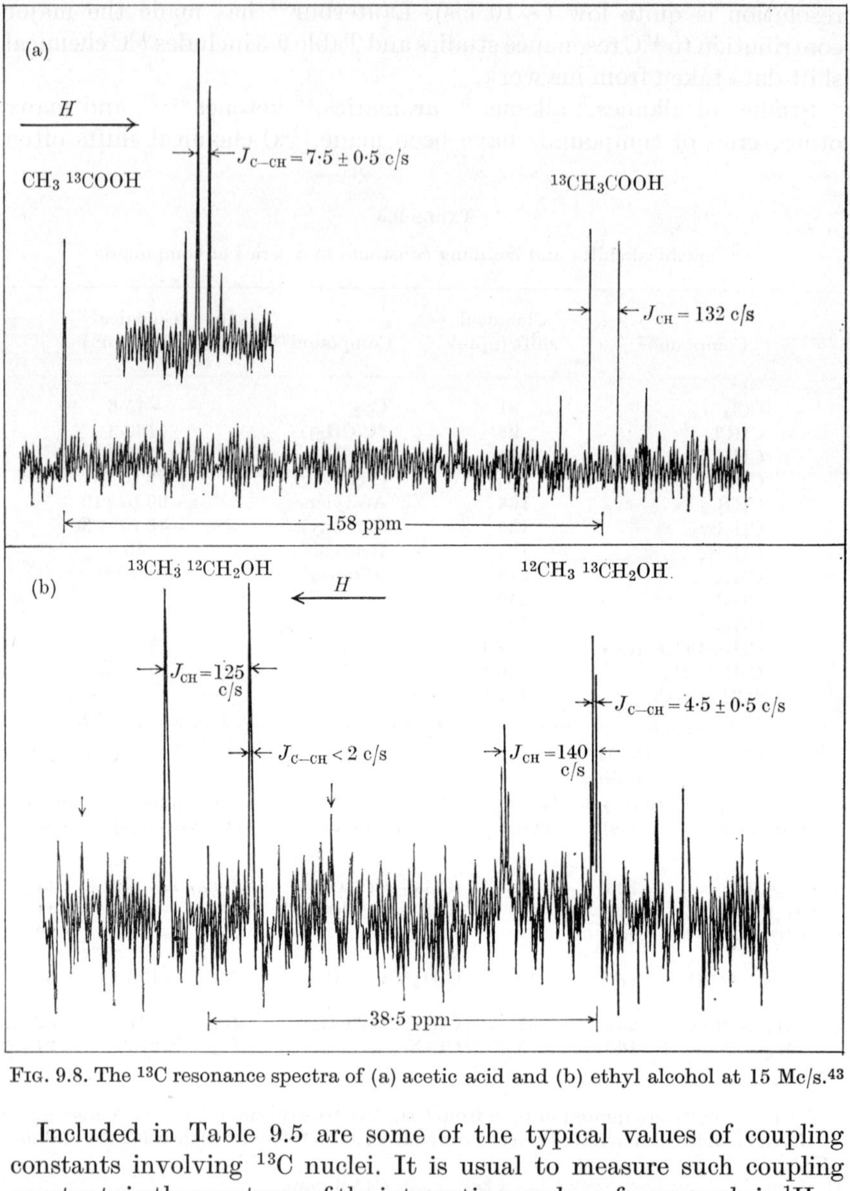

FIG. 9.8. The ^{13}C resonance spectra of (a) acetic acid and (b) ethyl alcohol at 15 Mc/s.[43]

Included in Table 9.5 are some of the typical values of coupling constants involving ^{13}C nuclei. It is usual to measure such coupling constants in the spectrum of the interacting nucleus, for example in ^{1}H or ^{19}F spectra where they appear in ^{13}CH and ^{13}CF satellite bands almost equally displaced about the signals from the nuclei attached to the corresponding ^{12}C atoms. Coupling constants between directly bonded carbon and hydrogen nuclei depend on the percentage *s* character in the bond.[40]

By using 15 mm O.D. spinning sample tubes, side-band modulation techniques, and operating in the absorption mode with fairly slow sweep rates, considerable improvement in resolution is possible. Figure 9.8 shows the ^{13}C resonance spectra of acetic acid and ethanol recorded under such conditions[43]; it is possible to resolve long range carbon, hydrogen spin-spin interactions (~ 2 c/s) between nuclei separated by two bonds. Such resolution is also useful in resolving bands with small chemical shift differences as observed in the ^{13}C spectra of aromatic molecules.[36]

The potential value of ^{13}C resonance studies to the organic chemist should stimulate a rapid increase in the amount of attention given to this nucleus in the future.

Nitrogen-14

^{14}N, the most commonly occurring isotope of nitrogen (99·6% abundance) has a spin number $I = 1$ and possesses a quadrupole moment. Its magnetic moment is small and thus it has a very low natural sensitivity to n.m.r. detection (0·101% that of hydrogen at constant field strength). Despite its unfavourable properties, ^{14}N chemical shifts have been measured for many compounds and a list of typical values is given in Table 9.6.[44]

TABLE 9.6

^{14}N chemical shifts of some common compounds[44]

Compound	Chemical shift (ppm)	Compound	Chemical shift (ppm)
NH_4^+	602	CN^-	380
$(C_3H_7)_2NH$	575	$C(NO_2)_4$	300
$(CH_3CH_2)_3N$	575	Pyridine	276
N_2H_4	566	N_2	268
NH_3	544	NO_3^-	254
$(NH_2)_2CO$	536	$C_6H_5NO_2$	252
CH_3CONH_2	498	n-$C_3H_7NO_2$	228
SCN^-	406	NO_2^-	0
CH_3CN	385		

Chemical shifts measured from NO_2^- ion; positive shifts are to high fields of the reference.

Oxygen-17

This isotope has a spin-number $I = 5/2$. It is difficult to study by n.m.r. because of its low natural abundance ($3{\cdot}7 \times 10^{-2}$ %), poor natural sensitivity to n.m.r. detection, and considerable quadrupole moment. Despite this, several ^{17}O studies have been reported, the spectra being observed usually in the dispersion mode[45–49]. The selection of ^{17}O chemical shifts listed in Table 9.7 indicates the observed range of ^{17}O chemical shifts.

The presence of a single absorption band in the ^{17}O spectra of the -COOH groups of formic, acetic and propionic acids, indicates that the two oxygen nuclei have a similar environment thus supporting the resonance picture of the molecules. Furthermore, the observed chemical shift is intermediate[46] between those expected for C—O and ketonic C═O types of oxygen. ^{17}O chemical shifts have been used to determine the extent of partial double bond character in C—O type bonds.[45]

Successful studies of slow keto–enol equilibria have been made using ^{17}O resonance measurements.[46]

TABLE 9.7

^{17}O chemical shifts of a series of compounds

Compound (Ref. 47)	Chemical shift (ppm)	Compound (Ref. 46)	Chemical shift (ppm)
H_2O	0	Dioxane	3
n-Butyl nitrite	−170, −380	CH_3OH	37
H_2O_2 (30%)	−190	C_2H_5OH	−5
CH_3COOH	−220	C_3H_7COOH	−250
HCOOH	−270	CH_3COCH_3	−568
HCOONa	−300	CH_3CHO	−595
Formamide	−320	Furan	−243
$C_2H_5NO_3$	−340, −470	H_2SO_4 (98%)	−138
$NaNO_3$	−430	HNO_3 (69%)	−409
SO_2 (liq.)	−540	D_2O (96%)	3
$C_2H_5NO_2$	−640	$HClO_4$ (60%)	−288
$NaNO_2$	−690		

Chemical shifts measured in ppm from H_2O reference; negative shifts are to low fields of the reference.

Although ^{13}C, ^{14}N and ^{17}O chemical shifts have been measured for numerous compounds, up to the present there have been but few applications of such measurements to determination of molecular structures. With improved instrument sensitivity, achieved by using higher magnetic field strengths and computer averaging techniques, more studies of these nuclei in molecules of unknown structure will be undertaken.

Phosphorus-31

^{31}P is the 100% naturally occurring isotope of phosphorus: it has a spin number $I=\frac{1}{2}$ and its relaxation times are sufficiently short to allow the nucleus to be studied using the absorption mode under fairly slow sweep rate conditions. It is an excellent nucleus for n.m.r. investigation and from the extensive studies which have been made there is an abundance of n.m.r. data available for ^{31}P.[1, 4–6] A selection of typical ^{31}P chemical shifts and coupling constants involving the ^{31}P nucleus is provided in Table 9.8. At the high field strengths now used for ^{31}P studies (23,400 gauss) high quality spectra can be obtained using samples contained in 5 mm. (outside diameter) spinning sample tubes. An example is the ^{31}P spectrum of $P(OCH_3)_3$ reproduced in Fig. 9.9. The well resolved fine structure components arise from spin–spin interaction with nine equivalent methyl hydrogen nuclei.[30].

TABLE 9.8

^{31}P chemical shifts and coupling constants

(a)

Compound	Chemical shift (ppm)	Compound	Chemical shift (ppm)
$POBr_3$	+102·9	$CH_3(C_2H_5O)P(O)H$	−32·0
$P(CH_3)_3$	+61·0	$C_6H_5P(O)Cl_2$	−34·5
$P(C_2H_5)_3$	+20·4	$CH_3(C_2H_5O)P(O)Cl$	−39·5
$P(CF_3)_2CN^+$	+40·7	$CH_3P(O)Cl_2$	−44·5
$P(n.C_4H_9)_3$	+32·3	$C_2H_5(C_2H_5O)P(O)Cl$	−45·0
$[(OH)_2P(O)]_2O$	+10·6	$(C_2H_5O)_2P(S)Cl$	−68·1
$P(C_6H_5)_3$	+5·9	$(C_2H_5O)_2P(S)H$	−69·0
$P(CF_3)_3$**	+2·6		
$POCl_3$	−1·9	$CH_3(C_2H_5)P(O)Cl$	−72·0
$(C_2H_5O)_2P(O)Cl$	−2·8	$CH_3P(S)Cl_2$	−79·8
$(CH_3O)_2P(O)H$	−11·3	$[CH_3(C_2H_5O)P(S)]_2O$	−85·4
$H_2(OH)P(O)$	−12·0	$(C_2H_5O)_2P(S)SH$	−85·7
$CH_3(i\text{-}C_3H_7O)P(O)F$	−16·1	$CH_3(C_2H_5O)P(S)OH$	−88·8
$[CH_3(C_2H_5O)P(O)]_2O$	−22·4	$CH_3(C_2H_5O)P(S)Cl$	−94·2
$(C_2H_5O)_2POSH$	−24·0	PF_3*	−97·0
CH_3POF_2	−27·4	$(CF_3)_2PF$**	−123·9
$CH_3(C_2H_5O)P(O)OH$	−28·5	PI_3*	−178·0
$PSCl_3$	−28·8	PCl_3	−219·4
$CH_3P(O)(OH)_2$	−31·1	PBr_3	−227·4

TABLE 9.8—*continued*

(b)

Compound or group	P—H Coupling constant (c/s)	Ref.	Compound or group	P—F Coupling constant (c/s)	Ref.
>P—H	180 to 210	5	$(CH_3)_2PF_3$	960 (equatorial) 772 (axial)	55
>P(O)H	500 to 700	5	$(CH_3)_3PF_2$	545	55
>P(S)H	490 to 630	5	$(CF_3)_3PF_2$	988	55
H—X—P(O)< (with bonds on P)	~10	4	$(CF_3)_2PCl$	85·1	51
H—X—P(S)< (with bonds on P)	15 to 20	4	$(CF_3)_2PCl_3$	193·3	51
H—X—Y—P(S) (with bonds on P)	15 to 20	4	$(CF_3)_3P(O)$	113·4	51
H—X—Y—P(O) (with bonds on P)	⩽5	4	$(CF_3)_2PH$	68·6	51
H—X—P<	15 to 20	4	PF_3	1410 ± 20	50
H—X—Y—P<	15 to 20	4	HPF_6	707 ± 20	50
$(CH_3O)_2P(O)H$	690·3	54	FP(O)R(Halogen)	~1100	56
$(C_2H_5O)_2P(S)H$	645	4			
YCH_2P (Halogen)$_2$	~19	56			

These ^{31}P chemical shifts have been extracted from a compilation made by H. Finegold.[4] The shifts were measured from 85% H_3PO_4 external reference: positive shifts are to high fields of the reference.

* Ref. 50 ** Ref. 51

The smaller marked bands on the spectrum arise from ^{13}C—^{31}P coupling which is present in 3·3% of the molecules.

It is interesting to note that in triethyl phosphorus, $P(C_2H_5)_3$, the phosphorus is coupled more strongly to the CH_3 hydrogen nuclei ($J_{P—CH_3} = 13·7$ c/s) than it is to the CH_2 hydrogen nuclei ($J_{P—CH_2} = 0·5$ c/s).[52] This is a feature common to alkyl derivatives of N, Pb, Tl, Hg and other heavy atoms. In trivinyl phosphine, $P(\overset{\alpha}{C}H=\overset{\beta}{C}H_2)_3$, the P—H coupling involving the α-CH hydrogen nucleus is much larger ($J_{PH} = 11·74$ c/s) than the corresponding coupling constant in $P(C_2H_5)_3$: the coupling

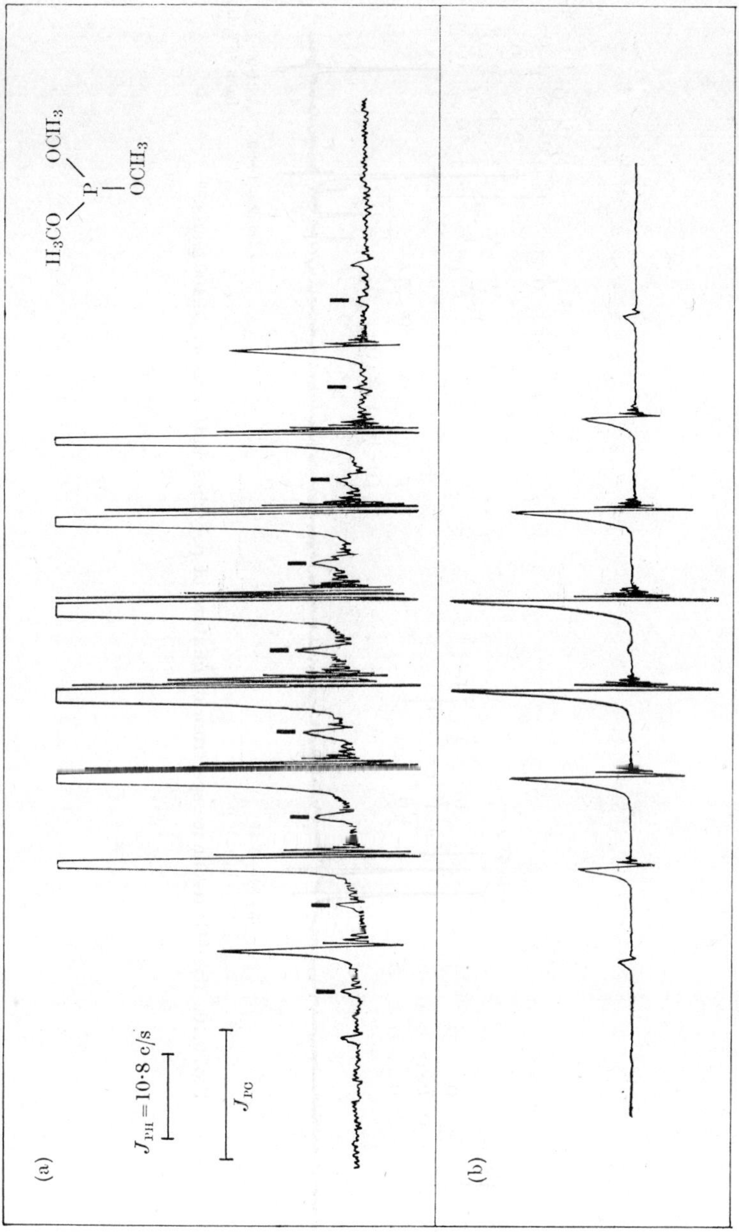

FIG. 9.9. The ^{31}P resonance spectrum of $P(OCH_3)_3$ at 23,400 gauss.[30]

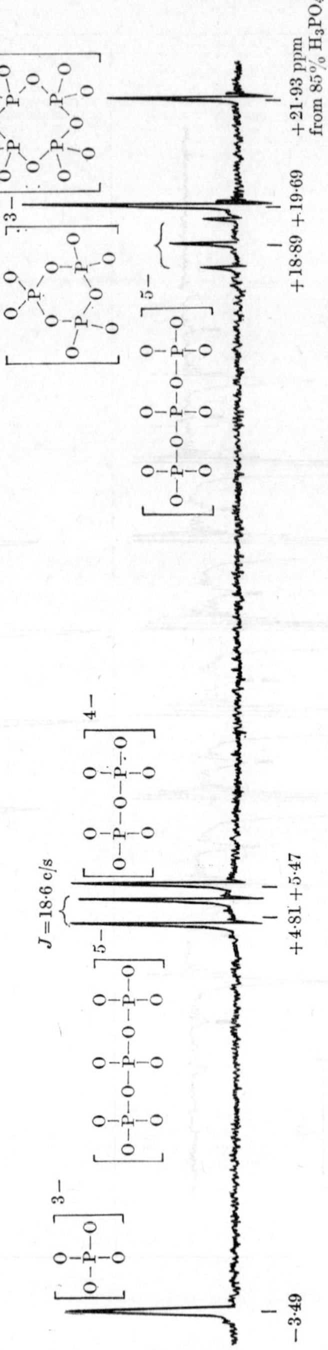

FIG. 9.10. The ^{31}P resonance spectrum of mixture of polyphosphate ions at 23,400 gauss.[30]

constants with the β-hydrogen nuclei are $J^{cis}_{P—H} = 13{\cdot}62$ c/s and $J^{trans}_{P—H} = 30{\cdot}21$ c/s.[53]

N.M.R. has provided a particularly useful tool for the study of polyphosphoric acids. Figure 9.10 indicates the ease with which a mixture of polyphosphate ions can be analysed using the technique.

Many phosphonitrilic compounds have been characterized by ^{31}P n.m.r. measurements and for certain derivatives it is possible to determine the nature of the substitution. Thus the ^{31}P spectrum of a mercaptophosphazene shown in Fig. 9.11 proves that two thioethyl groups are substituted on the same phosphorus atom.[57] The two absorption multiplets are in the intensity ratio 1:2, and in addition to the small P—P triplet splitting in the low field multiplet, there is a large quintet splitting ($J_{P—CH_2} = 18{\cdot}1$ c/s). This is consistent only with the two thioethyl groups being on the same phosphorus atom which thus interacts with four equivalent hydrogen nuclei in the two $-CH_2$ groups. The ^1H resonance spectrum confirms this assignment.

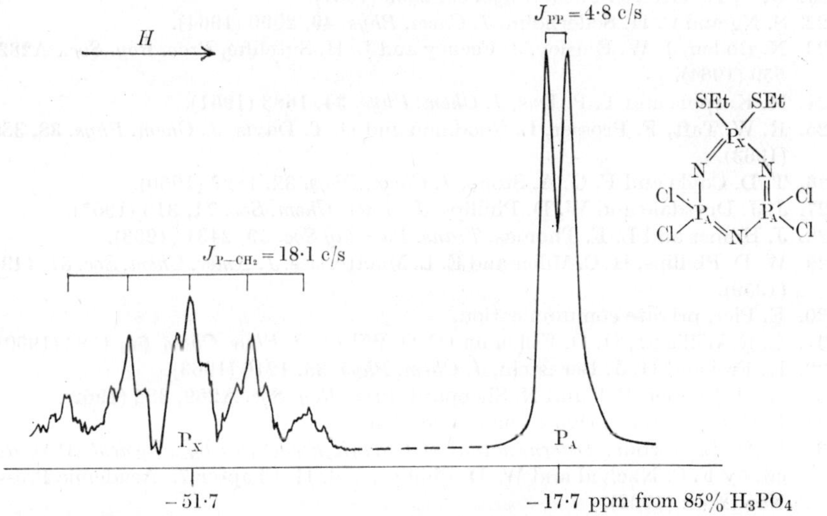

FIG. 9.11. The ^{31}P resonance spectrum at 16·2 Mc/s of a mercaptophosphazene $N_3P_3Cl_4(SC_2H_5)_2$.[57]

References

1. J. A. Pople, W. G. Schneider and H. J. Bernstein, *High-resolution Nuclear Magnetic Resonance*. McGraw-Hill, New York (1959).
2. W. D. Phillips, In *Determination of Organic Structures by Physical Methods*, ed. by F. C. Nachod and W. D. Phillips, Vol. II, Chapter 6. Academic Press, New York (1962).

3. P. C. Lauterbur, *Ann. N. Y. Acad. Sci.* **70**, 841 (1958).
4. H. Finegold, *Ann. N.Y. Acad. Sci.* **70**, 875 (1958).
5. R. A. Y. Jones and A. R. Katritzky, *Angew. Chem.* **74**, 60 (1962).
6. J. W. Emsley, J. Feeney and L. H. Sutcliffe, *High Resolution Nuclear Magnetic Resonance Spectroscopy*. Pergamon Press, Oxford (1965).
7. E. G. Brame, *Analyt. Chem.* **34**, 591 (1962).
8. N. F. Ramsey, *Phys. Rev.* **78**, 699 (1950).
9. A. Saika and C. P. Slichter, *J. Chem. Phys.* **22**, 26 (1954).
10. H. S. Gutowsky and C. J. Hoffman, *J. Chem. Phys.* **19**, 1259 (1951).
11. L. H. Meyer and H. S. Gutowsky, *J. Chem. Phys.* **57**, 481 (1953).
12. N. Muller and D. T. Carr, *J. Phys. Chem.* **67**, 112 (1963).
13. N. Boden, J. W. Emsley, J. Feeney and L. H. Sutcliffe, unpublished results.
14. N. Boden, J. W. Emsley, J. Feeney and L. H. Sutcliffe, *Mol. Phys.* **8**, 133 (1964).
15. A. Saika and H. S. Gutowsky, *J. Amer. Chem. Soc.* **78**, 4818 (1956).
16. J. Feeney and L. H. Sutcliffe, *Trans. Faraday Soc.* **56**, 1559 (1960).
17. J. Feeney, L. H. Sutcliffe and S. M. Walker, unpublished results.
18. L. Petrakis and C. H. Sederholm, *J. chem. Phys.* **35**, 1243 (1961).
19. D. F. Evans, *Discuss. Faraday Soc.* **34**, 139 (1962).
20. R. K. Harris and N. Sheppard, *Trans. Faraday Soc.* **59**, 606 (1963).
21. G. V. D. Tiers, *J. Chem. Phys.* **35**, 2263 (1961).
22. S. Ng and C. H. Sederholm, *J. Chem. Phys.* **40**, 2090 (1964).
23. N. Boden, J. W. Emsley, J. Feeney and L. H. Sutcliffe, *Proc. Roy. Soc.*, **A282**, 559 (1964).
24. M. Karplus and T. P. Das, *J. Chem. Phys.* **34**, 1683 (1961).
25. R. W. Taft, F. Prosser, L. Goodman and G. T. Davis, *J. Chem. Phys.* **38**, 380 (1963).
26. T. D. Coyle and F. G. A. Stone, *J. Chem. Phys.* **32**, 1892 (1960).
27. J. J. Drysdale and W. D. Phillips, *J. Amer. Chem. Soc.* **71**, **319** (1957).
28. J. Homer and L. F. Thomas, *Trans. Faraday Soc.* **59**, 2431 (1963).
29. W. D. Phillips, H. C. Miller and E. L. Muetterties. *J. Amer. Chem. Soc.* **81**, **4496** (1959).
30. E. Pier, private communication.
31. R. E. Williams, H. D. Fisher and C. O. Wilson, *J. Phys. Chem.* **64**, 1583 (1960).
32. K. Frei and H. J. Bernstein, *J. Chem. Phys.* **38**, 1216 (1963).
33. R. M. Lynden-Bell and N. Sheppard, *Proc. Roy. Soc.* **A269**, 385 (1962).
34. P. C. Lauterbur, private communication.
35. P. C. Lauterbur, *Determination of Organic Structures by Physical Methods*, ed. by F. C. Nachod and W. D. Phillips, Vol. II, Chapter 7. Academic Press, New York (1962).
36. P. C. Lauterbur, *J. Amer. Chem. Soc.* **83**, 1838 (1961).
37. P. C. Lauterbur, *J. Chem. Phys.* **38**, 1406 (1963).
38. R. A. Friedel and H. L. Retcofsky, *J. Amer. Chem. Soc.* **85**, 1300 (1963).
39. G. E. Maciel and G. C. Ruben, *J. Amer. Chem. Soc.* **85**, 3903 (1963).
40. J. N. Shoolery, *J. Chem. Phys.* **31**, 1427 (1959).
41. G. J. Karabatsos, *J. Amer. Chem. Soc.* **83**, 1230 (1961).
42. N. Muller and D. E. Pritchard, *J. Chem. Phys.* **31**, 1471 (1959).
43. J. N. Shoolery, private communication
44. B. E. Holder and M. P. Klein, *J. Chem. Phys.* **23**, 1956 (1955).
45. S. S. Dharmatti, K. H. Sundra Rao and R. Vijayaraghavan, *Nuovo Cimento* 656 (1959).

46. H. A. Christ, *Helv. Phys. Acta* **33**, 572 (1960).
47. H. E. Weaver, B. M. Tolbert and R. C. La Force, *J. Chem. Phys.* **23**, 1956 (1955).
48. R. Bramley, B. N. Figgis and R. S. Nyholm, *Trans. Faraday Soc.* **58**, 1893 (1962).
49. B. N. Figgis, R. G. Kidd and R. S. Nyholm, *Proc. Roy. Soc.* **A269**, 469 (1962).
50. H. S. Gutowsky and D. W. McCall, *J. Chem. Phys.* **22**, 162 (1954).
51. K. J. Packer, *J. Chem. Soc.* 960 (1963).
52. P. T. Narasimhan and M. T. Rogers, *J. Chem. Phys.* **34**, 1049 (1961).
53. W. A. Anderson, R. Freeman and C. A. Reilly, *J. Chem. Phys.* **39**, 1518 (1963).
54. H. Finegold, private communication.
55. E. L. Muetterties, W. Mahler and R. Schmutzler, *Inorg. Chem.* **2**, 613 (1963).
56. L. C. D. Groenweghe, L. Maier and K. Moedritzer, *J. Phys. Chem.* **66**, 901 (1962).
57. N. Boden, J. W. Emsley, J. Feeney and L. H. Sutclifie, *Chem. and Ind.* 1909 (1962).
58. H. S. Gutowsky, C. H. Holm, A. Saika and G. A. Williams, *J. Amer. Chem. Soc.* **79**, 4596 (1957).

APPENDIX

PROTON MAGNETIC RESONANCE CORRELATION TABLES

Compiled by

J. A. ELVIDGE

Tables of Chemical Shifts

The τ values in the following tables are mostly average values from several compounds and/or different sources (see Refs. 1–3, 5, 7). Discrepancies from values observed in particular cases may be as much as $\pm 0{\cdot}5$ ppm.

TABLE 1

(Approx.) τ values for the protons of CH_3, CH_2 and CH groups (attached to various groups X, and to saturated hydrocarbon residues, R, R′, etc.)

X	$CH_3.X$	$R'CH_2.X$	$R'R''CH.X$
—R	9·10	8·75*	8·50
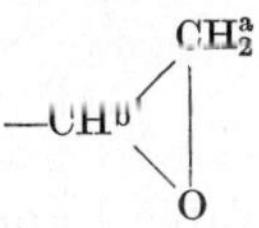	8·68	a 7·6, 7·3	b 7·0
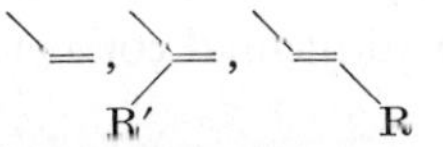	8·31	8·05	7·4
 etc. (end-of-chain position)	8·18	—	—
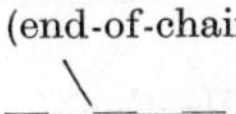 etc. (in-chain position)	8·03	7·8	—
—C≡C—R	8·2	—	—
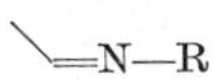	8·0	—	—
—CO.OR	8·00	7·90	—
—CN	8·00	7·52	7·3
$—CO.NH_2$, $—CO.NR_2$	7·98	7·95	—

TABLE 1—*continued*

X	$CH_3.X$	$R'CH_2.X$*	$R'R''CH.X$
—CO.OH	7·93	7·66	7·43
—CO.R	7·90	7·60‡	7·5
—SH, —SR	7·90	7·60	6·9
$—NH_2$, $—NR_2$	7·85	7·50	7·13
—I	7·84	6·85	5·8
—CHO	7·83	7·8	7·6
—Ph†	7·66	7·38	7·13
—COPh	7·38	—	6·42
—Br	7·35	6·66	5·9
—NH.CO.R, —NR.CO.R‴	7·1	6·7	6·5
—Cl	6·98	6·56	5·98
—OR	6·70	6·64	6·2
$—N^+R_3$	6·67	6·60	6·5
—OH	6·62	6·44	6·15
$—O.SO_2.OR$	6·42	—	—
—O.CO.R	6·35	5·85	4·99
—O.Ph	6·27	6·10	6·0
—O.CO.Ph	6·10	5·77	4·88
—F	5·74	5·65	—
$—NO_2$	5·67	5·60	5·4

* If present in 3-membered ring +1·05
If present in 5-membered ring −0·25
If present in 6-membered ring −0·20
If present in 7-membered ring −0·30

‡ Except for a CH_2 next to >C=O, when the sign is changed.

† *o*-, *m*, *p*-Substituents: Electron-withdrawing, 0 to −0·2; Electron-donating, 0 to +0·1

TABLE 2

(a) CH_3, CH_2, CH positioned β to the X groups of Table 1 appear at 0·05 to 1·0 *lower* values than shown above. The additional corrections are as follows:

X′	β-Shifts $CH_3—C—X'$	$CH_2—C—X'$	$CH—C—X'$
—C=C	−0·1	−0·05	
—CO.OH, —CO.OR	−0·25		
—CN	−0·43	−0·4	
$—CO.NH_2$	−0·23		
—CO.R, —CHO	−0·2		

TABLE 2—*continued*

X	β-Shifts CH_3—C—X	CH_2—C—X	CH—C—X
—SH, —SR	—0·45	—0·3	
—NH_2, —NR_2	—0·1	—0·05	
—I	−1·0	—0·5	−0·4
—Ph	−0·35	−0·3	
—Br	−0·8	−0·6	−0·25
—NH.CO.R	−0·1		
—Cl	−0·6	−0·4	−0·02
—OH, —OR	−0·27	−0·1	
—O.CO.R	−0·37		
—O.Ph	−0·47	−0·35	
—F		−0·2	
—NO_2	−0·65	−0·8	
—C—Y*		0·0 to −0·15	

* Y is any common function (—OH, —CO, —N<, Halogen, etc.)

(b) CH_3, CH_2, CH attached to naphthyl and higher hydrocarbons and to α-pyridyl, pyrimidyl, furyl, thienyl, thiazolyl, etc., appear at values 0·05 to 0·65 *lower* than in toluene (7·66). CH_3, CH_2, CH attached to pyrryl, β- or γ-pyridyl come at values 0·05 to 0·3 *higher*

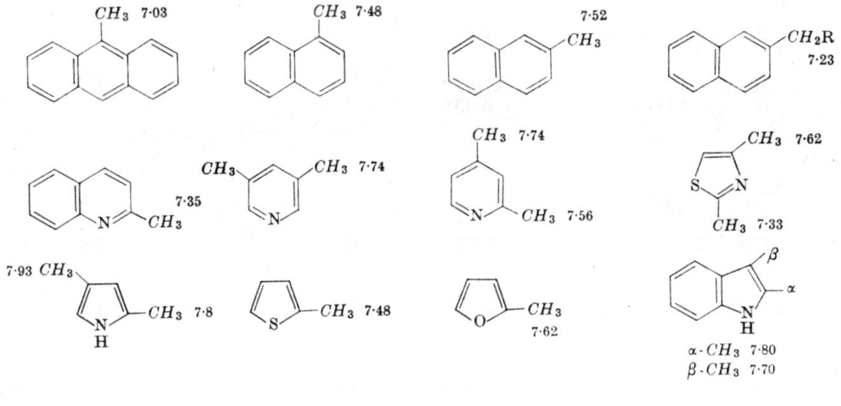

Table 3

(Modified Shoolery Rules)

Approximate τ values for the protons of CH_2 (and CH) groups bearing more than one functional substituent

$$\text{For } CH_2\langle^{X^1}_{X^2} \qquad \tau\, CH_2 = 8{\cdot}75 - \Sigma a$$

$$\left[\text{Less accurate for } CH{-}X^1,\,X^2,\,X^3\right]$$

X	a	X	a
—=	0·75	—Ph	1·3
—≡	0·90	—Br	1·9
—CO.OH, —CO.OR	0·7	—Cl	2·0
—CN, —CO.R	1·2	—OR, —OH	1·7
—SR	1·0	—O.CO.R	2·7
$—NH_2$, $—NR_2$	1·0	—OPh	2·3
—I	1·4		

Table 4

(a) (Approx.) τ values for H attached to unsaturated groups

H—≡—	8·20*	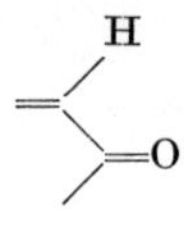	4·2
H—≡—O—	7·60*		
H—≡—=— etc.	7·2*		
H—≡—Ph	7·07*	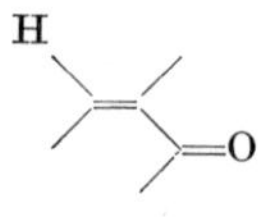	4·0
H—≡—CO—	6·83*		
$H_2C{=}C(R)(R')$	5·35	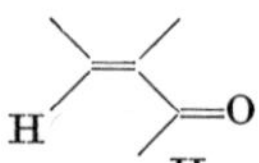	3·8
H_2C=—=— etc.	5·1	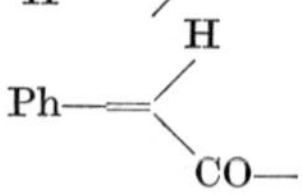 (cis or trans)	3·4

* Signals moved to lower-field by a trace of pyridine.

TABLE 4—*continued*

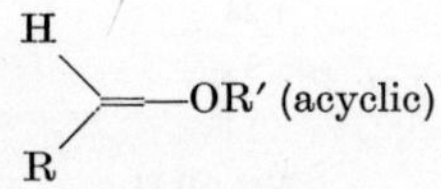	5·0	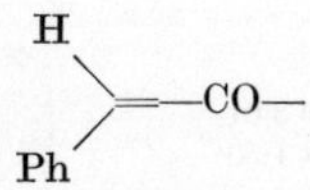	2·2
$=C=CH$ (allenic)	4·8		
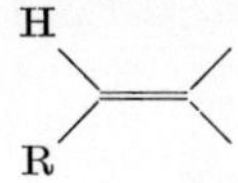	4·7	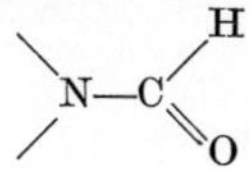	2·15
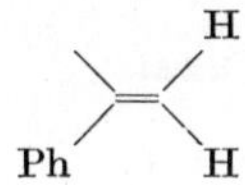	4·95 4·65	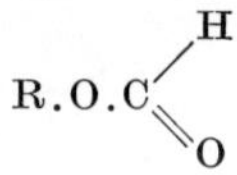	1·97
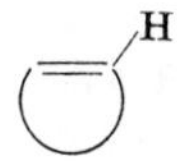	4·4	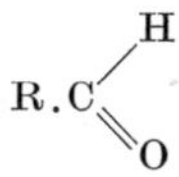	0·35
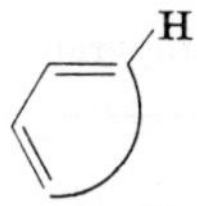	4·3	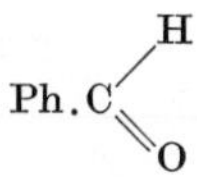	0·1
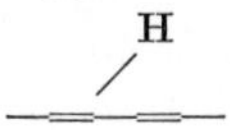 (in-chain)	3·8		
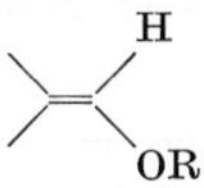 (acyclic)	3·2		
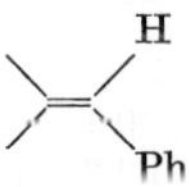	3·0		

(b) (Approx.) τ values for H attached to aromatic groups

 Benzene	(Ph—H)	2·73	*o*-substituents, 0 to ±1·0 *m*-substituents, 0 to ±0·4 *p*-substituents, 0 to ±0·6	+ for electron donating group; − for electron withdrawing group

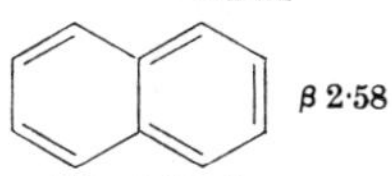

Naphthalene

TABLE 4—*continued*

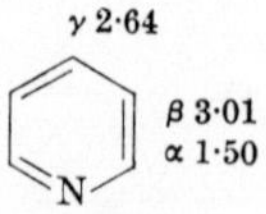

Pyridine

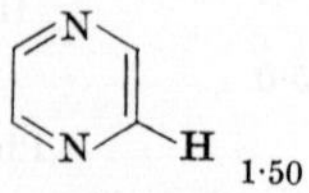

Pyrazine

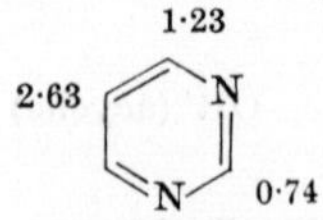

Pyrimidine

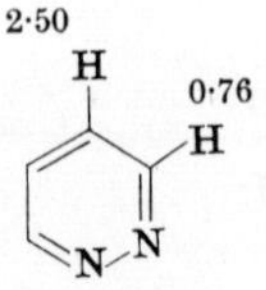

Pyridazine

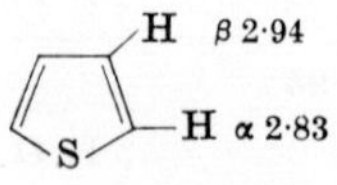

Thiophen

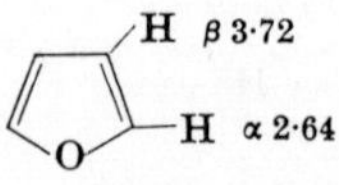

Furan

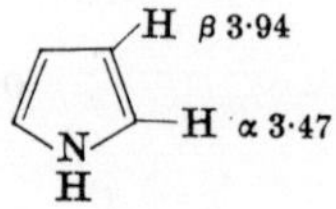

Pyrrole

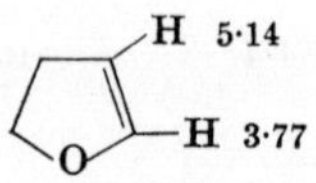

2,3-Dihydrofuran

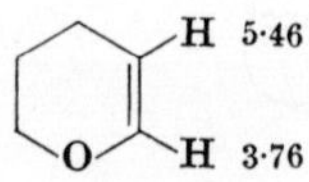

2,3-Dihydropyran

TABLE 5

(Approx.) Shifts in the position of benzene protons (τ 2·73) caused by substituents

Substituent	*ortho*	*meta*	*para*
—CH_3	0·15	0·1	0·1
—═	−0·2	−0·2	−0·2
—CO.OH, —CO.OR	−0·8	−0·15	−0·2
—CN	−0·3	−0·3	−0·3
—CO.NH_2	−0·5	−0·2	−0·2
—CO.R	—0·6	—0·3	—0·3
—SR	−0·1	0·1	0·2
—NH_2	0·8	0·15	0·4
—$N(CH_3)_2$	0·5	0·2	0·5
—I	−0·3	0·2	0·1
—CHO	−0·7	−0·2	−0·4
—Br	0	0	0
—NH.CO.R	−0·4	0·2	0·3
—Cl	0	0	0
—NH_3^+	−0·4	−0·2	−0·2
−OR	0·2	0·2	0·2
—OH	0·4	0·4	0·4
—O.CO.R	−0·2	0·1	0·2
—NO_2	−1·0	−0·3	−0·4

TABLE 6

(Approx.) τ values for the protons of –OH, –NH, and –SH groups. All signals removable by deuteration with D_2O, with D_2O/DCl (trace) in the case of amines, and with NaOD/D_2O in the case of amides and imides.

R.OH	9·5 to 6·0, lower for enols; H-bonded enols usually −1 to −6; lines not always sharp.
Ār.OH	5·5, lowered by H bonding to 1·0 or so.
R.CO_2H	0 to −3 or so.
R.NH_2, R.NHR′	8 to 5 } lines usually broadened.
Ār.NH_2, Ār.NHR′	6·5 to 4 or lower } lines usually broadened.
R.CO.NH_2	5 to 1·5; lines often very broad (even unobservable).
R.CO.NH.CO.R′	1 to − 2 or so; lines broadened.
R.SH	9 to 8; lower for enthiols (e.g. ca. 5), and lines may then be broadened.
Ār.SH	ca. 6·5.
=NOH	ca. 0 to −2; may be broad.

Signals from the above groups show special characteristics

R.OH *pure*	OH signal should be a multiplet because of coupling with adjacent protons, but usually appears as a slightly broadened singlet.
with a trace of acid	Signal is a sharp line, between the expected positions for ROH and H_2O (ca. 5·2) because of catalysed rapid exchange of the OH proton with water. Position of line is concentration-dependent.
R.OH and Ār.OH and thiols	Positions of signals depend on the concentration and temperature of the solution, and the nature of the solvent. A thiol proton is often coupled to adjacent protons.
R.CO_2H	Position of OH signal, in non-polar solvents, is independent of concentration (at least over 5 to 10% range). Signal is shifted by a trace of pyridine.
R.NH_2, RR′NH } *pure*	Single line because of exchange, usually fairly sharp, but may be broadened. Position depends on solvent.
with a trace of acid	Line is shifted to a position between that for R.NH_2 and H_2O.
R.NH_2+conc. acid, R.NH_3^+	Complete protonation suppresses exchange, and the signal then disappears because the protons, being coupled to N (which has $I = 1$), give a triplet signal in which the splitting is large J_{NH} = ca. 50 c/s)
RR′NH_2^+, R_3NH^+	Sharp low-field line. In protonated s- and t-amines, the nitrogen relaxation is fast (i.e. the spin-state changes of the N are fast) so that there is effectively no coupling to the attached proton(s).
R.CO.NH_2	Broad signal, sometimes virtually unobservable. Can be sharpened by alkaline catalysis of proton exchange. Position depends on solvent.
R.CO.NHR′	Usually sharper than primary amide signals.

TABLE 7

Approximate line positions of some protic solvents
(The positions vary a little with solute)

Solvent	τ	Solvent	τ
$CF_3.CO_2H$	0·17	Dioxan	6·32
$H.CO.NH_2$	2·15 (CH)	$CH_3.OH$	6·53 (CH_3)
$CHCl_3$	2·73	$(CH_3)_2SO$	7·40
Benzene	2·73	$(CH_3)_2CO$	7·83
$CHBrCl_2$	2·80	$CH_3.CO_2H$	7·90 (CH_3)
$CHBr_3$	3·15	$CH_3.CN$	8·00
CH_2Cl_2	4·70	Cyclohexane	8·57
H_2O	ca. 5·2	$(CH_3)_3COH$	8·73 (CH_3)
$CH_3.NO_2$	5·67		

The intense solvent-signal is always accompanied by spinning side-bands which occur in a region ~ 30 to 60 c/s on each side. Nitrobenzene and pyridine obscure a region τ 3 to 1.

For solutions in H_2O and D_2O, dioxan (τ 6·32) and acetonitrile (τ 8·00) are fairly good internal standards (better than t-butanol),[4] and DSS, sodium dimethyl-4-silapentane-1-sulphonate (Me_3Si-$[CH_2]_3.SO_3Na$), with τ 10·00 is highly convenient (Refs. 8, 2a, No. 481).

Table of Proton–Proton Spin Coupling Constants

TABLE 8

Function		J_{AB} c/s
Methanes 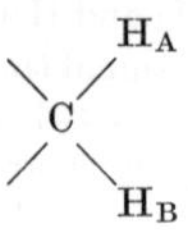	Geminal	− 10 to − 18, depending on the electronegativities of attached groups (Ref. 5).
Ethanes $CH_3.CH_2X$ (A B)	Vicinal	5 to 10 (free rotation) $J_{av} = 8·4 - 0·4\,E$ (Ref. 5a) where E is the Huggins electronegativity of X (Ref. 6).
(Except $CH_3.CHO$ (A B)	where J_{AB} =	2·9)

TABLE 8—*continued*

Function		J_{AB} c/s
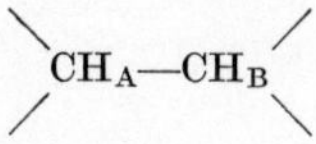	Vicinal	0 to 12 (no rotation) depending on the dihedral angle: for example 8 c/s at 0°; 0 at 85°; 9·2 c/s (Ref. 2e) to 11 c/s (Ref. 1d) at 180°. Ref. 2e gives $J=8{\cdot}5\cos^2\phi-0{\cdot}28$ for $0°\leqslant\phi\leqslant 90°$ and $J=9{\cdot}5\cos^2\phi-0{\cdot}28$* for $90°\leqslant\phi\leqslant 180°$ (cf. Fig. A.1)
Vinyl derivatives		
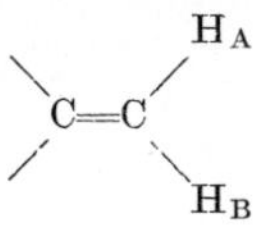	Geminal	−3 to +7 (Ref. 5): −3·2 in vinyl fluoride −2·2 in vinyl methyl ether −1·4 in vinyl chloride +1·0 in vinyl cyanide +1·1 in styrene +2·0 in ethylene itself +1·9 in 1-butene
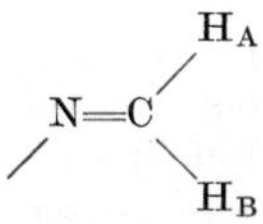		8 to 16
Ethylenes		
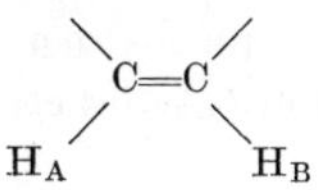	*cis*	acyclic 5 to 14 cyclic 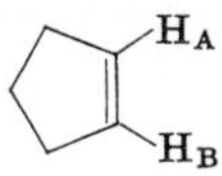 3 to 6, 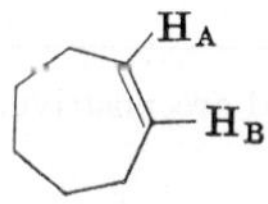 9 to 14
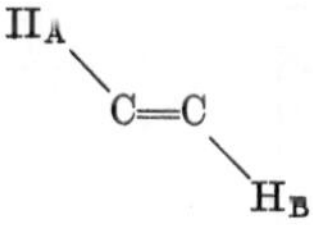	*trans*	11 to 19
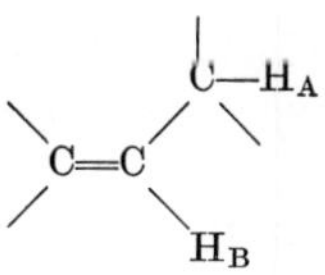		4 to 10
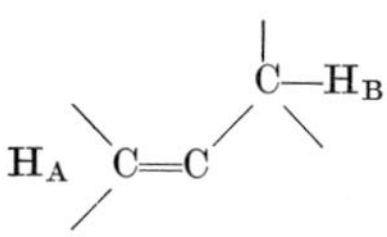	allylic (*cis* or *trans*)	0 to −2 (0 to +1 when the *sp*2 carbon atoms form part of an aromatic ring)

* Compare Chapter 8, p. 141. More detailed tables of J values will be found on pages 132, 134, 142, 145 and 148.

TABLE 8—*continued*

Function		J_{AB} c/s			
H_A—C—C=C—C—H_B	homo-allylic	0 to 2			
>C=CH_A—CH_B=C<		10 to 13			
>CH_A—C≡CH_B		−2 to −3			
>CH_A—C≡C—CH_B<		2 to 3			
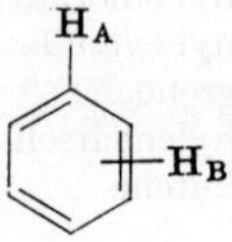	Benzene	*ortho* 7 to 10; *meta* 2 to 3; *para* 1 or less, often unobservable.			
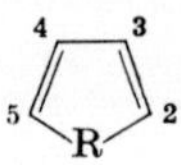	R=O, Furan	$J_{2,3}$ 1·8	$J_{3,4}$ 3·5	$J_{2,4}$ 0·8 (Ref. 7)	$J_{2,5}$ 1·6
	R=NH, Pyrrole	2·6	3·4	1·4	2·1
	R=S, Thiophen	5·2	3·6	1·3	2·7
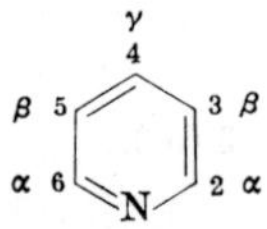	Pyridine	5·5	7·5	1·9	0·9
		in addition $J_{3,5}=1·6$; $J_{2,6}=0·4$ c/s (Ref. 3a)			

Note: Signs unspecified are positive.

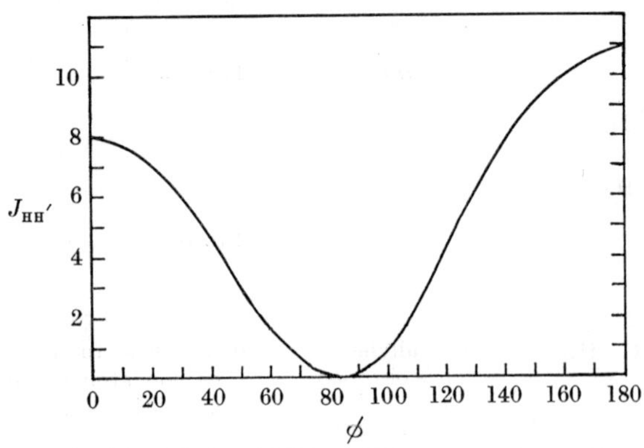

FIG. A.1. The relationship of the vicinal coupling constant to ϕ the dihedral angle (Ref. 1d).

References

1. (a) L. M. Jackman, *Applications of Nuclear Magnetic Resonance Spectroscopy in Organic Chemistry*. Pergamon Press, London (1959). (b) J. D. Roberts, *Nuclear Magnetic Resonance. Applications to Problems in Organic Chemistry*. McGraw-Hill, New York (1959). (c) J. D. Roberts, *J. Chem. Education* **37**, 581 (1961). (d) H. Conroy, "Nuclear Magnetic Resonance in Organic Structural Elucidation". In *Advances in Organic Chemistry*, ed. by R. A. Raphael, E. C. Taylor and H. Wynberg, Vol. 2, p. 265. Interscience, New York (1960).
2. (a) Varian N.M.R. Spectra Catalogs 1 and 2, Varian Assocs., Palo Alto, California (1962, 1963). (b) N. F. Chamberlain, *Analyt. Chem.* **31**, 56 (1959). (c) L. H. Meyer, A. Saika and H. S. Gutowsky, *J. Amer. Chem. Soc.* **75**, 4567 (1953). (d) H. S. Gutowsky, M. Karplus and D. M. Grant, *J. Chem. Phys.* **31**, 1278 (1959). (e) M. Karplus, *J. Chem. Phys.* **30**, 11 (1959). (f) R. H. Martin, Tetrahedron **20**, 897 (1964).
3. (a) J. A. Pople, W. G. Schneider and H. J. Bernstein, *High-resolution Nuclear Magnetic Resonance*, McGraw-Hill, New York (1959). (b) J. D. Roberts, *An Introduction to the Analysis of Spin-Spin Splitting in High Resolution Nuclear Magnetic Resonance Spectra*. Benjamin, New York (1961). (c) Corio, *Chem. Rev.* **60**, 363 (1960).
4. R. A. Y. Jones, A. R. Katritzky, J. N. Murrell and N. Sheppard, *J. Chem. Soc.* 2576 (1962).
5. (a) C. N. Banwell and N. Sheppard, *Discuss. Faraday Soc.* **34**, 115 (1962). (b) H. J. Bernstein and N. Sheppard, *J. Chem. Phys.* **37**, 3012 (1962). (c) C. N. Banwell and N. Sheppard, *Mol. Phys.* **3**, 351 (1960).
6. M. L. Huggins, *J. Amer. Chem. Soc.* **75**, 4123 (1953).
7. R. J. Abraham and H. J. Bernstein, *Canad. J. Chem.* **39**, 216 (1961).
8. G. V. D. Tiers and R. I. Coon, *J. Org. Chem.*, **26**, 2097 (1961).

CONVERSION TABLES

(c/s at 60 Mc/s to τ values)

c/s	τ	c/s	τ	c/s	τ	c/s	τ
1	9·98	44	9·27	87	8·55	130	7·83
2	9·97	45	9·25	88	8·53	131	7·82
3	9·95	46	9·23	89	8·52	132	7·80
4	9·93	47	9·22	90	8·50	133	7·78
5	9·92	48	9·20	91	8·48	134	7·77
6	9·90	49	9·18	92	8·47	135	7·75
7	9·88	50	9·17	93	8·45	136	7·73
8	9·87	51	9·15	94	8·43	137	7·72
9	9·85	52	9·13	95	8·42	138	7·70
10	9·83	53	9·12	96	8·40	139	7·68
11	9·82	54	9·10	97	8·38	140	7·67
12	9·80	55	9·08	98	8·37	141	7·65
13	9·78	56	9·07	99	8·35	142	7·63
14	9·77	57	9·05	100	8·33	143	7·62
15	9·75	58	9·03	101	8·32	144	7·60
16	9·73	59	9·02	102	8·30	145	7·58
17	9·72	60	9·00	103	8·28	146	7·57
18	9·70	61	8·98	104	8·27	147	7·55
19	9·68	62	8·97	105	8·25	148	7·53
20	9·67	63	8·95	106	8·23	149	7·52
21	9·65	64	8·93	107	8·22	150	7·50
22	9·63	65	8·92	108	8·20	151	7·48
23	9·62	66	8·90	109	8·18	152	7·47
24	9·60	67	8·88	110	8·17	153	7·45
25	9·58	68	8·87	111	8·15	154	7·43
26	9·57	69	8·85	112	8·13	155	7·42
27	9·55	70	8·83	113	8·12	156	7·40
28	9·53	71	8·82	114	8·10	157	7·38
29	9·52	72	8·80	115	8·08	158	7·37
30	9·50	73	8·78	116	8·07	159	7·35
31	9·48	74	8·77	117	8·05	160	7·33
32	9·47	75	8·75	118	8·03	161	7·32
33	9·45	76	8·73	119	8·02	162	7·30
34	9·43	77	8·72	120	8·00	163	7·28
35	9·42	78	8·70	121	7·98	164	7·27
36	9·40	79	8·68	122	7·97	165	7·25
37	9·38	80	8·67	123	7·95	166	7·23
38	9·37	81	8·65	124	7·93	167	7·22
39	9·35	82	8·63	125	7·92	168	7·20
40	9·33	83	8·62	126	7·90	169	17·8
41	9·32	84	8·60	127	7·88	170	7·17
42	9·30	85	8·58	128	7·87	171	7·15
43	9·28	86	8·57	129	7·85	172	7·13

c/s	τ	c/s	τ	c/s	τ	c/s	τ
173	7·12	228	6·20	283	5·28	338	4·37
174	7·10	229	6·18	284	5·27	339	4·35
175	7·08	230	6·17	285	5·25	340	4·33
176	7·07	231	6·15	286	5·23	341	4·32
177	7·05	232	6·13	287	5·22	342	4·30
178	7·03	233	6·12	288	5·20	343	4·28
179	7·02	234	6·10	289	5·18	344	4·27
180	7·00	235	6·08	290	5·17	345	4·25
181	6·98	236	6·07	291	5·15	346	4·23
182	6·97	237	6·05	292	5·13	347	4·22
183	6·95	238	6·03	293	5·12	348	4·20
184	6·93	239	6·02	294	5·10	349	4·18
185	6·92	240	6·00	295	5·08	350	4·17
186	6·90	241	5·98	296	5·07	351	4·15
187	6·88	242	5·97	297	5·05	352	4·13
188	6·87	243	5·95	298	5·03	353	4·12
189	6·85	244	5·93	299	5·02	354	4·10
190	6·83	245	5·92	300	5·00	355	4·08
191	6·82	246	5·90	301	4·98	356	4·07
192	6·80	247	5·88	302	4·97	357	4·05
193	6·78	248	5·87	303	4·95	358	4·03
194	6·77	249	5·85	304	4·93	359	4·02
195	6·75	250	5·83	305	4·92	360	4·00
196	6·73	251	5·82	306	4·90	361	3·98
197	6·72	252	5·80	307	4·88	362	3·97
198	6·70	253	5·78	308	4·87	363	3·95
199	6·68	254	5·77	309	4·85	364	3·93
200	6·67	255	5·75	310	4·83	365	3·92
201	6·65	256	5·73	311	4·82	366	3·90
202	6·63	257	5·72	312	4·80	367	3·88
203	6·62	258	5·70	313	4·78	368	3·87
204	6·60	259	5·68	314	4·77	369	3·85
205	6·58	260	5·67	315	4·75	370	3·83
206	6·57	261	5·65	316	4·73	371	3·82
207	6·55	262	5·63	317	4·72	372	3·80
208	6·53	263	5·62	318	4·70	373	3·78
209	6·52	264	5·60	319	4·68	374	3·77
210	6·50	265	5·58	320	4·67	375	3·75
211	6·48	266	5·57	321	4·65	376	3·73
212	6·47	267	5·55	322	4·63	377	3·72
213	6·45	268	5·53	323	4·62	378	3·70
214	6·43	269	5·52	324	4·60	379	3·68
215	6·42	270	5·50	325	4·58	380	3·67
216	6·40	271	5·48	326	4·57	381	3·65
217	6·38	272	5·47	327	4·55	382	3·63
218	6·37	273	5·45	328	4·53	383	3·62
219	6·35	274	5·43	329	4·52	384	3·60
220	6·33	275	5·42	330	4·50	385	3·58
221	6·32	276	5·40	331	4·48	386	3·57
222	6·30	277	5·38	332	4·47	387	3·55
223	6·28	278	5·37	333	4·45	388	3·53
224	6·27	279	5·35	334	4·43	389	3·52
225	6·25	280	5·33	335	4·42	390	3·50
226	6·23	281	5·32	336	4·40	391	3·48
227	6·22	282	5·30	337	4·38	392	3·47

c/s	τ	c/s	τ	c/s	τ	c/s	τ
393	3·45	445	2·58	497	1·72	549	·85
394	3·43	446	2·57	498	1·70	550	·83
395	3·42	447	2·55	499	1·68	551	·82
396	3·40	448	2·53	500	1·67	552	·80
397	3·38	449	2·52	501	1·65	553	·78
398	3·37	450	2·50	502	1·63	554	·77
399	3·35	451	2·48	503	1·62	555	·75
400	3·33	452	2·47	504	1·60	556	·73
401	3·32	453	2·45	505	1·58	557	·72
402	3·30	454	2·43	506	1·57	558	·70
403	3·28	455	2·42	507	1·55	559	·68
404	3·27	456	2·40	508	1·53	560	·67
405	3·25	457	2·38	509	1·52	561	·65
406	3·23	458	2·37	510	1·50	562	·63
407	3·22	459	2·35	511	1·48	563	·62
408	3·20	460	2·33	512	1·47	564	·60
409	3·18	461	2·32	513	1·45	565	·58
410	3·17	462	2·30	514	1·43	566	·57
411	3·15	463	2·28	515	1·42	567	·55
412	3·13	464	2·27	516	1·40	568	·53
413	3·12	465	2·25	517	1·38	569	·52
414	3·10	466	2·23	518	1·37	570	·50
415	3·08	467	2·22	519	1·35	571	·48
416	3·07	468	2·20	520	1·33	572	·47
417	3·05	469	2·18	521	1·32	573	·45
418	3·03	470	2·17	522	1·30	574	·43
419	3·02	471	2·15	523	1·28	575	·42
420	3·00	472	2·13	524	1·27	576	·40
421	2·98	473	2·12	525	1·25	577	·38
422	2·97	474	2·10	526	1·23	578	·37
423	2·95	475	2·08	527	1·22	579	·35
424	2·93	476	2·07	528	1·20	580	·33
425	2·92	477	2·05	529	1·18	581	·32
426	2·90	478	2·03	530	1·17	582	·30
427	2·88	479	2·02	531	1·15	583	·28
428	2·87	480	2·00	532	1·13	584	·27
429	2·85	481	1·98	533	1·12	585	·25
430	2·83	482	1·97	534	1·10	586	·23
431	2·82	483	1·95	535	1·08	587	·22
432	2·80	484	1·93	536	1·07	588	·20
433	2·78	485	1·92	537	1·05	589	·18
434	2·77	486	1·90	538	1·03	590	·17
435	2·75	487	1·88	539	1·02	591	·15
436	2·73	488	1·87	540	1·00	592	·13
437	2·72	489	1·85	541	·98	593	·12
438	2·70	490	1·83	542	·97	594	·10
439	2·68	491	1·82	543	·95	595	·08
440	2·67	492	1·80	544	·93	596	·07
441	2·65	493	1·78	545	·92	597	·05
442	2·63	494	1·77	546	·90	598	·03
443	2·62	495	1·75	547	·88	599	·02
444	2·60	496	1·73	548	·87	600	·00

THE INTERPRETATION OF PROTON MAGNETIC RESONANCE SPECTRA

The following notes provide a useful summary of important points necessary for solving the problems which follow (p. 201).

(1) *The positions of signals*

The chemical shifts give clues to the structural environment of the protons concerned. In using the correlation Tables (p. 179) it must be remembered that the values are not absolute and there may be deviations of up to $\pm 0{\cdot}5$ ppm from the observed values on occasion.

(2) *The intensities of signals*

Divide the total rise in the integral trace over the whole spectrum, by the number of protons, and so obtain the size of a one-proton step. With this information and from the sizes of the individual steps, the numbers of protons responsible for the various signals can be derived.

(3) *Multiplicities of signals*

The multiplicity of a signal indicates the number of protons to which there is coupling i.e. the number of protons on *adjacent* carbon atoms. It may also indicate the presence of an adjacent magnetic nucleus of another kind. The magnitude of the coupling (J c/s) often gives stereochemical information or information concerning the interrelationships of protons or groups.

The number of lines in a multiplet is $(n+1)$ where n is the number of adjacent protons, provided that

(i) the shift between the coupled groups (Δ c/s) is large compared to J;
(ii) all the couplings are equal (see Chapter 4).

If more than one J is involved in a multiplet, then it is best to consider the effect of the several couplings in separate stages and to draw a splitting diagram (see Problem 13). The above simple splitting rule is readily extended to the case where a proton or group of protons is coupled either to a single magnetic nucleus of another kind with spin quantum number I, or to a group (n) of such nuclei. Then, provided only one coupling is involved, the multiplicity of the proton signal is $(2nI+1)$.

(4) *Operating Conditions*

It is necessary to know how the spectrum has been recorded. The line positions cannot be read off correctly unless we take account of the undernoted points (i) to (v) below.

(i) The operating frequency. All spectra in the following problems were run at 60 Mc/s. Line positions are given by

$$\tau = (10-\delta) = 10-\Delta/60$$

where Δ is the position of the line in c/s from TMS.

Conversion tables (c/s to τ) are given on pp. 191–193.

(ii) The sweep width. This gives the operative scale on the chart. The charts used here have five scales printed across the top, 0–50, 0–100, 0–250, 0–500, and 0–1000 c/s. When the 0–500 c/s scale is operative, then the δ and τ scales, printed above and below the chart, can be used to obtain the chemical shifts in ppm directly.

(iii) The offset. This denotes a displacement of the position of the reference line. Thus 60 c/s offset means that the TMS line is now 60 c/s to the *right-hand side*, off the high-field end of the chart. To obtain the *true* line positions, the offset figure (60 c/s) must be added to the *apparent* line position on the chart.

A negative offset would imply the reverse.

(iv) The position of the reference line. This may not be positioned accurately on the appropriate ordinate of the graticule. If the reference line lies to the high-field side (i.e. to the right) then the discrepancy has to be *added* to the apparent line position. If the reference line lies to the low-field side of the particular ordinate, then the discrepancy has to be *substracted* from the apparent line positions in order to obtain the correct values.

(v) The solvent used must be noted in case there are lines in the spectrum arising from this—obviously these must be disregarded.

(5) *Spin coupling patterns*

Two spin-coupled nuclei (Chapter 4, p. 52, Chapter 5, p. 70).

AX spectrum. Consists of four lines grouped as two doublets: components of each doublet are of equal intensity in the first-order limit. ν_A and ν_X are the mid-points of each doublet. J_{AX} is the separation between the lines within each doublet.

AB spectrum. Consists of two distorted doublets as below

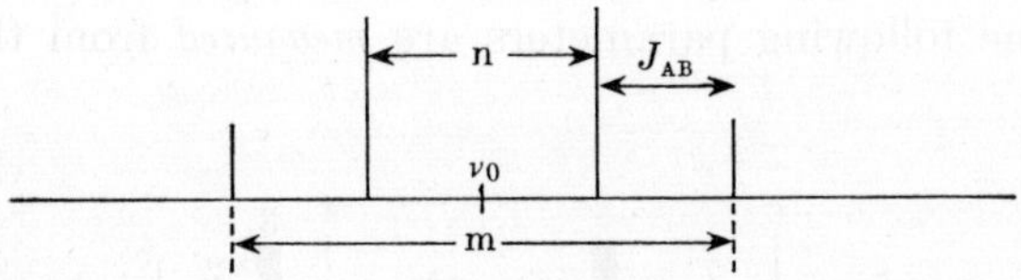

$\nu_A = \nu_0 + \frac{1}{2}\sqrt{m.n}$

$\nu_B = \nu_0 - \frac{1}{2}\sqrt{m.n}$

J_{AB} is the separation between the lines within each doublet.

Three spin-coupled nuclei

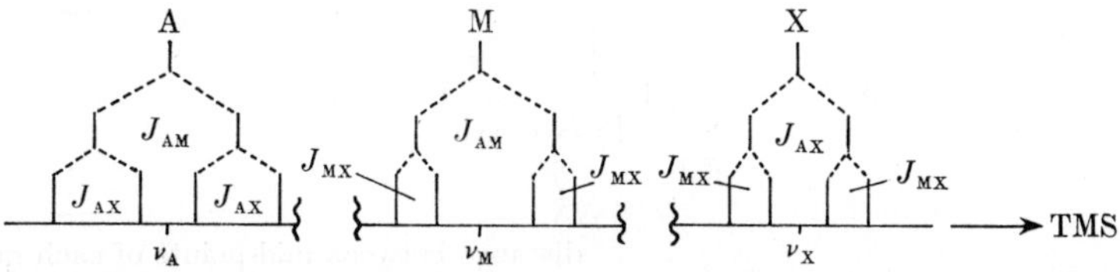

AMX spectrum. (Chapter 5, p. 73). There are 12 lines—4 quartets, when $J_{AX} \neq J_{MX} \neq J_{AM}$. If two of the coupling constants are equal, then one signal appears as a triplet, and the spectrum consists of 11 *lines*. Thus if $J_{AM} = J_{AX}$, the A signal appears as a 1:2:1 triplet:

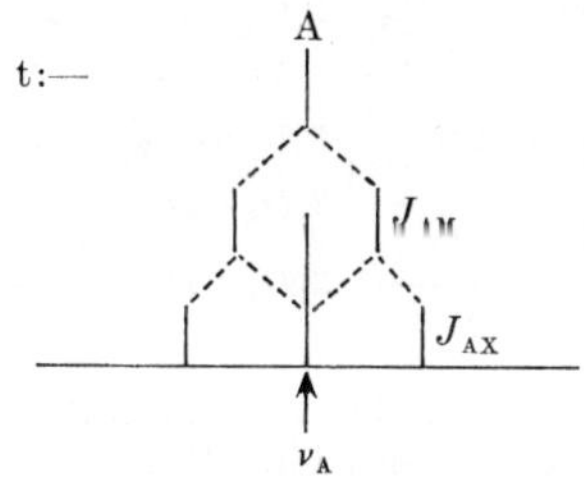

If $J_{AM} = J_{AX} = J_{MX}$, then *each* signal becomes a triplet, and the spectrum contains 9 lines only.

In every case, the origin position of the signals and the coupling constants are directly observable.

ABX spectrum (Chapter 6). This consists of two parts, the AB part (two interleaved quartets) and the X part (four *or* six lines).

For the AB portion of the spectrum, the four higher field lines are designated B lines, the lower field lines, A lines. Thus $\nu_A - \nu_B$ (measured from TMS), is positive.

Where overlapping of lines occurs the number of lines may be less than given above and the picture has to be suitably modified (cf. Chapter 6, p. 96). The following parameters are *measured* from the spectrum

AB part of spectrum

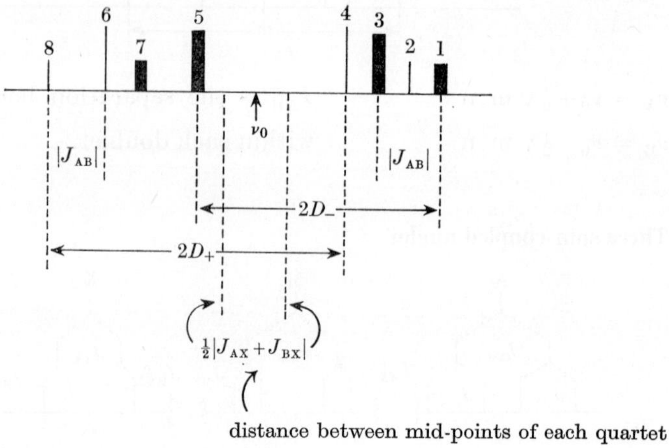

X part of spectrum (symmetrical about ν_X):

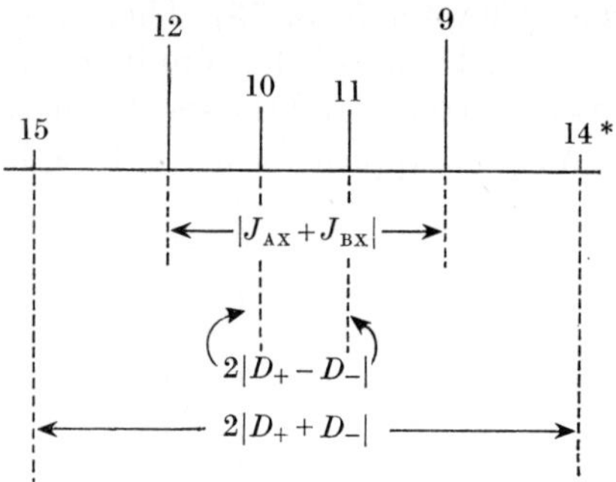

* Note that line 13 is forbidden in ABX cases; lines 13, 14 and 15 are forbidden in the AMX case. All three transitions are allowed in ABC.

(D_+ and D_- are defined as positive quantities and are strictly measured in a direction away from the TMS line.)

The measured values are substituted in the following equations

$$(\nu_A - \nu_B) + \tfrac{1}{2}|J_{AX} - J_{BX}| = (4D_+^2 - J_{AB}{}^2)^{\frac{1}{2}}$$

$$(\nu_A - \nu_B) - \tfrac{1}{2}|J_{AX} - J_{BX}| = (4D_-^2 - J_{AB}{}^2)^{\frac{1}{2}}$$

which are then solved for $(\nu_A - \nu_B)$ and $|J_{AX} - J_{BX}|$

It follows that

$$\nu_A = \nu_0 + \tfrac{1}{2}(\nu_A - \nu_B)$$

$$\nu_B = \nu_0 - \tfrac{1}{2}(\nu_A - \nu_B) \quad \text{whilst}$$

$|J_{AX}|$ and $|J_{BX}|$ are likewise simply obtained.

Absolute signs of the three J values cannot be extracted from the spectrum: the *relative* signs of J_{AX} and J_{BX} may be found by calculation of line intensities (see Chapter 6, p. 98 and Table 6.1).

In an ABX system where $J_{AB} \gg (\nu_A - \nu_B) + \frac{1}{2}|J_{AX} - J_{BX}|$, lines 10 and 11 merge (transitions 14 and 15 are forbidden) so that the X part of the spectrum appears as a 1:2:1 triplet. Although this might suggest that $J_{AX} = J_{BX}$, this may not be so: it means only that $D_+ \approx D_-$ (and in consequence that $\phi_+ \approx \phi_-$). Values for J_{AX} and J_{BX} cannot then be extracted from the spectrum.

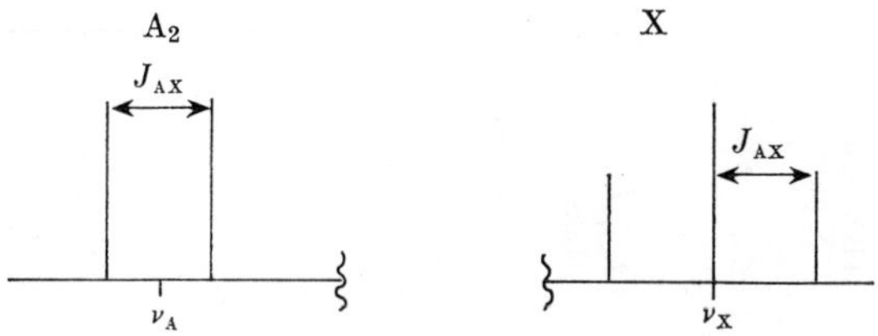

As will be obvious, the main difficulty lies in assigning correctly, the lines in the original spectrum. Where, however, as in a series of very closely related compounds, this can be done with certainty, the following method may be used. The eight AB lines are treated as two separate quartets one arising from each orientation of the X spin in the external field. The splitting diagrams for the AB and the X parts are shown below (Fig. A.2). The method for extracting the information is self explanatory.

A_2X (or *AX_2*)

As expected from the simple splitting rule this gives a 5 line spectrum. The origin positions and coupling constant J_{AX} are directly obtainable.

A_2B (or AB_2). The spectrum contains 9 lines, one of which may be small or non-existent. The appearance of the group depends on the magnitude of $J/\nu_A - \nu_B) = J/\Delta\nu$. See Chapter 5, Fig. 5.6.

Origin positions of the protons and the coupling constant is extracted as follows (cf. Chapter 5, p. 83).

$$\nu_B = \nu_3$$

$$\nu_A = \tfrac{1}{2}(\nu_5 + \nu_7)$$

$$J_{AB} = \tfrac{1}{3}(\nu_4 + \nu_8 - \nu_1 - \nu_6)$$

where ν indicates the line positions in c/s from TMS

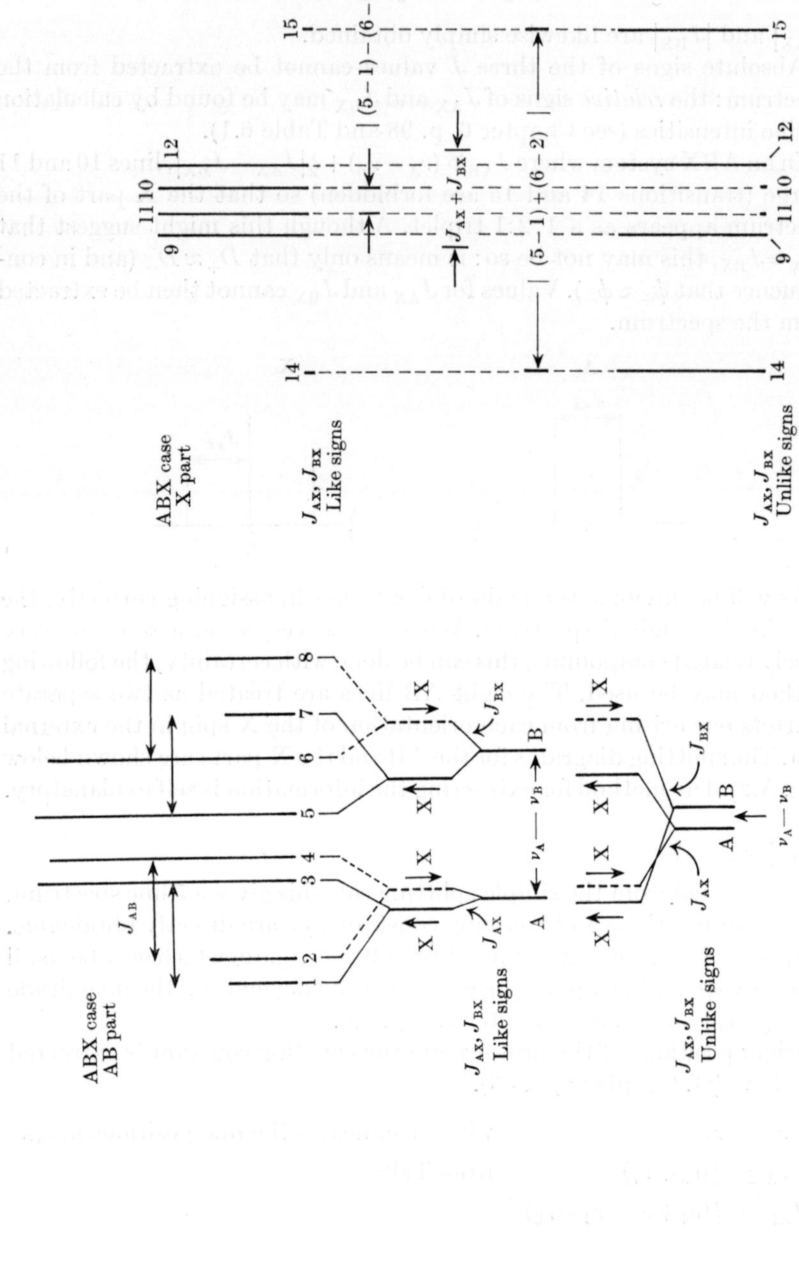
ABX case
AB part
J_{AB}
1 2 3 4 5 6 7 8
X X X X
J_{AX}
J_{BX}
A — $\nu_A - \nu_B$ — B
J_{AX}, J_{BX}
Like signs
J_{AX}, J_{BX}
Unlike signs
$\nu_A - \nu_B$
ABX case
X part
9 1110 12 15 14
$J_{AX} + J_{BX}$
$(5-1) - (6-2)$
$(5-1) + (6-2)$
J_{AX}, J_{BX}
Like signs
J_{AX}, J_{BX}
Unlike signs

FIG. A.2.

EXAMPLES IN THE INTERPRETATION OF N.M.R. SPECTRA

PROBLEMS

Spectrum No. 1

Description of sample:	$C_7H_7O_2N$
Objective:	Identify the compound
Instrumental conditions:	Solvent—carbon tetrachloride
	Sweep width—500 c/s
	Sweep offset—zero

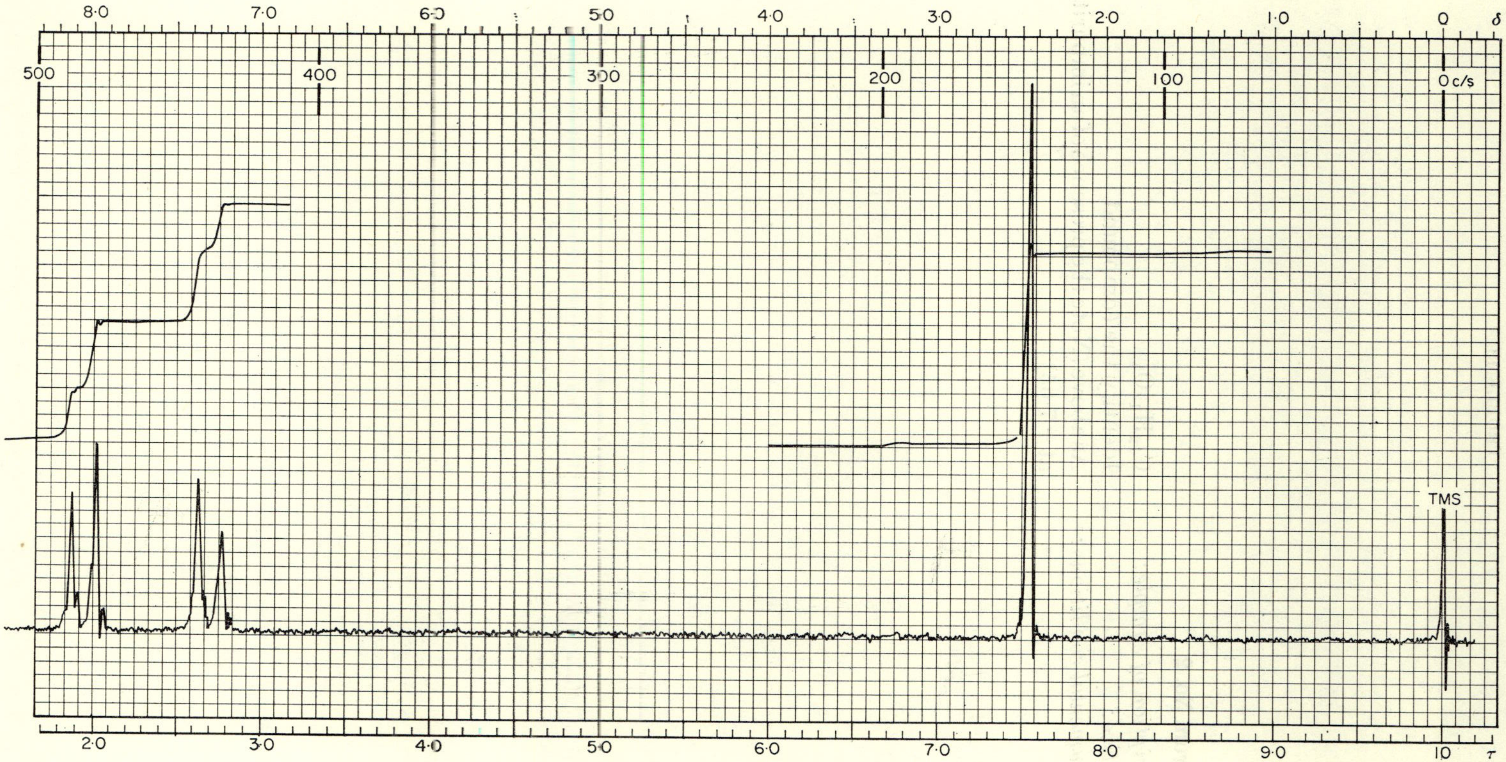

Spectrum No. 1.

Spectrum No. 2

Description of sample:	$C_{12}H_{14}O_4$
Objective:	Identify the compound
Instrumental conditions:	As above (Note the presence of an impurity at τ 7·65)

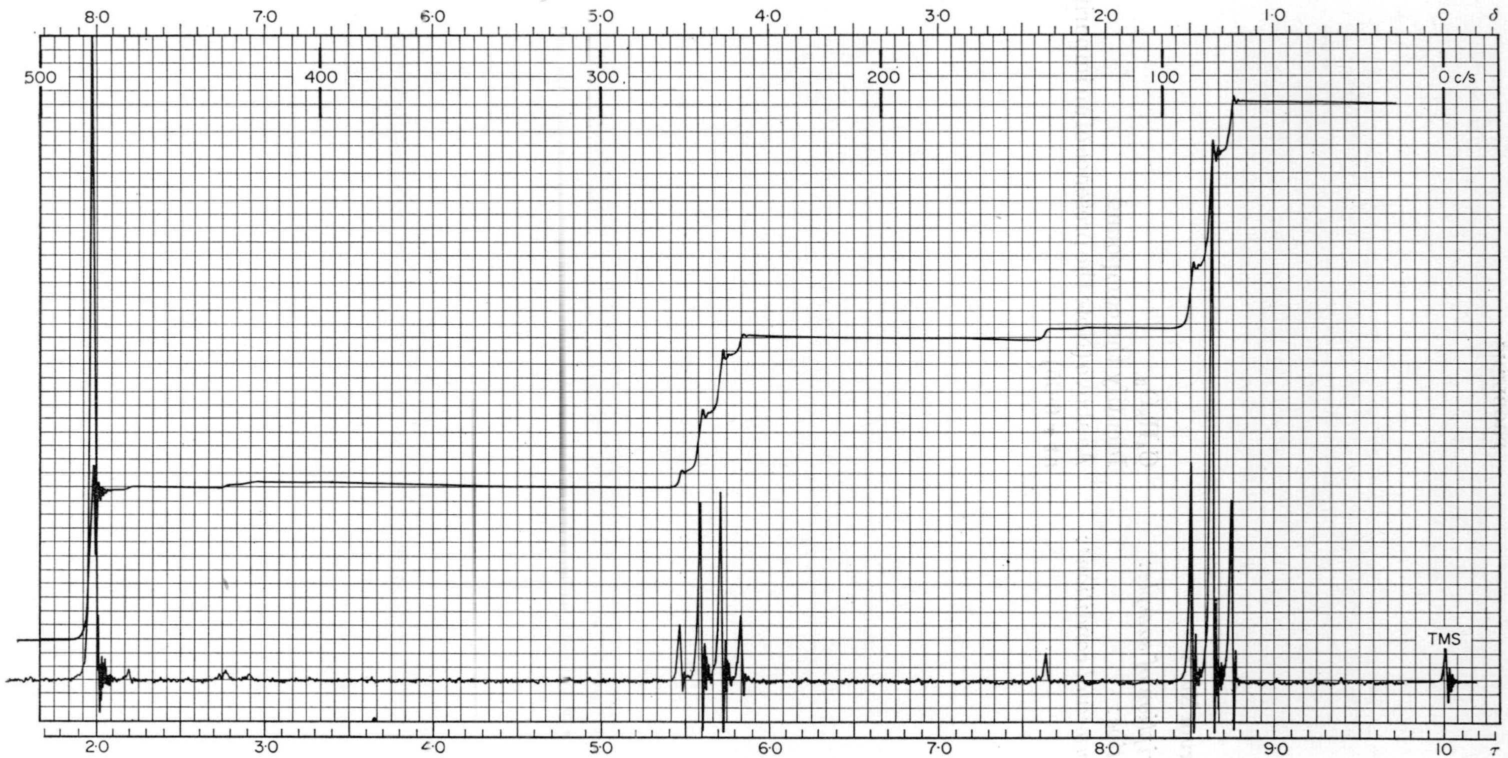

Spectrum No. 2.

Spectrum No. 3

Description of sample:	$C_4H_8O_3$
Objective:	Identify the compound
Instrumental conditions:	As above (Note the presence of an impurity at τ 6·65)

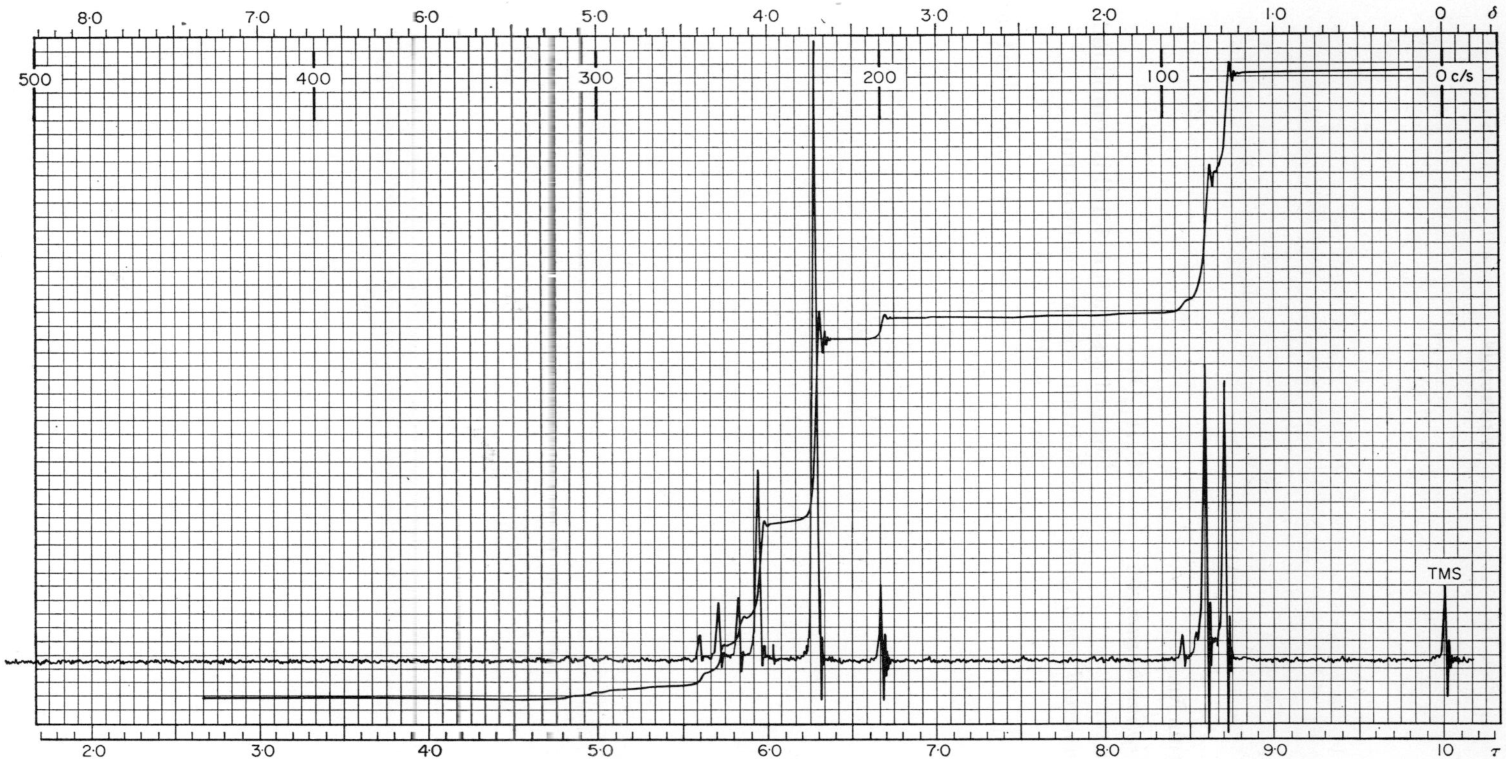

Spectrum No. 3.

Spectrum No. 4

Description of sample:	$C_9H_{10}O_3$
Objective:	Suggest a structure for this compound
Instrumental conditions:	Solvent—deuterochloroform Sweep width—500 c/s Sweep offset—zero for main chart and 240 c/s for inset A

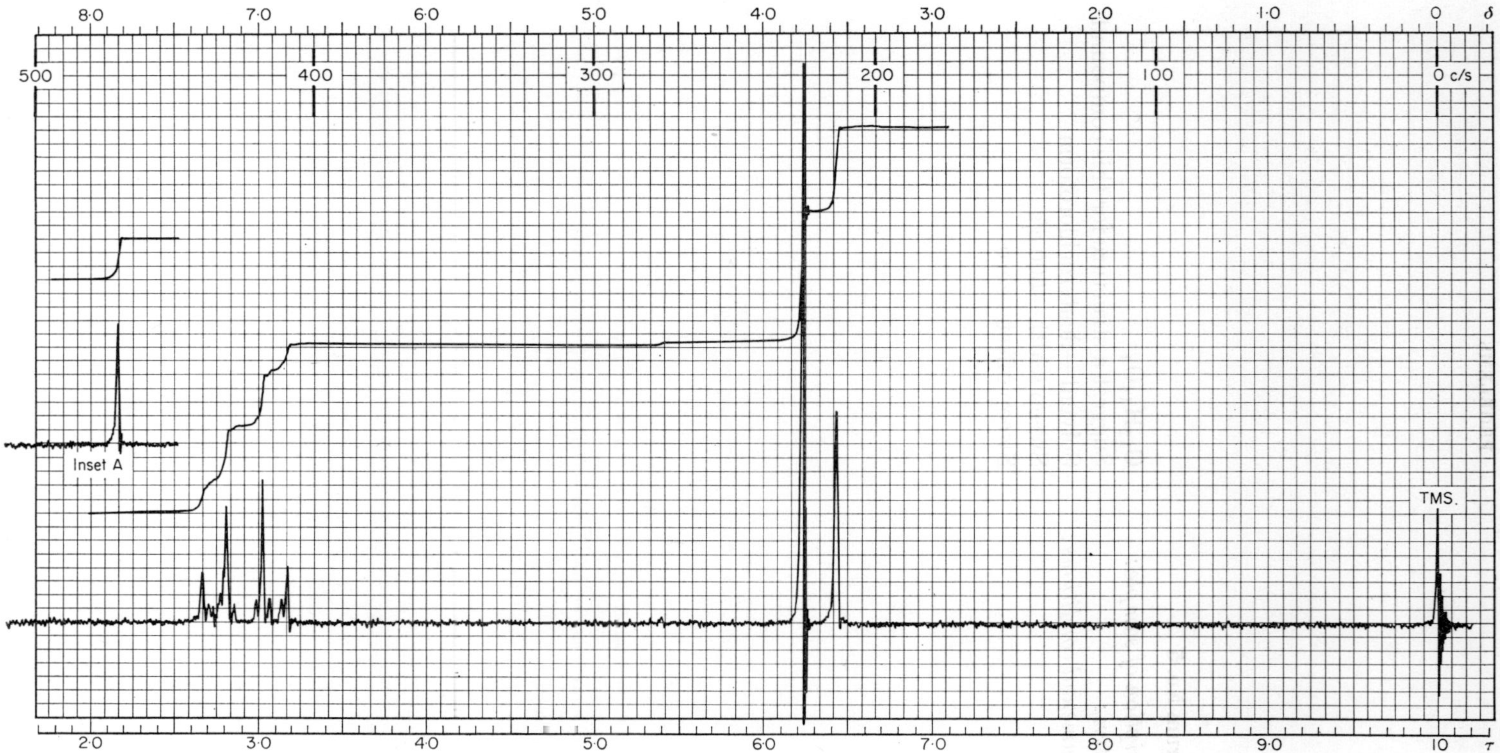

Spectrum No. 4.

Spectrum No. 5

Description of sample: $C_{12}H_{14}O_4$
Objective: Suggest a structure for this compound
Instrumental conditions: Solvent—deuterochloroform
Sweep width—500 c/s
Sweep offset—zero

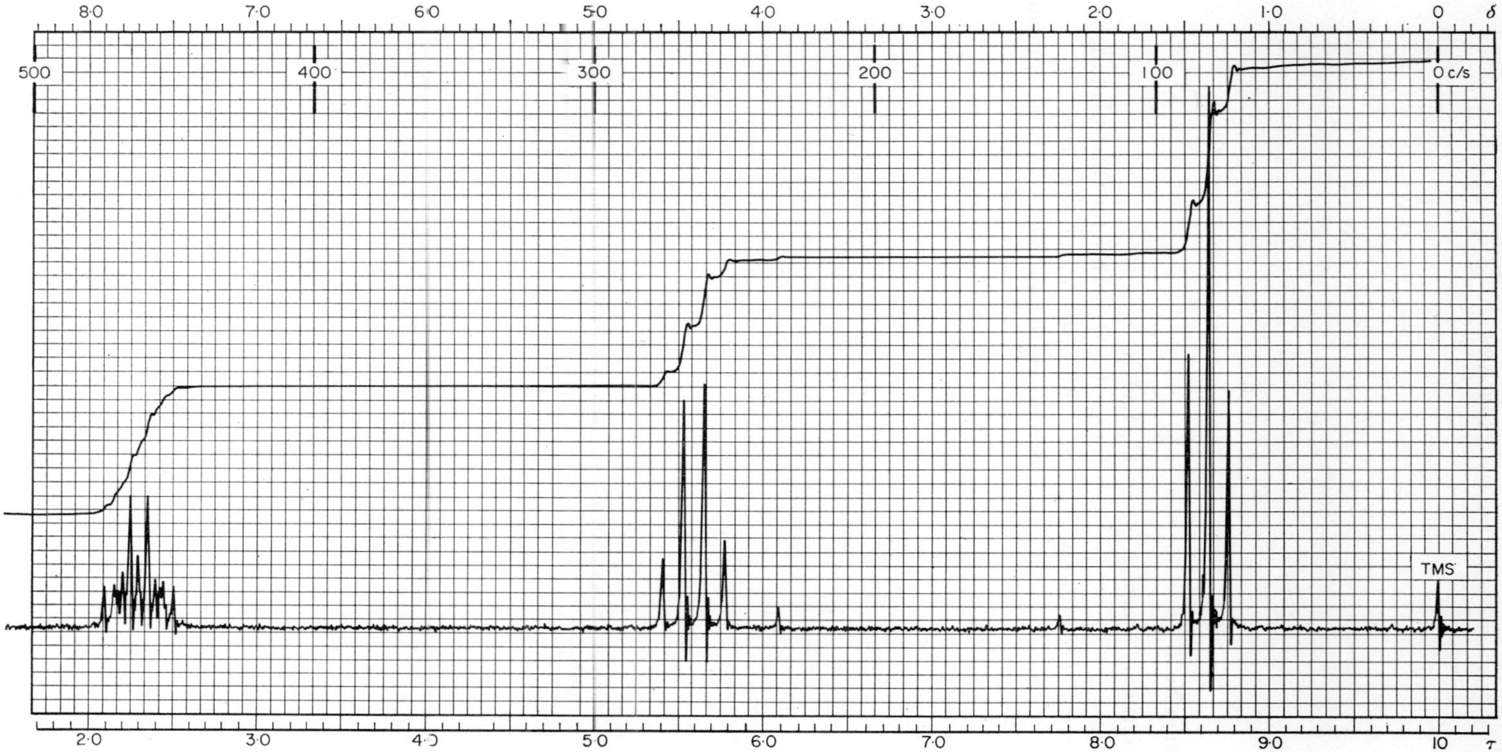

Spectrum No. 5.

Spectrum No. 6

Description of sample:	$C_5H_9O_4N$
Objective:	Suggest a structure for this substance
Instrumental conditions:	Solvent—carbon tetrachloride
	Sweep width and offset as in Spectrum No. 5

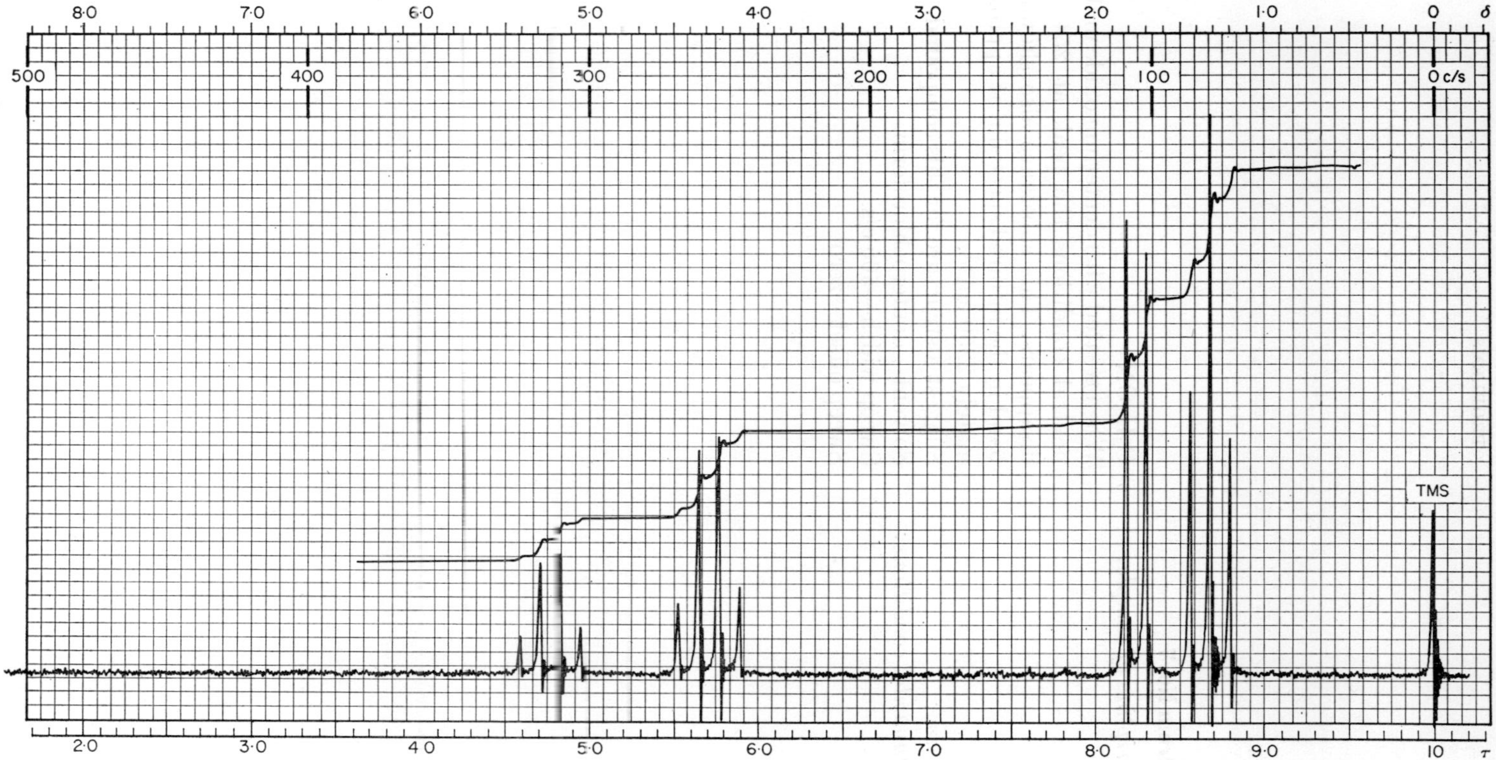

Spectrum No. 6.

Spectrum No. 7

Description of sample: C_3H_5ON

When treated with deuterium oxide under neutral conditions the peak at 254 c/s disappears. This substance also shows a band in the infrared spectrum at approximately 2250 cm^{-1}.

Objective: Suggest a structure for this compound.

Instrumental conditions: Solvent—deuterochloroform

Sweep width—500 c/s

Sweep offset—zero

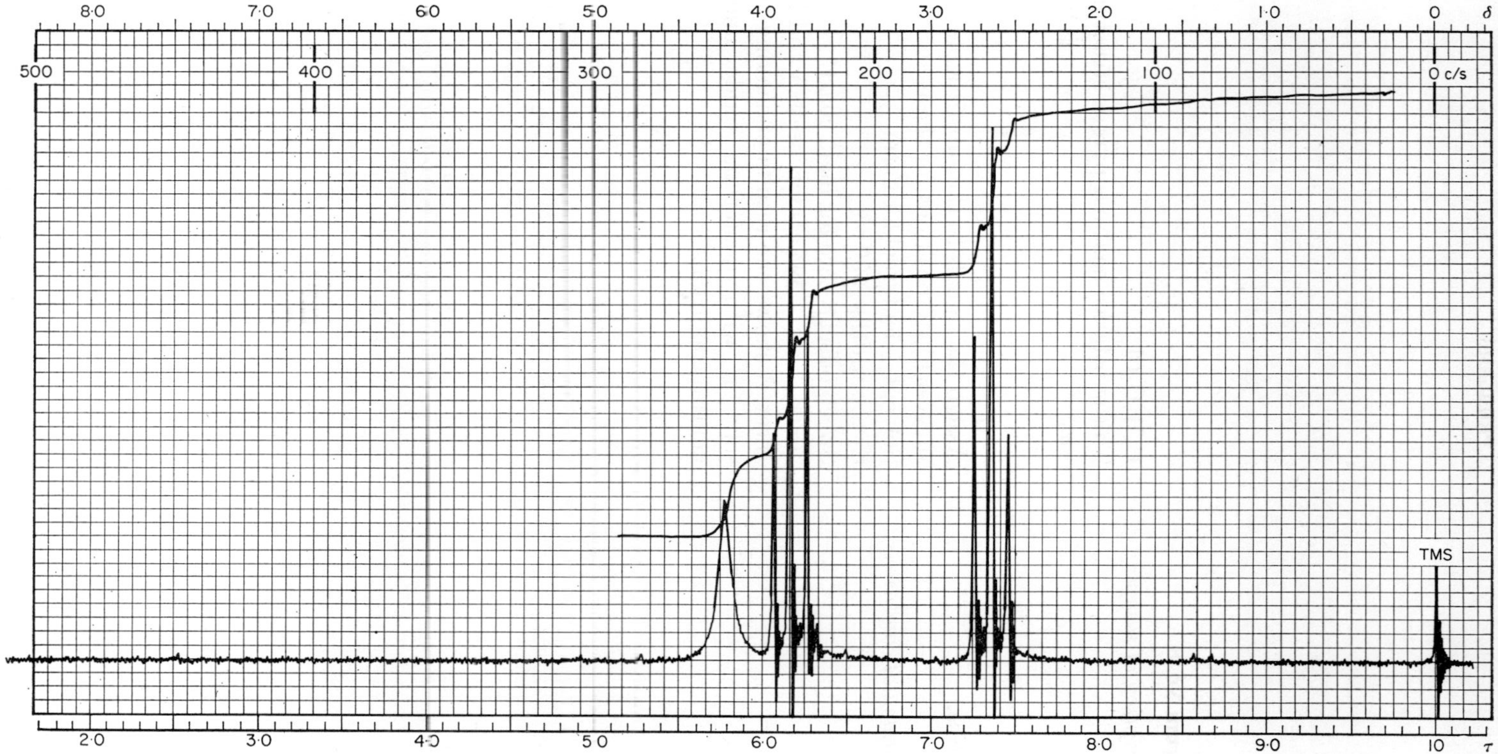

Spectrum No. 7.

Spectrum No. 8

Description of sample: $C_8H_7O_5Cl$

This substance has an α-pyrone structure, viz.

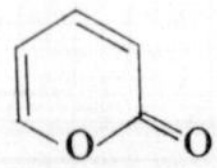

Objective: Assign the signals to the substituents and suggest likely positions for these.

Instrumental conditions: Solvent—deuterochloroform

Sweep width—500 c/s

Sweep offset—zero for main chart

395 c/s for insets A and B

Inset B shows the results of shaking the solution with deuterium oxide under neutral conditions

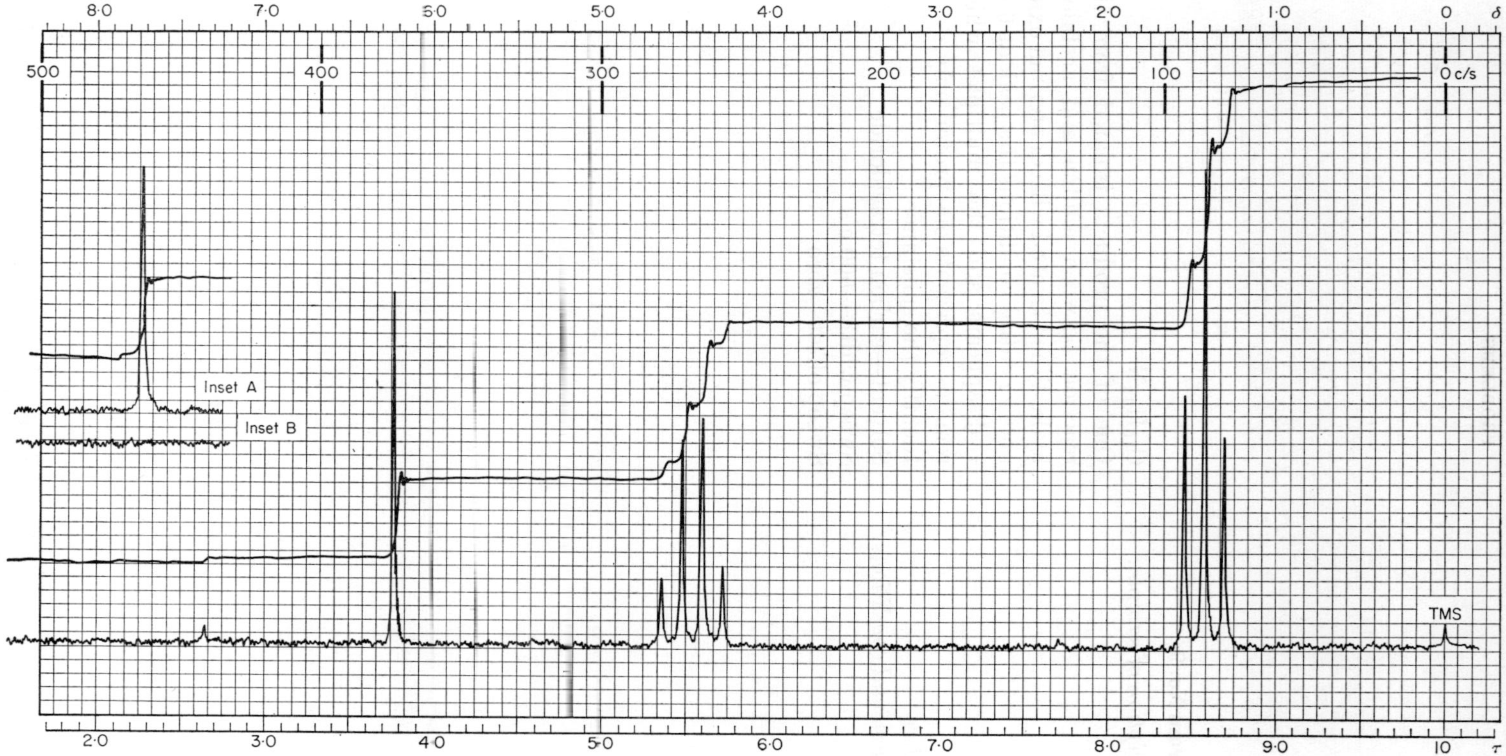

Spectrum No. 8.

Spectrum No. 9

Description of sample: $C_{10}H_{13}N$

One of the products isolated from the reaction between aniline and acetaldehyde. When treated with deuterium oxide under neutral conditions the peak at 6·65 τ disappears and the multiplet which remains represents one proton only.

Objective: Identify the substance

Instrumental conditions: Solvent—carbon tetrachloride

Sweep width—500 c/s

Sweep offset—zero

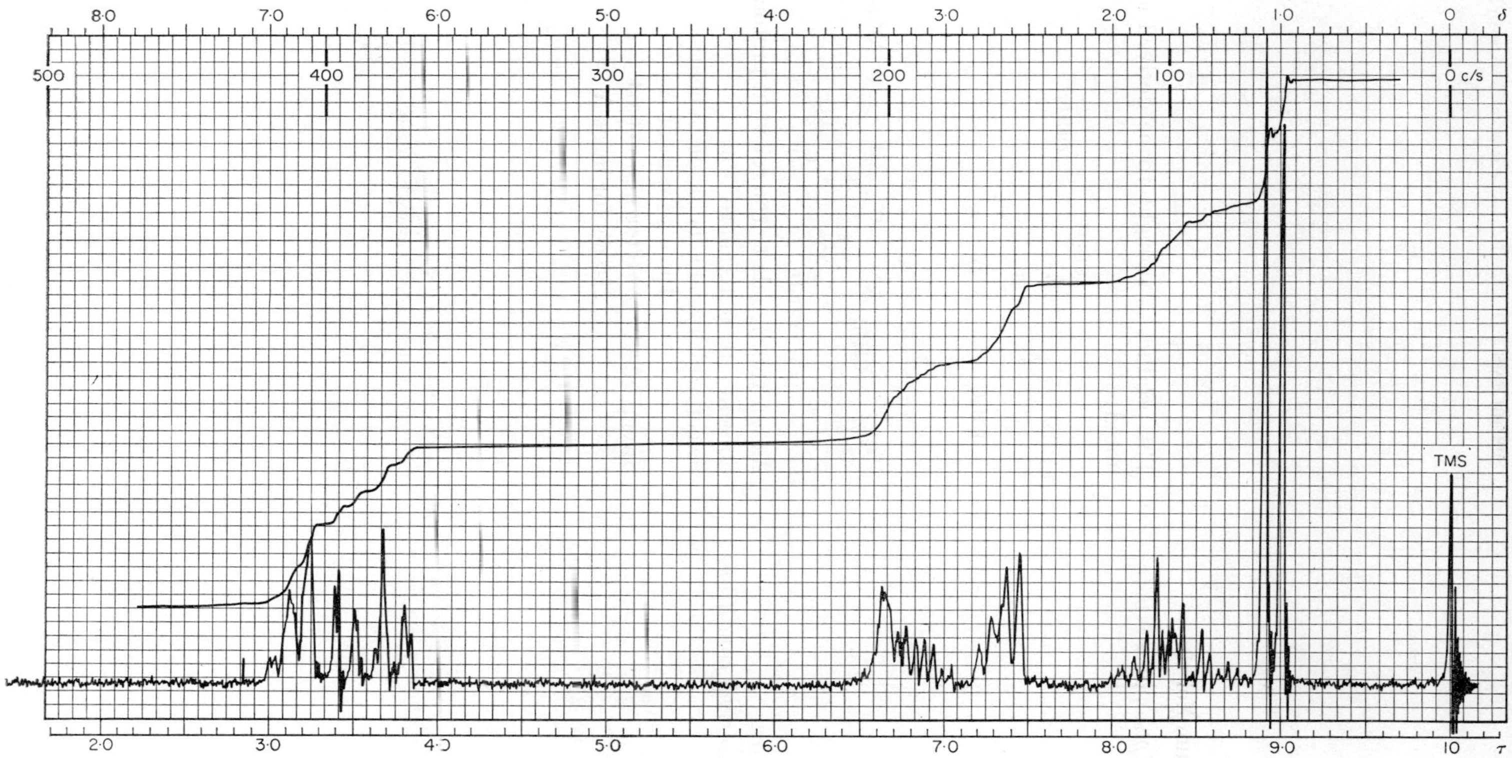

Spectrum No. 9.

Spectrum No. 10

Description of sample: The product is thought to be derived as noted:

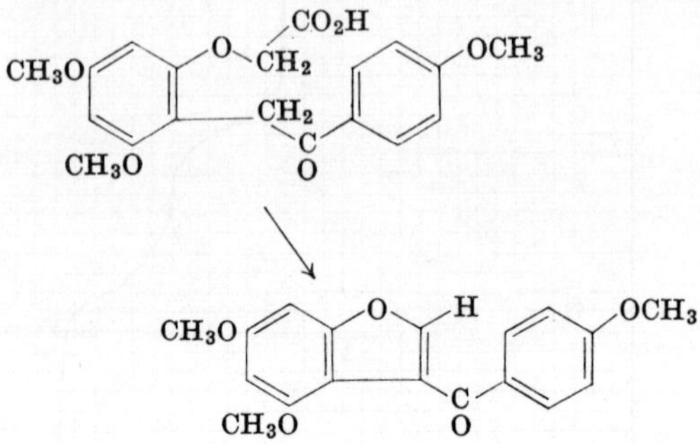

Objective: Assign the signals to the relevant protons

Instrumental conditions: Solvent—deuterochloroform

Sweep width—500 c/s for main chart
100 c/s for inset A

Sweep offset—zero for main chart
210 c/s for inset A

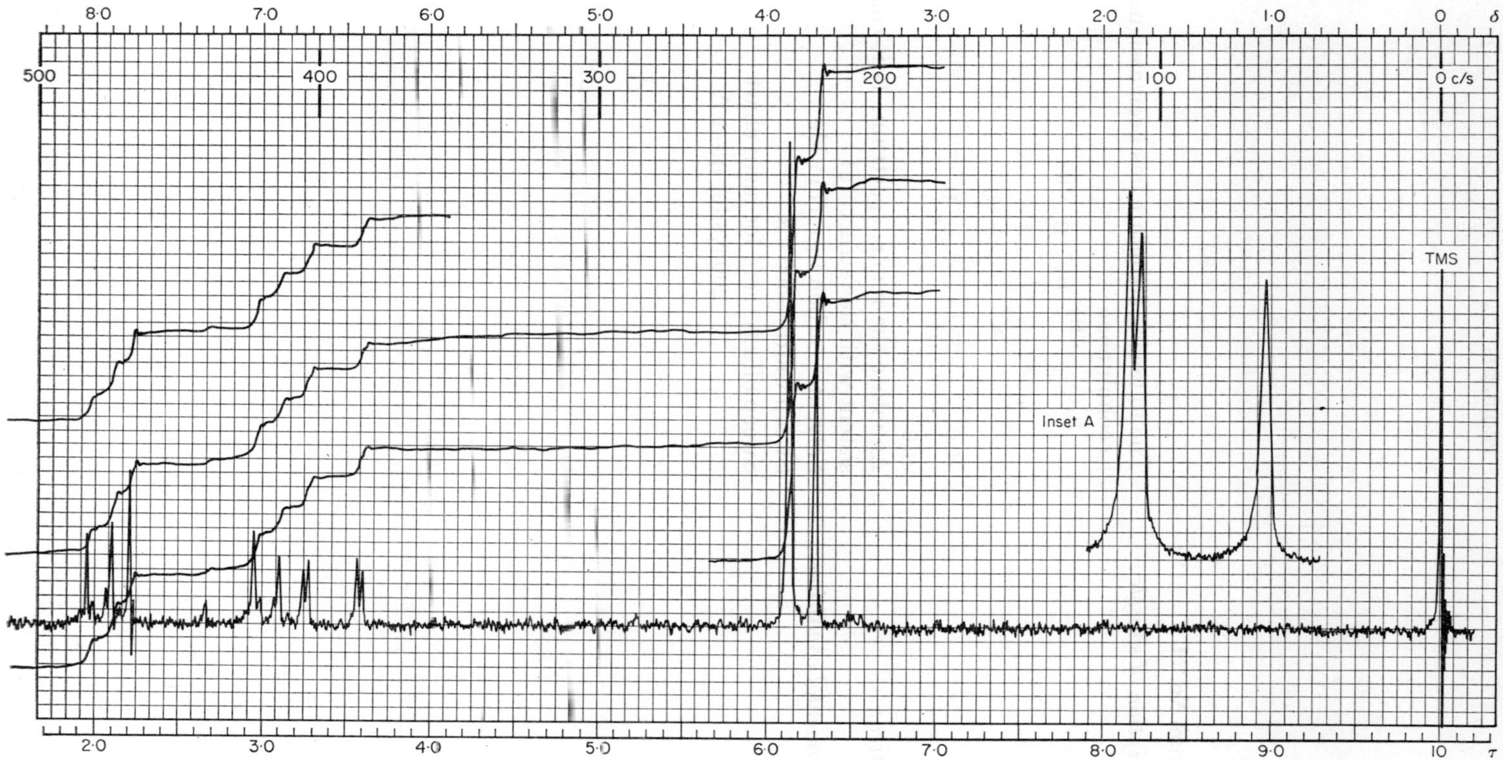

Spectrum No. 10.

Spectrum No. 11

Description of sample:	The following structure has been assigned to cohulupone

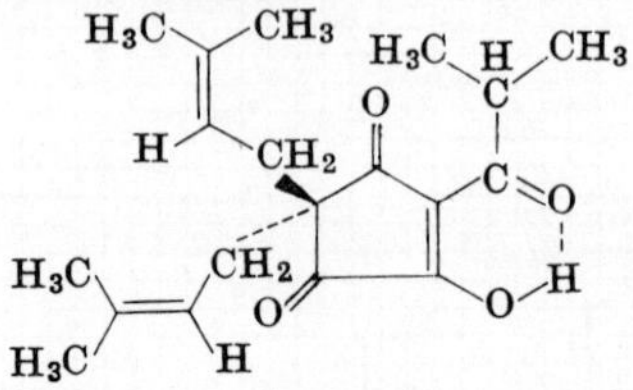

Objective:	Annotate the n.m.r. spectrum. Is it consistent with this structure?
Instrumental conditions:	Solvent—carbon tetrachloride Sweep width—500 c/s Sweep offset—zero for main chart 200 c/s for inset A

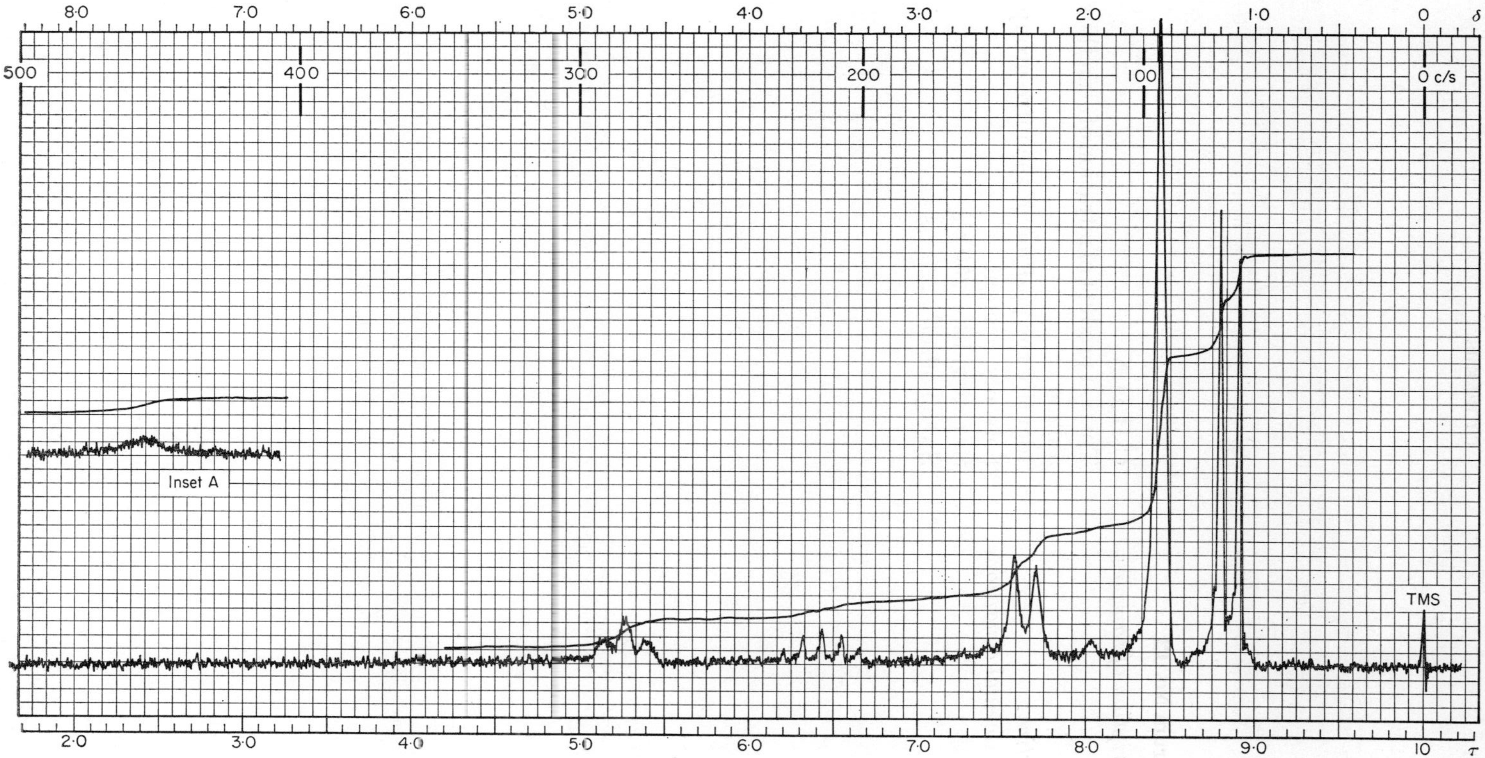

Spectrum No. 11.

Spectrum No. 12

Description of sample:	$C_7H_8O_3N_2$ This is a product of nitration of *o*-methoxy-aniline
Objective:	Identify the position of the nitro group
Instrumental conditions:	Solvent—acetone Sweep width—250 c/s Sweep offset—220 c/s

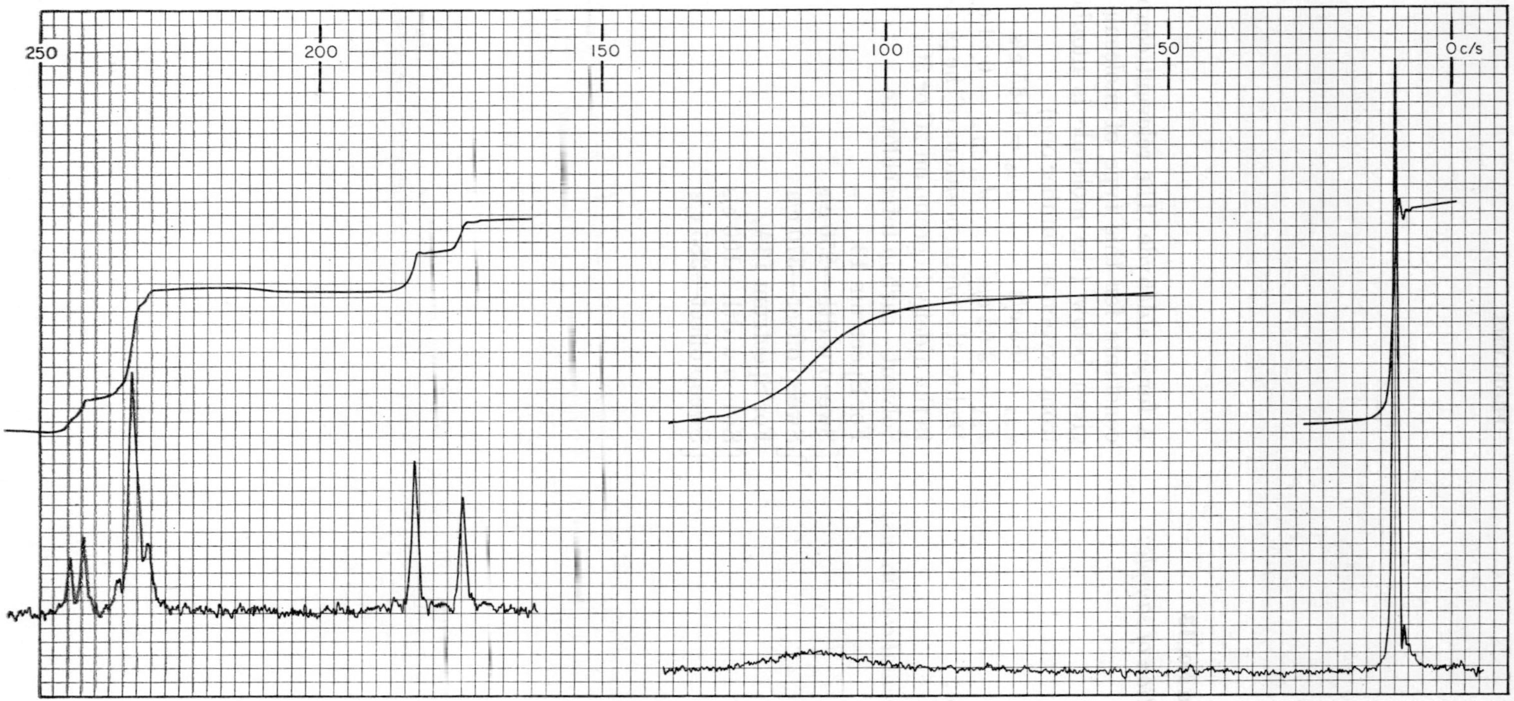

Spectrum No. 12.

Spectrum No. 13

Description of sample:	A furan aldehyde. The relevant coupling constants for this type of molecule are given below (the molecule is numbered from oxygen as number one)
	$J_{35} = 0{\cdot}8$ c/s $\quad J_{45} = 1{\cdot}65$ c/s
	$J_{34} = 3{\cdot}62$ c/s $\quad J_{5, \mathrm{CH}(2)} = 0{\cdot}81$ c/s
Objective:	Deduce the origin positions of the protons. Is the aldehyde group on 2C or 3C?
Instrumental conditions:	Solvent—carbon tetrachloride
	For Spectrum 13,
	Sweep width—250 c/s; Sweep offset—370 c/s
	For Spectrum 13A,
	Sweep width—100 c/s; Sweep offset—385 c/s

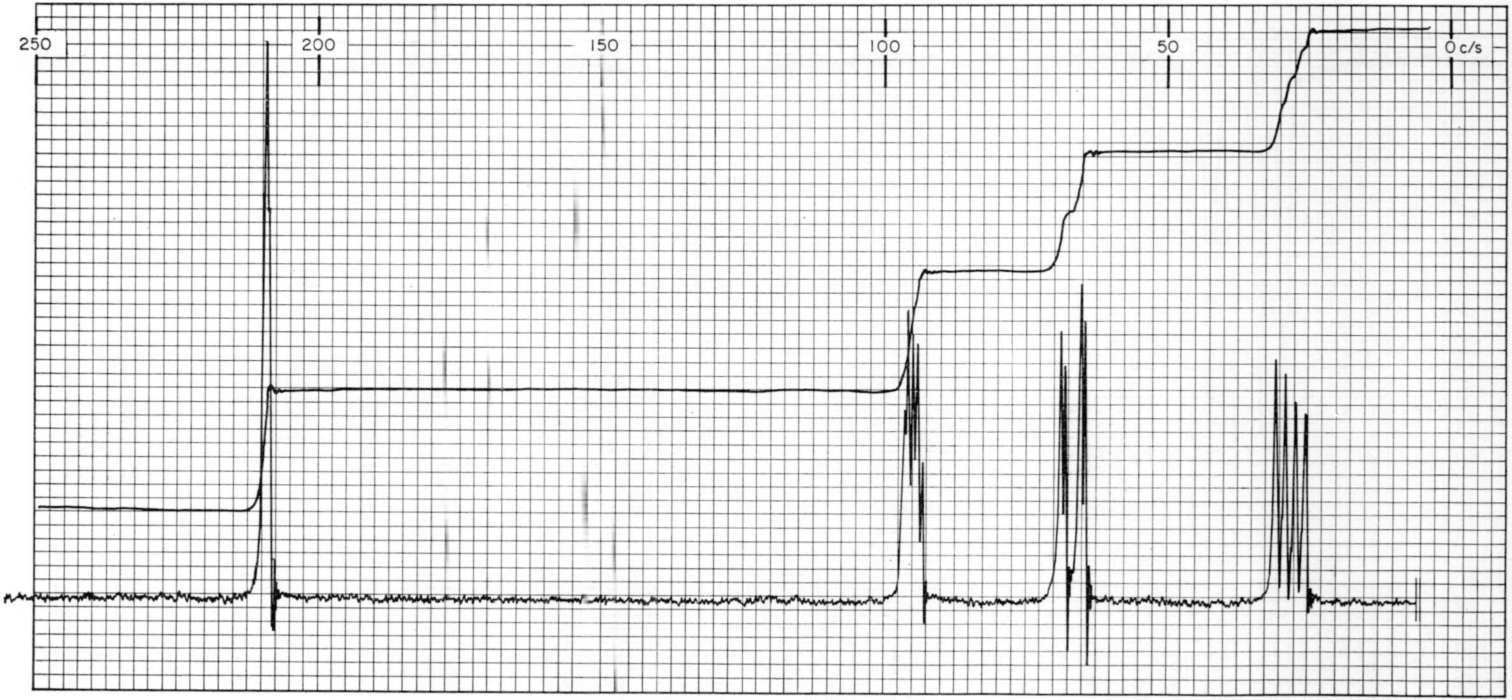

Spectrum No. 13.

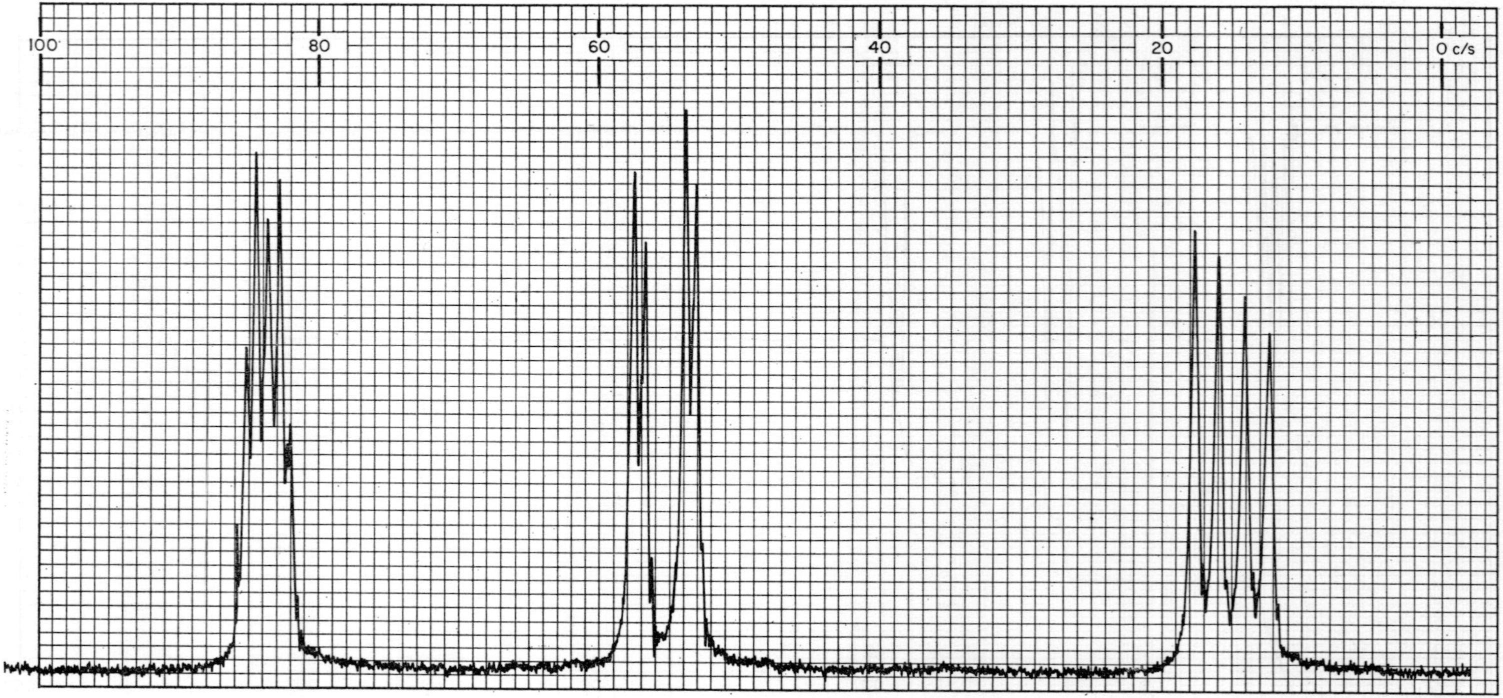

Spectrum No. 13A.

Spectrum No. 14

Description of sample: $C_7H_6O_4$ An α-pyrone with a methoxy-carbonyl substituent.

Objective: Deduce the origin positions of the proton signals.
What can you say about the structure of this compound?

Instrumental conditions: Solvent—deuterochloroform
For Spectrum 14,
Sweep width—500 c/s
Sweep offset—zero
For Spectrum 14A,
Sweep width—100 c/s
Sweep offset—380 c/s
A general indication of the relative chemical shifts in the α-pyrone nucleus is as follows:

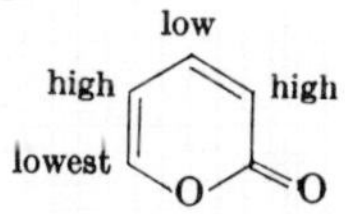

This pattern is largely independent of the substituents.

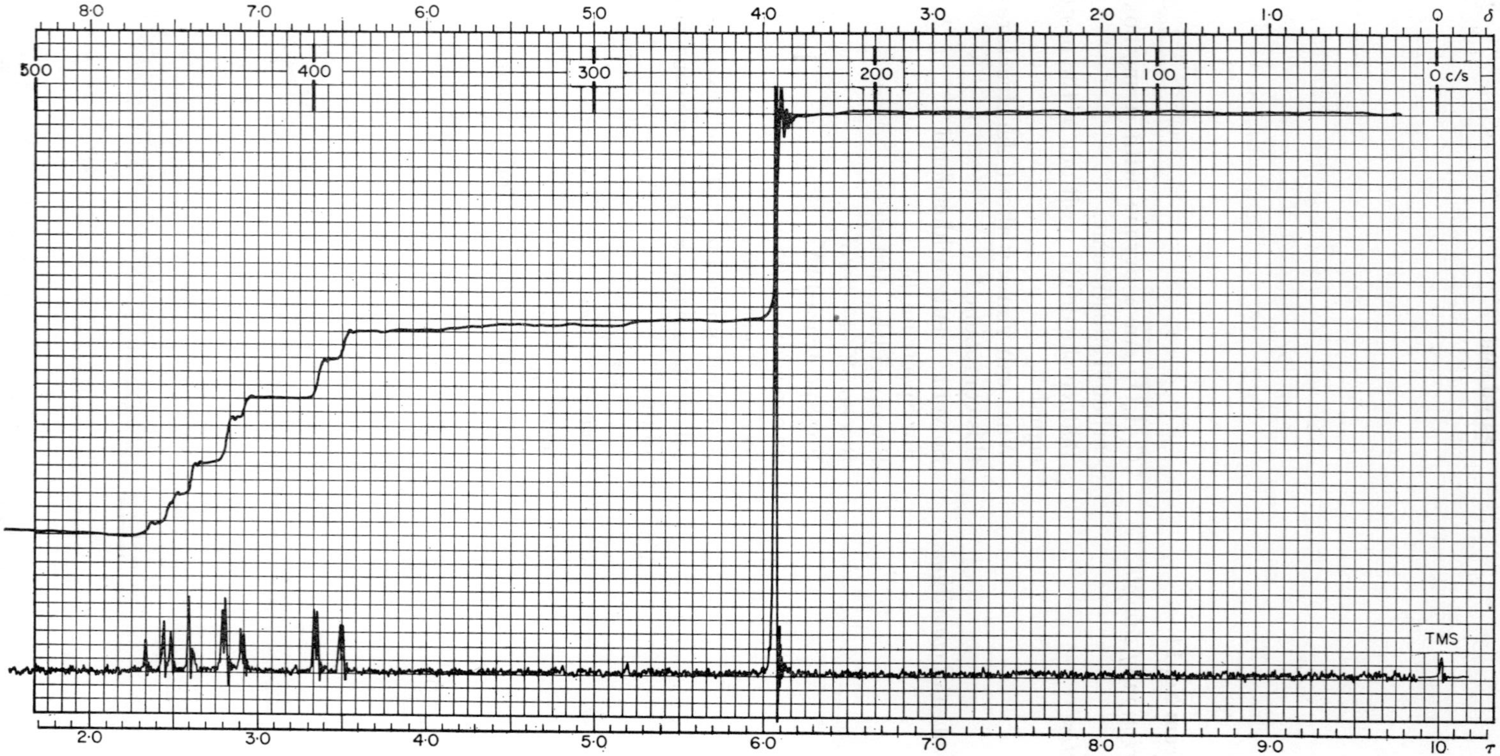

Spectrum No. 14.

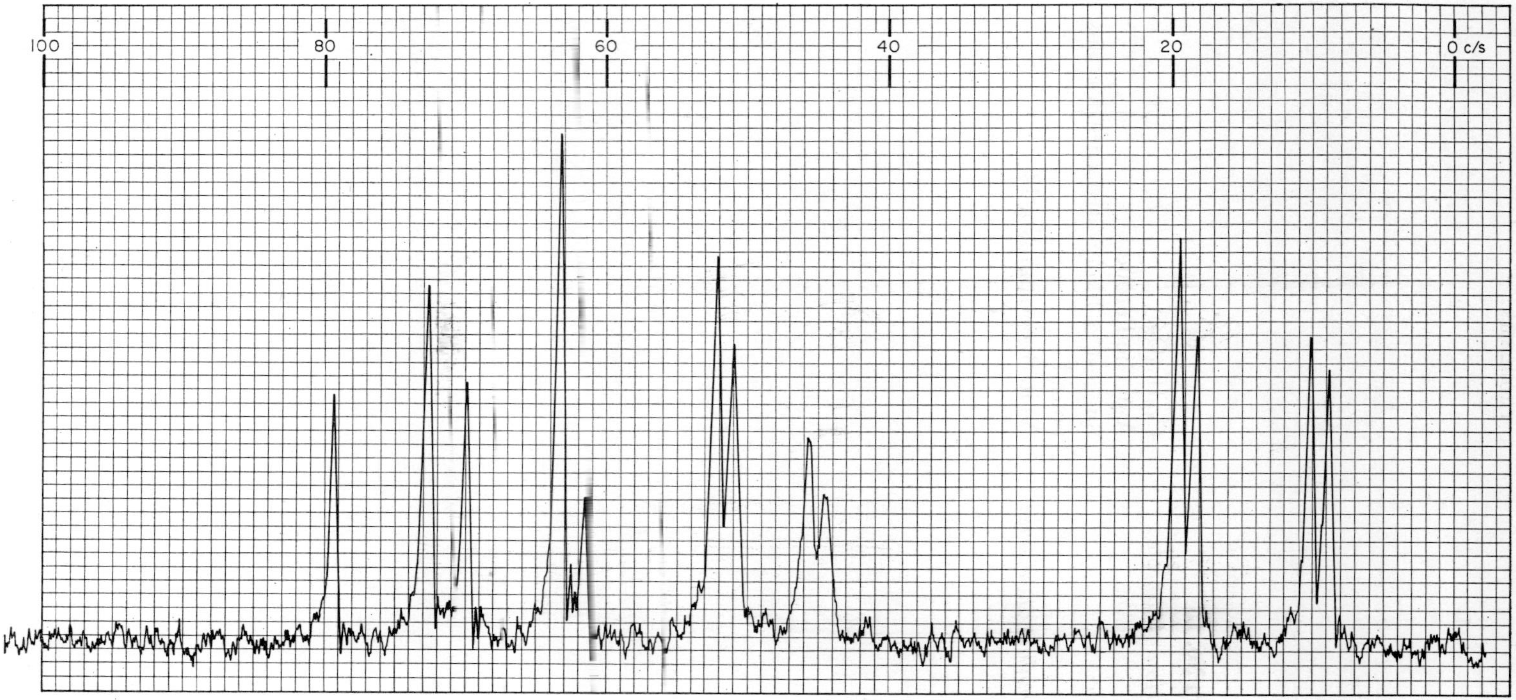

Spectrum No. 14A.

Spectrum No. 15

Description of sample: $C_{10}H_{14}O_3$
A cyclic ketone

Objective: Suggest a likely structure for this compound

Instrumental conditions: As in Spectrum No. 14

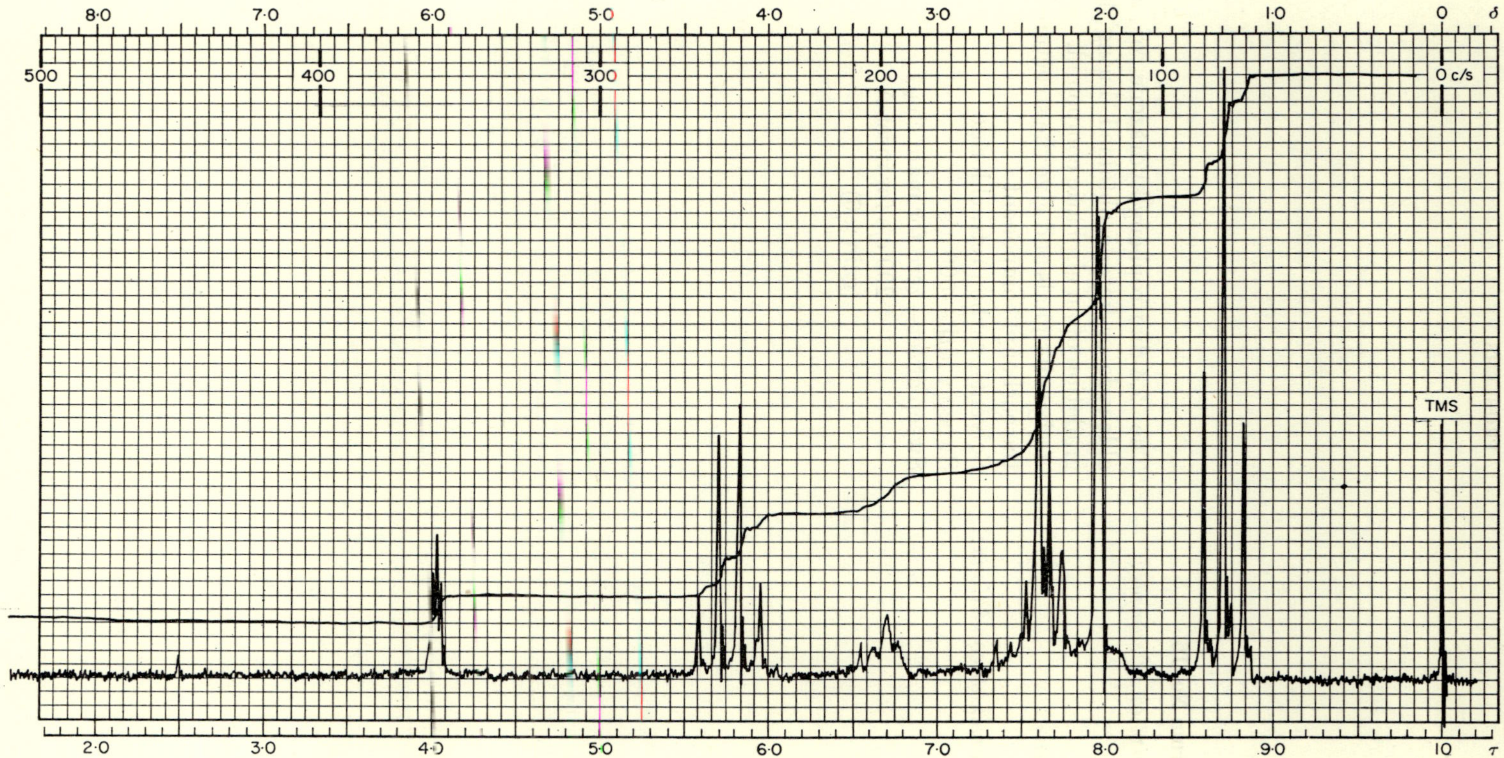

Spectrum No. 15.

Spectrum No. 16

Description of sample:	The 2,4-dinitrophenylhydrazone of an aldehyde or ketone obtained during degradation of an aliphatic compound.
Objective:	Assign all signals and give the structure of the carbonyl component
Instrumental conditions:	Solvent—carbon tetrachloride Sweep width—500 c/s Sweep offset—zero for main chart 200 c/s for inset A (gain × 6)

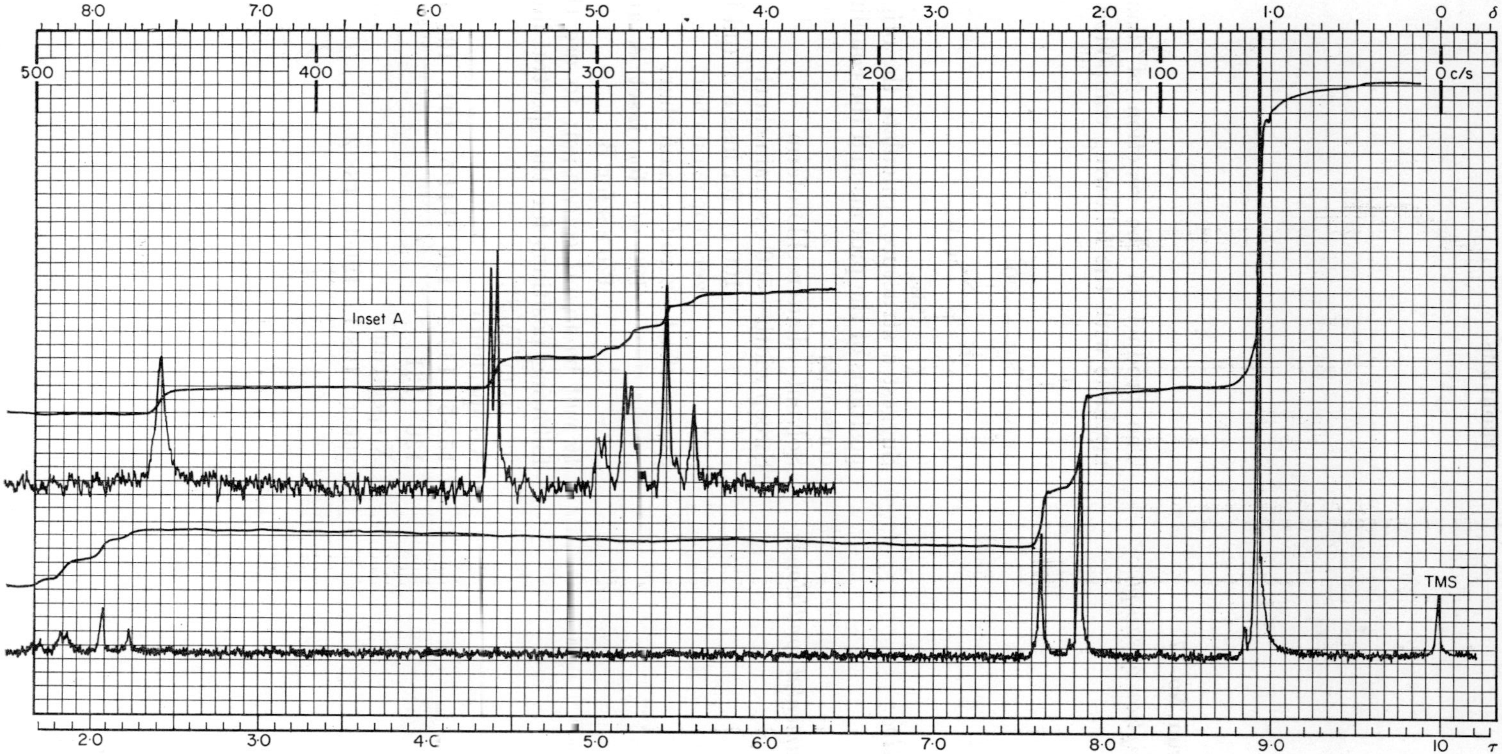

Spectrum No. 16.

Spectrum No. 17

Description of sample: The product of the reaction of diazomethane with the following compound

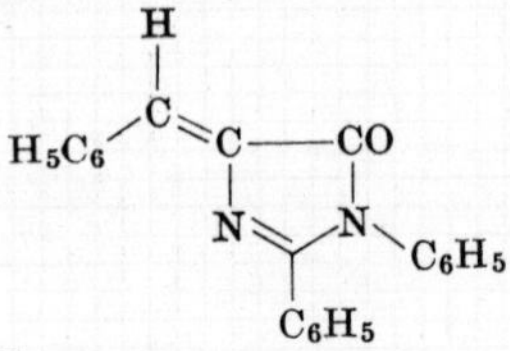

Objective: Suggest a structure: calculate the origin positions of the protons and the coupling constants

Instrumental conditions: Solvent—deuterochloroform

Sweep width—500 c/s main chart
100 c/s for insets A and B
zero for main chart

Sweep offset—zero for main chart
130 c/s for inset A and
120 c/s for B

It will be found convenient to use the following positions for the spectral lines (c/s from TMS).

AB spectrum

152·5 147·5 146 144·5 141 139·5 136 131

X spectrum

206·4 198·1 196·4 188·2

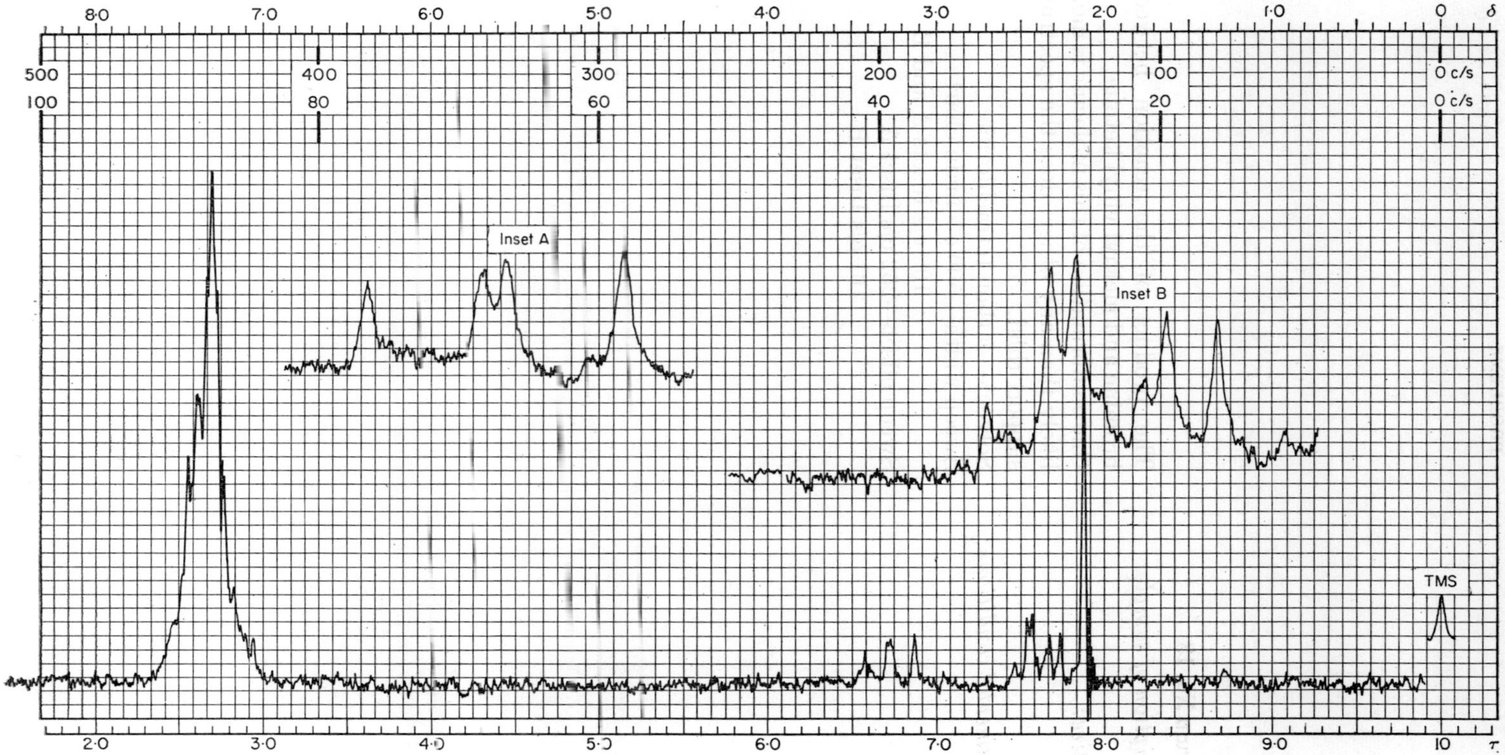

Spectrum No. 17.

Spectrum No. 18

Description of sample:	The product of the reaction between bromine and methyl vinyl ketone
Objective:	One conformation of this molecule is more likely than any other: does the spectrum support this?
Instrumental conditions:	Solvent—carbon tetrachloride Sweep width—500 c/s main chart 100 c/s Insets A and B Sweep offset—252 c/s Inset A 161 c/s Inset B

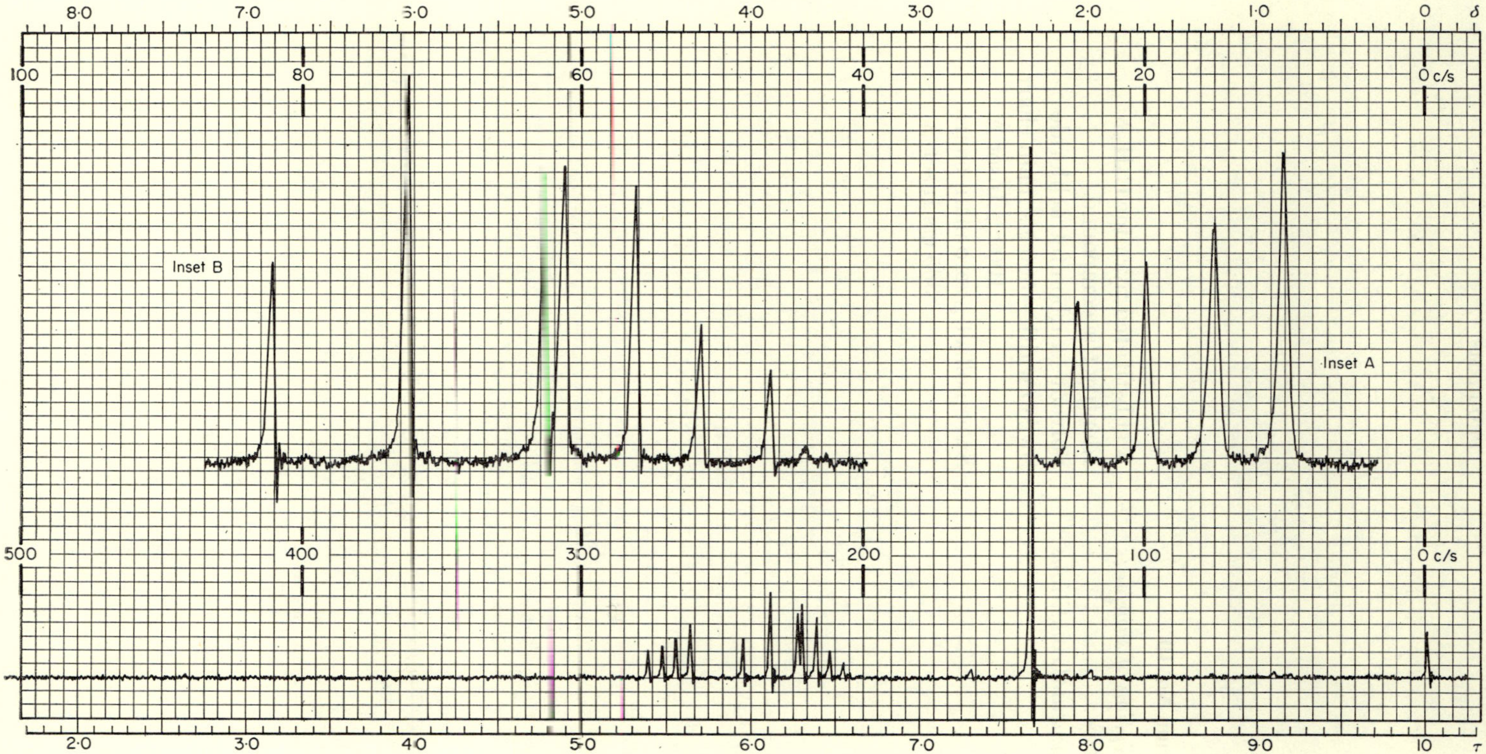

Spectrum No. 18.

Spectrum No. 19

Description of sample:	A substituted pyridine C_7H_9N
Objective:	Identify the compound
Instrumental conditions:	Solvent—carbon tetrachloride
	Sweep width—500 c/s main chart
	100 c/s Inset A
	Sweep offset—zero main chart
	365 c/s Inset A

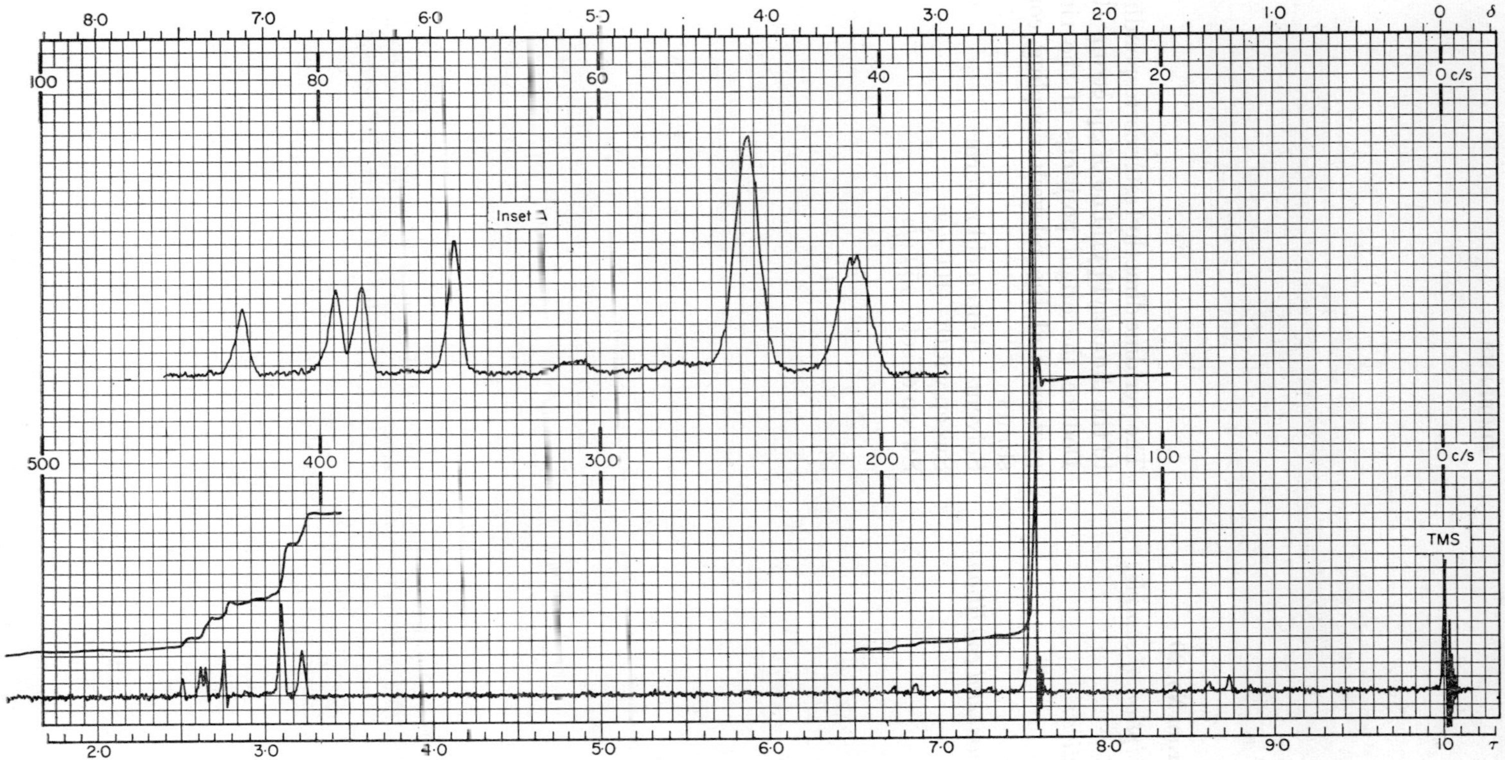

Spectrum No. 19.

Spectrum No. 20

Description of sample:	The pyridine dimethyl ester corresponding to the dimethyl pyridine of Spectrum 19
Objective:	Deduce the origin positions of the ring protons and the relevant coupling constants
Instrumental conditions:	Solvent—deuterochloroform Sweep width—100 c/s Sweep offset—450 c/s

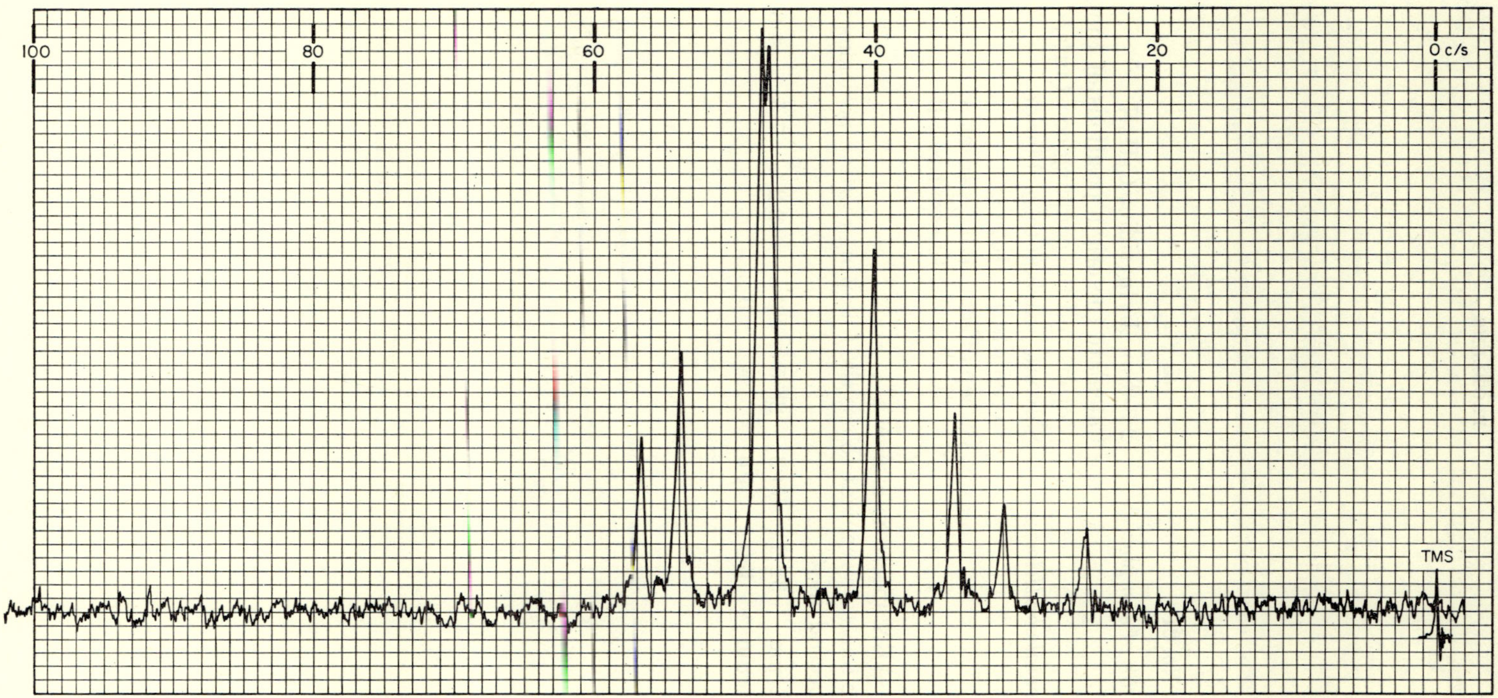

Spectrum No. 20.

Spectrum No. 21

Description of sample:	2,3-dimethylpyridine C_7H_9N
Objective:	Calculate the origin positions of the ring protons and the coupling constants.
Instrumental conditions:	Solvent—carbon tetrachloride Sweep width—100 c/s Sweep offset—410 c/s Inset A 400 c/s Inset B

Only the ABX spectrum is shown
It will be found convenient to use the following relative positions in c/s for the spectral lines

AB spectrum	32·2	30·8	24·5	23·1	12·2	7·5	4·8	0
X spectrum	6·0	4·6	1·4	0				

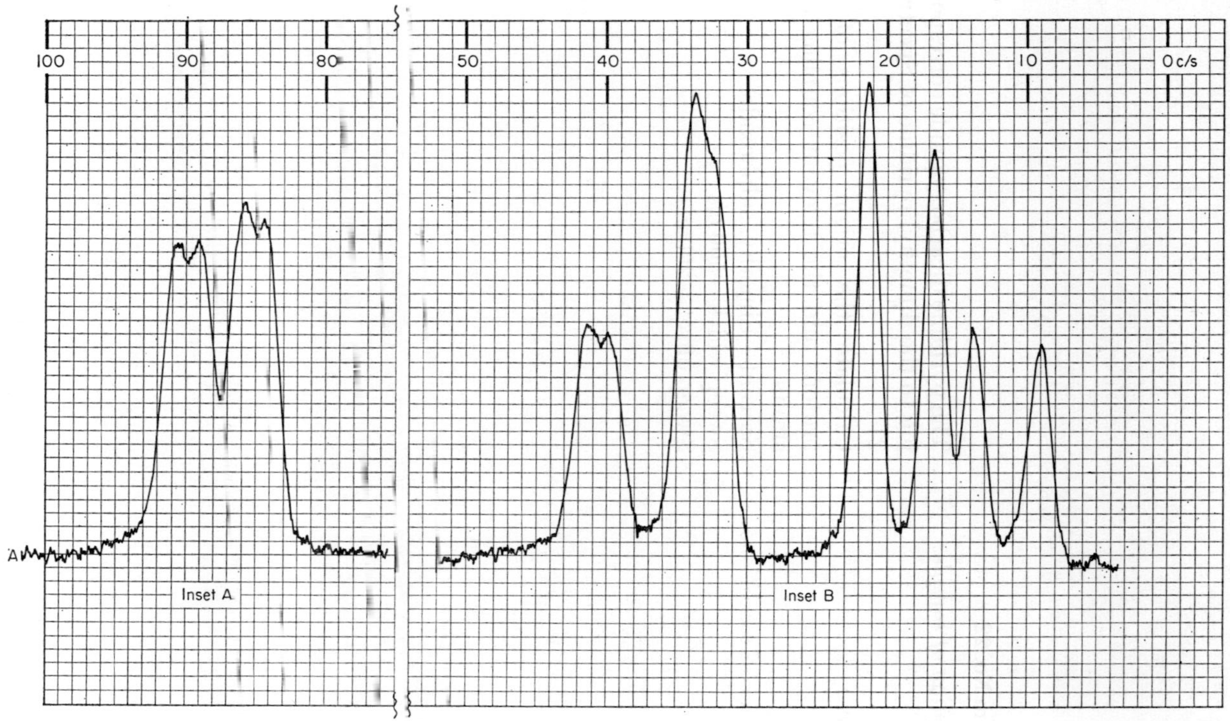

Spectrum No. 21.

Spectrum No. 22

Description of sample:	The pyridine dicarboxyamide corresponding to the dimethyl pyridine of Spectrum 21
Objective:	Calculate the origin positions of the protons and the relevant coupling constants
Instrumental conditions:	Solvent—dimethyl sulphoxide Sweep width—500 c/s Sweep offset—200 c/s

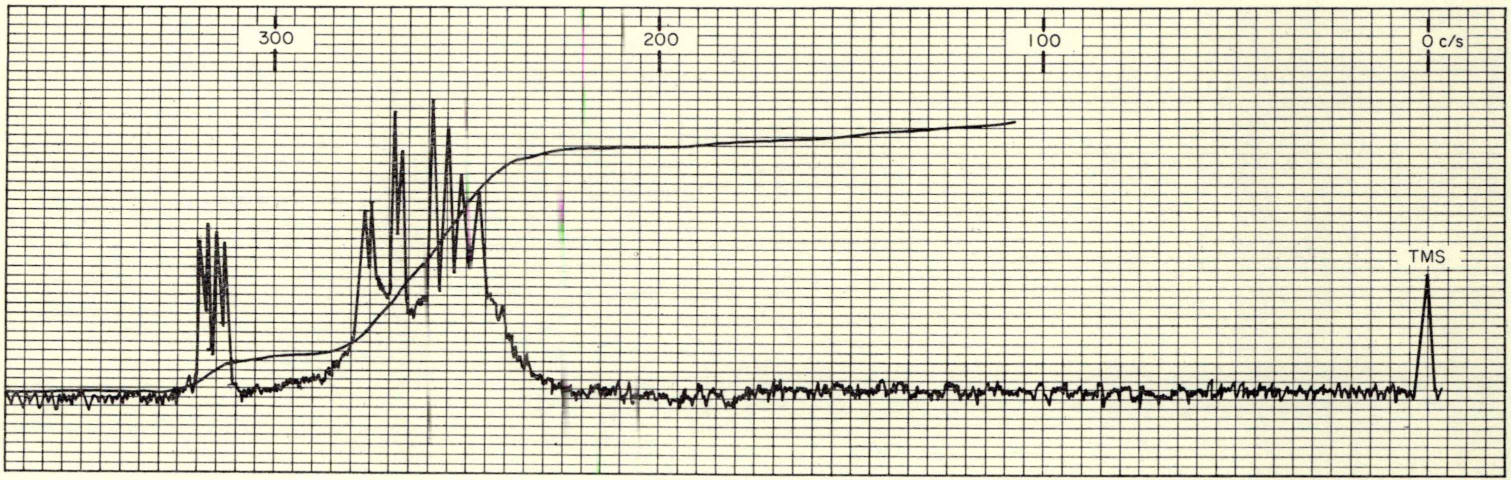

Spectrum No. 22.

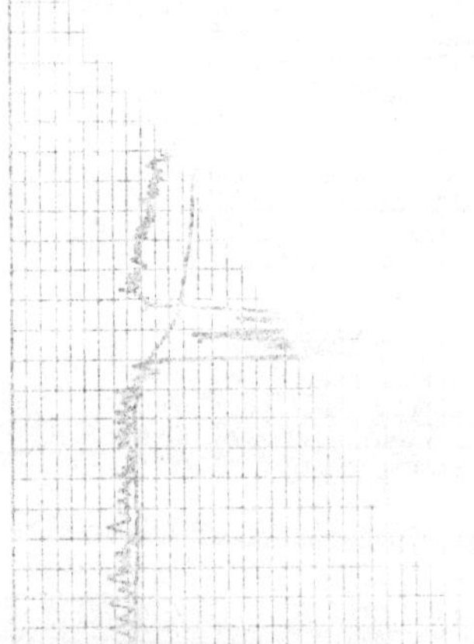

ANSWERS TO PROBLEMS

Statements of τ and J values given in the answers were taken from the full size spectra whilst the smaller size of the reproductions in this book slightly restricts such accuracy, this does not affect the general validity of the solutions.

Spectrum No. 1

The TMS line is accurately on zero, there is no offset, and the sweep width is 500 c/s. Hence the line positions can be read directly from the 0–500 c/s scale, or from the τ and δ scales. As the solvent was CCl_4, all lines must arise from the solute.

The spectrum of the compound $C_7H_7NO_2$ shows 5 lines. Two pairs of lines at low field form what is apparently an AB quartet pattern: at higher field is a singlet. To find the numbers of protons responsible for the individual signals, we take the total rise in the integral trace, $17 + 14 = 31$ units, and divide by the total number of protons, 7. This gives a step of $31/7 = 4{\cdot}4$ units/proton, from which it follows that the singlet must arise from a group of 3 equivalent protons, and each of the low-field pair of lines from 2 protons.

None of these last 4 lines corresponds individually to an integral number of protons.

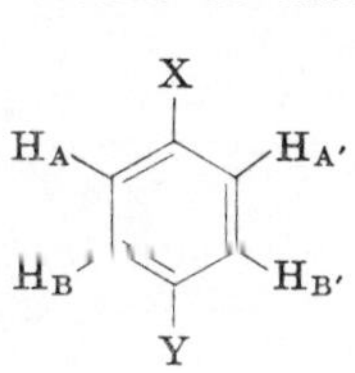

We conclude that there are two superimposed AB quartets. These lines appear in the aromatic region of the spectrum. If we assume, then, that the compound is benzenoid, we arrive at a *p*-substituted ring, as inset.

The separation within each doublet is 8 c/s, correct for ortho-coupling. One of the substituents is evidently a methyl group: Table 1 gives τ 7·66 for the methyl-group protons in toluene, and this is close to the observed position, τ 7·54, of the high field three-proton line in the spectrum. Subtraction of C_6H_4—CH_3 from the molecular formula leaves NO_2. The second substituent is then most likely to be a nitro group, so that the compound is probably *p*-nitrotoluene (I). The nitro substituent is strongly electron-withdrawing and this gives a reason for the methyl resonance line being at a lower-field position than in toluene. The lower-field half of the AB quartet must come from the two protons, (A), in (I) which flank the nitro group and the higher-field half from those, (B), which flank the methyl group. The predicted line positions for the ring-protons (from Table 5) are $2{\cdot}73 - 1 + 0{\cdot}1 = 1{\cdot}83$ and $2{\cdot}73 + 0{\cdot}15 - 0{\cdot}3 = 2{\cdot}58\ \tau$, in very reasonable agreement with the observed positions of the mid-points of the doublets, at $1{\cdot}95$ and $2{\cdot}70\ \tau$. Because $\frac{\Delta_{AB}}{J_{AB}} = \frac{45}{8} = 5{\cdot}6$ we are justified in taking the mid-points as the origin positions of the A and B protons (see Chapter 4).

The protons in the structure (I) do not strictly constitute two independent AB systems (although this is a reasonable first approximation) because there is coupling round the ring. The system is really an A_2B_2 type (strictly AA′BB′ because there are two different AB couplings, $J_{AB} = J_{A'B'}$ and $J_{AB'} = J_{A'B}$).* Consequently more than 4 lines of an AB pattern arise; such extra lines are just discernible in spectrum No. 1.

Summary

τ	Intensity	Multiplicity	J (c/s)	Assignment
7·54	3	Singlet	–	(C) CH_3
2·70	2	Doublet (half of AB)	8	(B) Aromatic protons *ortho* coupled to (A)
1·95	2	Doublet (half of AB)	8	(A) Aromatic protons *ortho* coupled to (B)

This substance is

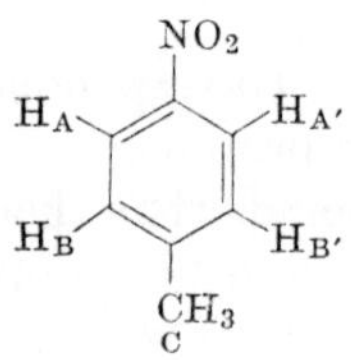

(I)

Spectrum No. 2

The spectrum has a sweep width of 500 c/s, there is no offset, and the TMS line is at zero. Hence the τ and δ scales can be used directly. Striking features in the spectrum of the compound ($C_{12}H_{14}O_4$) are the triplet and quartet signals, centred at τ 8·62 and 5·65, which have a common line separation of 7 c/s. These lines constitute an A_2X_3 spin-spin pattern and so could arise from an ethyl group $-CH_2-CH_3$. A strongly deshielding function must be attached to the methylene to account for its chemical shift of τ 5·65. The only reasonable possibility, from Table 1 (p. 179), second column, is an arylcarbonyloxy group, Ar.CO.O–, which may be expected

*See p. 112.

to give an attached methylene group a shift of about τ 5·77. In making the A_2X_3 assignment we note that the intensities of the quartet and triplet signals are in the ratio of 2:3, and the intensities of the individual lines within each multiplet are approximately 1:3:3:1 and 1:2:1, respectively These are essential requirements.

The spectrum also shows a strong line at τ 1·98 which is presumably from several equivalent aromatic-type protons. From left to right, the integral trace shows steps of 11 + 11 + 16 = 38 units. Dividing by the total number of protons, 14, we obtain a step of 2·7 units/proton. Hence the signals in the spectrum, from left to right, are from groups of 4, 4 and 6 protons. There must therefore be *two* equivalent ethyloxycarbonyl substituents on an aryl ring, which, evidently, also carries four equivalent hydrogen atoms: these groupings $(C_2H_5.O.CO)_2$ and C_6H_4 together make up the molecular formula. To account for the equivalence of the four aromatic protons, the ester groups must be *para*-disposed

Summary

τ	Intensity	Multiplicity	J (c/s)	Assignment
8·62	6	Triplet	7	(A) 2 × CH_3 of ester coupled to (B)
5·65	4	Quartet	7	(B) 2 × CH_2 of ester coupled to (A)
1·98	4	Singlet	–	(C) Aromatic protons

The compound is therefore diethyl terephthalate (II).

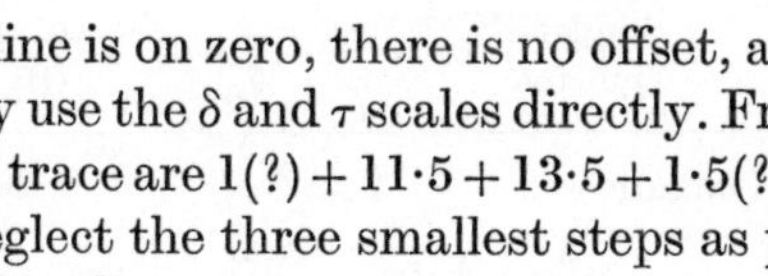

(II)

Spectrum No. 3

Again, the TMS line is on zero, there is no offset, and the sweep width is 500 c/s, so we may use the δ and τ scales directly. From left to right, the steps in the integral trace are 1(?) + 11·5 + 13·5 + 1·5(?) + 1(?) + 17·4 units, or 41 units if we neglect the three smallest steps as possibly being from impurity present. Dividing each total by the eight protons of the molecular formula, $C_4H_8O_3$, we obtain steps of 5·56 or 5·1 units/proton. Obviously, the three weakest signals can safely be ignored and we proceed, using the value of 5·1 units/proton.

At highest field there is a doublet, with $J = 7$ c/s, of total intensity corresponding to three protons. These could be on a methyl group and coupled to one other proton, so giving a doublet signal. This other proton would give a signal split as a 1:3:3:1 quartet with $J = 7$ c/s (in the absence of further coupling). Such a signal can be discerned at lowest field: the high-field line of the quartet is hidden, however, by another signal, at τ 5·94. The quartet and this signal together have an intensity of 11·5 units, corresponding to two protons. From the size of the steps in the integral trace over the three visible components of the quartet, it is clear that the quartet signal does indeed come from a single proton, and that the line at τ 5·94 comes from another. The quartet is centred at τ 5·77. So we conclude that the molecule is built from the fragments H, CH_3—CH–. The chemical shifts of the two single protons are considered later.

At τ 6·28, there is a singlet with an intensity of 13·5 units, which is somewhat low for a 3-proton signal. The signal is very sharp, however, and it is possible that the integration was done too rapidly to obtain a correct response: in any case the signal is too large to be attributed to only 2 protons. The first column of Table 1 indicates that the protons of a methyl ester group can have τ 6·35. We are therefore inclined to the view that the compound is a methyl ester and so we arrive at the partial structure CH_3—CH—CO—OMe, which can readily be completed with the remaining fragments O and H to give the molecule of methyl lactate (III). The hydroxyl proton could well have a chemical shift of τ 5·94 (Table 6, p. 185). Table 2 indicates that the β-methyl group would have a shift of $9{\cdot}1 - 0{\cdot}27 - 0{\cdot}25 = 8{\cdot}58\ \tau$ in reasonable agreement with the observed value of 8·65. Less accurately predictable is the chemical shift of the α-proton. Table 3 suggests a value of $8{\cdot}75 - 0{\cdot}7 - 1{\cdot}7 = 6{\cdot}35\ \tau$ which is rather higher than that observed.

Summary

τ	Intensity	Multiplicity	J (c/s)	Assignment
8·65	3	Doublet	7	(A) CH_3 coupled to (B)
6·28	3	Singlet	–	(D) CH_3 of ester
5·94	1	Singlet	–	(C) OH
5·77	1	Quartet	7	(B) CH coupled to (A)

This substance is methyl lactate.

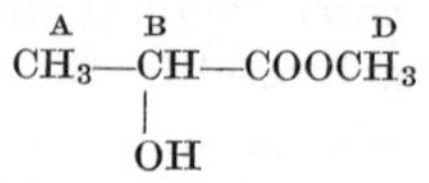

(III)

Spectrum No. 4

τ	H	Multiplicity	J (c/s)	Assignment
6·43	2	Singlet	–	(E) CH_2 (cf. value from Shoolery's rules, Table 3, Appendix, p. 182
6·23	3	Singlet	–	(D) CH_3O
3·06	2	Doublet (half of AB)	9	(C) Aromatic protons* ortho coupled to (B)
2·74	2	Doublet (half of AB)	9	(B) Aromatic protons* ortho coupled to (C)
−1·83	1	Singlet	–	(A) Carboxyl proton

* Two superimposed "AB" patterns (strictly AA'BB' see Chapter 7, p. 112)

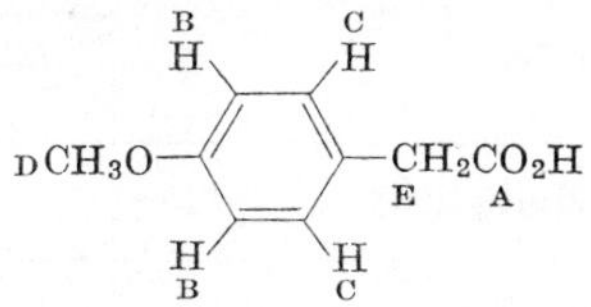

(IV)

Spectrum No. 5

τ	H	Multiplicity	J (c/s)	Assignment
8·63	6	Triplet	7·5	(C) 2 × CH_3 of ethyl ester coupled to (B)
5·60	4	Quartet	7·5	(B) 2 × CH_2 of ethyl ester coupled to (C)
centre at 2·30	4	Multiplet AA'BB'	–	(A) Aromatic protons—note symmetry about mid-point—a characteristic of the A_2B_2 cases (cf. Chapter 7, p. 107) (Wiberg and Nist (Interpretation of NMR Spectra, p. 310)

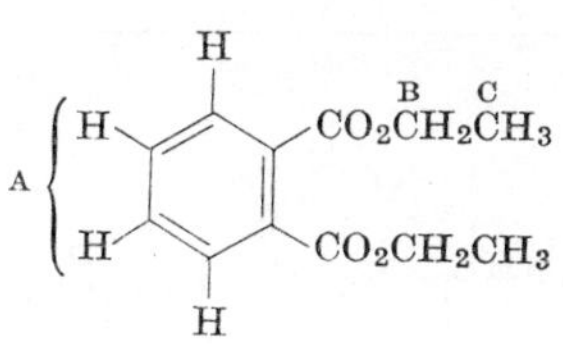

(V)

Compare the aromatic proton signals with those in spectrum No. 2. What pattern would you predict for the aromatic protons of the diethyl ester of benzene 1,3-dicarboxylic acid?

Spectrum No. 6

τ	H	Multiplicity	J (c/s)	Assignment
8·67	3	Triplet	7	(D) CH_3 coupled to (B)
8·22	3	Doublet	7	(C) CH_3 adjacent to (A)
5·71	2	Quartet	7	(B) CH_2 adjacent to —O.COR coupled to (D)
4·78	1	Quartet	7	(A) CH coupled to (C

$$\underset{C}{CH_3}.\underset{A}{\overset{\overset{NO_2}{|}}{CH}}.CO_2\underset{B}{CH_2}\underset{D}{CH_3} \qquad \text{(VI)}$$

Spectrum No. 7

τ	H	Multiplicity	J (c/s)	Assignment
7·37	2	Triplet	6	(C) CH_2 coupled to (B)
6·17	2	Triplet	6	(B) CH_2 coupled to (C)
5·78	1	Singlet	–	(A) OH removed by deuteration

$$\underset{A}{HO}.\underset{B}{CH_2}.\underset{C}{CH_2}.CN$$ (VII) (cf. Chapter 7, p. 113)

Spectrum No. 8

τ	H	Multiplicity	J (c/s)	Assignment
8·57	3	Triplet	7	(D) CH_3 of ethoxyl group coupled to (C)
5·55	2	Quartet	7	(C) CH_2 of ethoxyl group coupled to (D)
3·77	1	Singlet	–	(B) Proton directly attached to ring
−4·32	1	Singlet	–	(A) Enolic OH

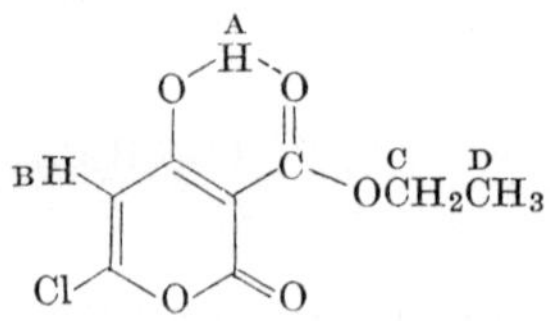

(VIII)

This arrangement of substituents does not necessarily represent the only solution.

Spectrum No. 9

τ	H	Multiplicity	J (c/s)	Assignment
8·95	3	Doublet	6	(E) Methyl group split by adjacent hydrogen (B)
~ 8·35	2	Multiplet	–	(D) CH_2 group surrounded by alkyl residues
~ 7·35	2	Multiplet	–	(C) CH_2 adjacent to aromatic ring
~ 6·75	2	Multiplet	–	(B) One proton exchangeable under neutral conditions, therefore NH. The second must be that causing the splitting of the methyl signal (E)
3–4	4	Multiplet	–	(A) Four aromatic protons. The high field position of this multiplet is due to the attached electron-donating substituent –NH

This is an example of a spectrum which has to be interpreted almost entirely on the basis of chemical shifts and the integral trace.

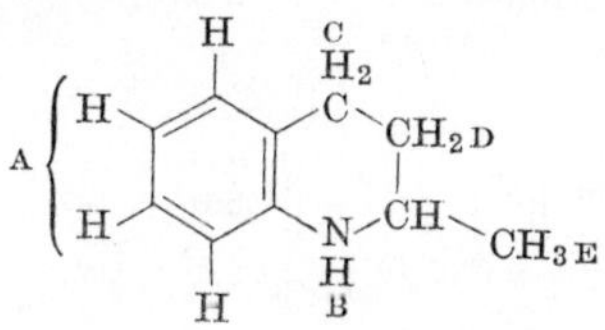

(IX)

Spectrum No. 10

τ	H	Multiplicity	J (c/s)	Assignment
6·29	3	Singlet	–	(E) Methoxyl group
6·13	6	Singlet	–	(D) 2 methoxyl groups
3·60	1	Doublet	2	(C) Aromatic protons *meta* splitting
3·27	1	Doublet	2	
3·03	2	Doublet	9	(A) Two superimposed AB patterns: *ortho* splitting (strictly AA′BB′)
2·04	2	Doublet	9	
2·22	1	Singlet	–	(B) Olefinic proton

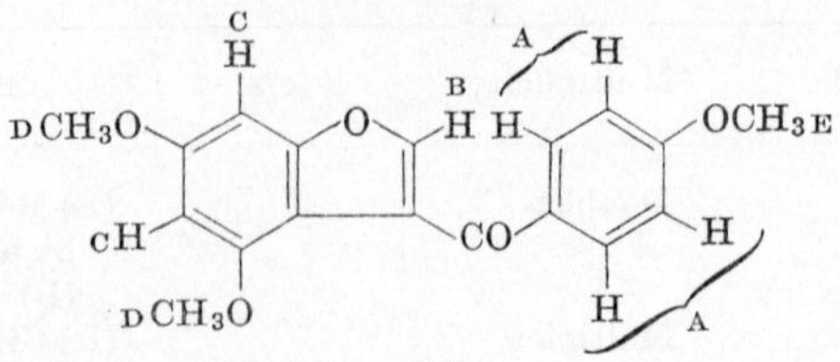

(X)

Spectrum No. 11

τ	H	Multiplicity	J (c/s)	Assignment
8·86	6	Doublet	6·5	(F) Two methyl groups
8·43	12	Singlet	–	(E) Four methyl groups on double bonds. Note broad nature of peak and shoulder on low field side: suggestion of splitting due to coupling to proton (B)
7·63	4	Doublet	8·0	(D) Two CH_2 groups attached to double bond: split by coupling to (B)
6·43	1	7 lines	6·5	(C) Hydrogen atom of *iso* propyl group: shifted downfield by adjacent C═O
5·27	2	Triplet	8·0	(B) Ethylenic protons split by coupling with adjacent methylene (D)
~ −0·9	1	Broad peak	–	(A) Hydrogen bonded enolic OH

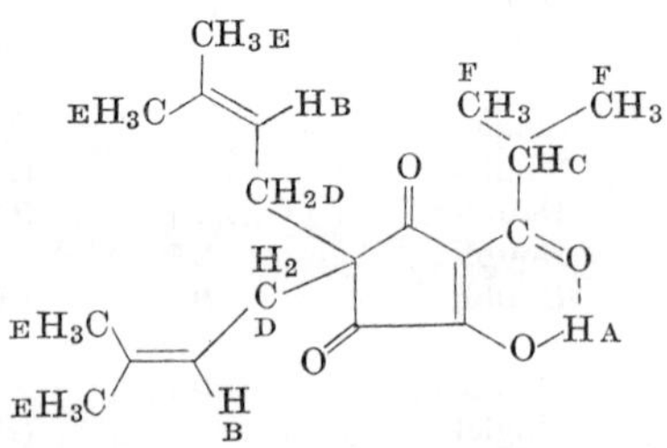

(XI)

Spectrum No. 12

There are four mononitro derivatives of *o*-methoxyaniline and using Table 5, p. 184 the approximate position of the aromatic protons in each of these may be calculated.

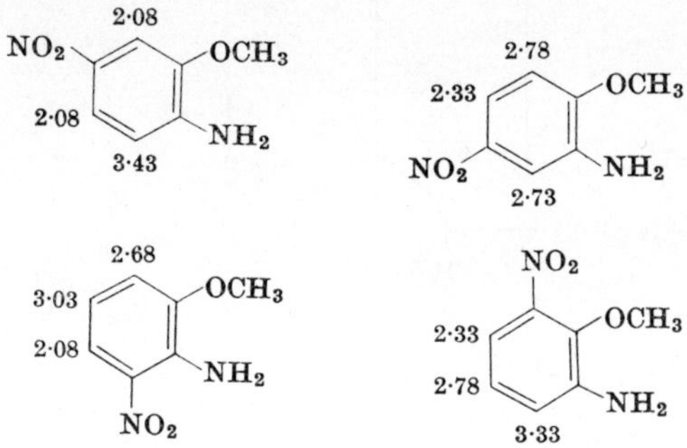

The last two cases seem contraindicated by the observed splitting pattern of the aromatic protons which approximates to an AB spectrum.

Working from the high field end of the spectrum, the OCH_3 (6·17 τ) and broad NH_2 signals (4·45 τ) are easily identified. The high field doublet of the AB system has lines as under

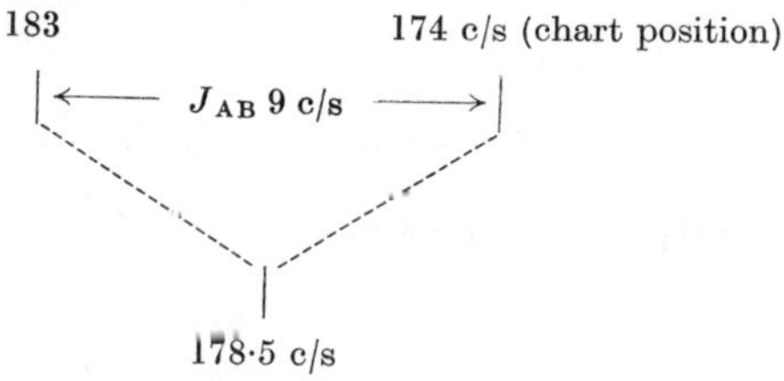

178·5 c/s

Offset 220

398·5 c/s

$\nu_B - 3{\cdot}36\tau$

This value agrees only with the first of the above formulae and this resonance may be assigned to the proton *ortho* to the amino group (B in the formula below).

The low field doublet of the AB system is complicated by having a second proton signal superimposed on it. First order analysis is as follows;

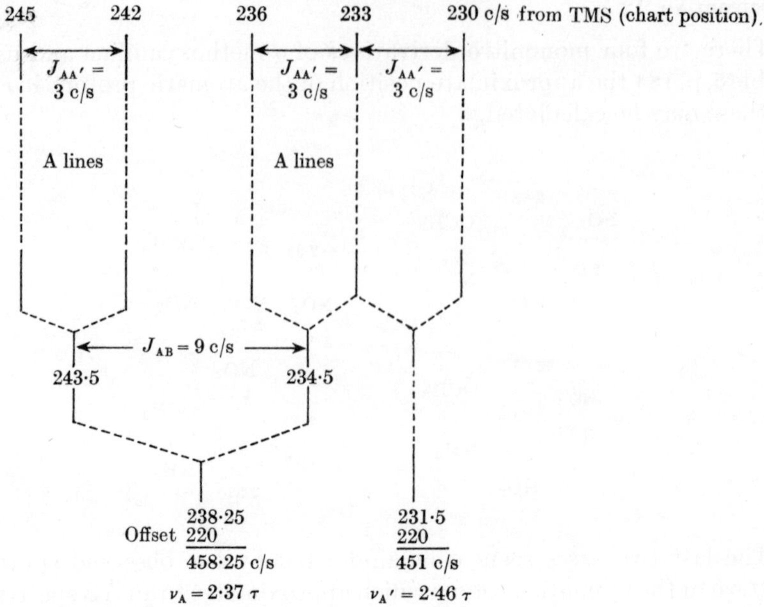

The unusually intense nature of the line at 233 c/s (chart position) which distorts the normal AB picture is thus explained.

Thus the nitration product of *o*-anisidine is

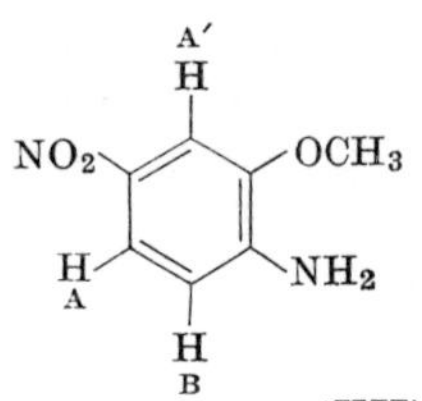

$\nu_{A'} = 2{\cdot}46\ \tau$ $\quad$ $J_{AA'} = 3$ c/s

$\nu_A = 2{\cdot}37\ \tau$

$\nu_B = 3{\cdot}36\ \tau$ $\quad$ $J_{AB} = 9$ c/s

(XII)

Spectrum No. 13 (AMX spectrum)

The middle multiplet of the three shown in Spectrum 13A, most obviously yields two coupling constants. Thus if we note the positions of the individual peaks in c/s we have,

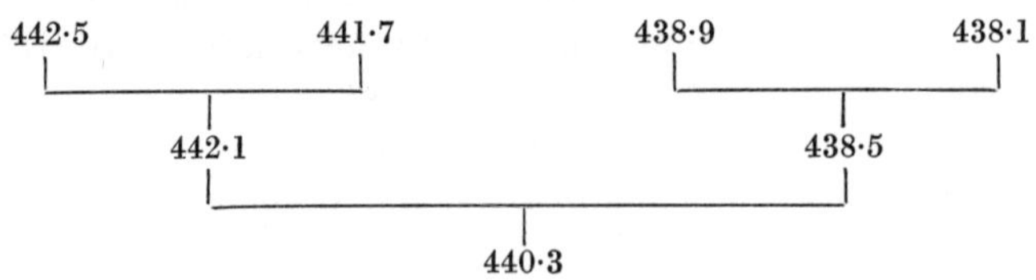

$J = 0{\cdot}8$ c/s $(J_{3,5} \equiv J_{2,4})$

$J = 3{\cdot}6$ c/s $(J_{3,4})$

Of these J values, the figure of 3·6 c/s can only be $J_{3,4}$ and this multiplet represents the hydrogen on C-3 or C-4. This same figure of 3·6 c/s is also seen in the highest field multiplet which also shows a splitting of 1·65 c/s ($J_{4,5}$) thus:—

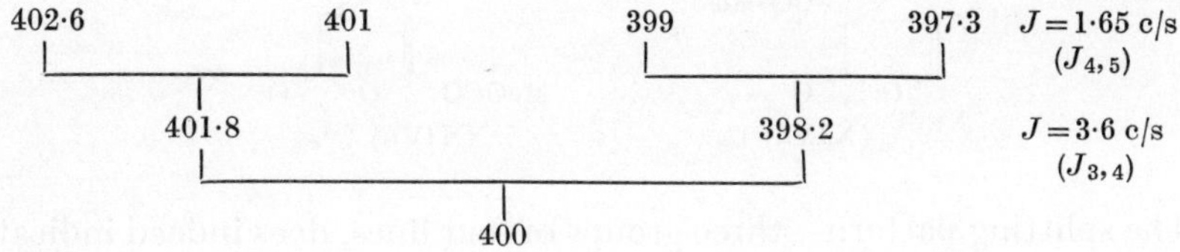

The high field multiplet may be assigned therefore to the hydrogen atom on C-4 and the middle multiplet thus must arise from the hydrogen on C-3.

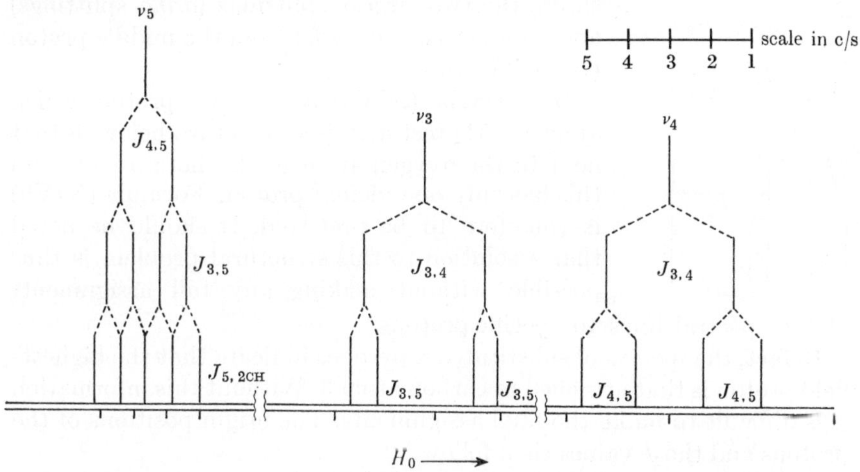

The low field multiplet is that adjacent to the oxygen atom of the ring and the splitting shows the requisite values for 4,5, 3,5 and 5,2 CH coupling. This is shown graphically on the above diagram.

This substance is furan-2-aldehyde.

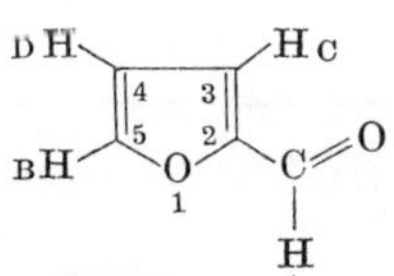

(XIII)

Proton	c/s	τ	Proton	c/s	τ
(D)	400	3·33	(B)	468·5	2·19
(C)	440·3	2·66	(A)	579·5	0·34

Spectrum No. 14 (AMX spectrum)

The low-field lines in this spectrum appear in three distinct groups each representing one proton. If the methoxycarbonyl group is on C-4 or C-5 then an AB type of spectrum might be predicted with one of the doublets split further by long-range coupling. The observed pattern is more complex and would suggest the CO_2Me to be on C-3 or C-6.

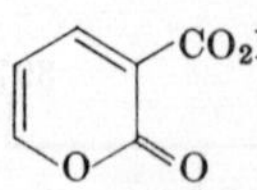

(XIVa)

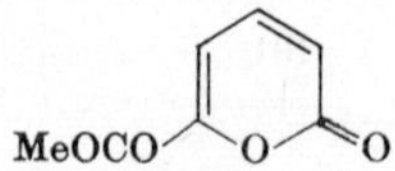

(XIVb)

The splitting pattern—three groups of four lines, does indeed indicate three adjacent protons each of which is coupled unequally to two other protons. Thus, each proton gives rise to two doublets. The "double doublet" at lowest field shows the two largest splittings (*ortho* splittings) and so must be the signal from the middle proton (H_A) of the three.

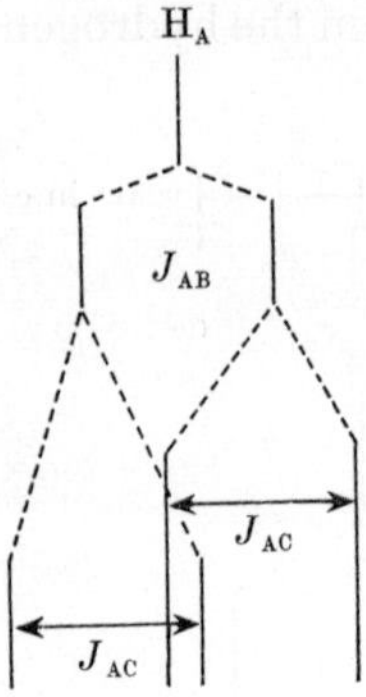

In formula (XIVa) above the proton which appears at lowest field (see structure below) is that next to the oxygen atom of the hetero ring, and this has only one vicinal proton. Formula (XIVb) is therefore to be preferred. It should be noted that a solution to this structural problem is thus possible without making any full assignments of the spectral lines to specific protons.

In fact, the spectra of substituted α pyrones indicate that the highest-field proton is that attached to carbon atom 3. Without this information it is difficult to make the final assignments. The origin positions of the protons and the J values then follow.

	A	B	C	J_{AB}	J_{BC}	J_{AC}
c/s	451·3	428·4	394	6·5	1·2	9·4
τ	2·48	2·86	3·43			

The signal at 6·08 τ = 3H arises from the methyl ester (D).

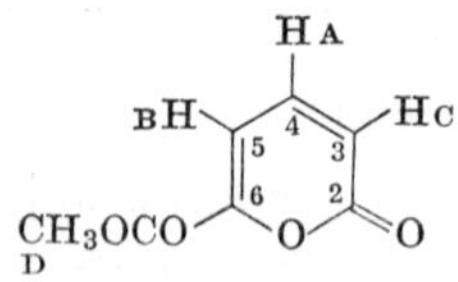

(XIVc)

Spectrum No. 15

τ	H	Multiplicity	J (c/s)	Assignment
8·7	3	Triplet	7	(F) CH_3 of ester coupled to (B)
7·95	3	"Singlet" (possibly double doublet)	1	(E) CH_3 on d.b. coupled to proton (A) and (C)
7·3–7·85	4	Multiplet	–	(D) $2 \times CH_2$ groups (cf. τ values from Table 1 footnote*
6·45–6·85	1	Multiplet	–	(C) proton α to COOEt and to d.b.
5·78	2	Quartet	7	(B) CH_2 of ester coupled to (F)
4·03	1	Multiplet	1	(A) olefinic hydrogen α to C═O and coupled to (E)

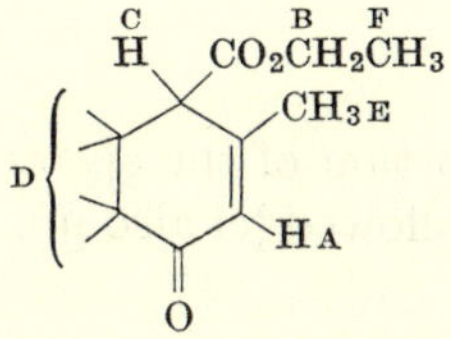

(XV)

Spectrum No. 16

τ	H	Multiplicity	J (c/s)	Assignment
8·91	9	Singlet	—	(G) $3 \times CH_3$ (no splitting: ∴ no adjacent H)
7·88	3	Singlet		(F) CH_3 next to C═O (C═N in derivative)
7·05	2	Singlet	—	(E) CH_2 next to C═O (C═N in derivative)
2·17	1	Doublet (half of AB)	9·5	(D) Aromatic H, *ortho* splitting to (C)
1·75	1	Doublet (half of AB) (peaks further split)	9·5 (2·5)	(C) Aromatic H, *ortho* splitting to (D) and *meta* splitting to (B)
1·06	1	Doublet	2·5	(B) Aromatic H, *meta* splitting
−0·92	1	Singlet	—	(A) Aryl NH

Structure of ketone:

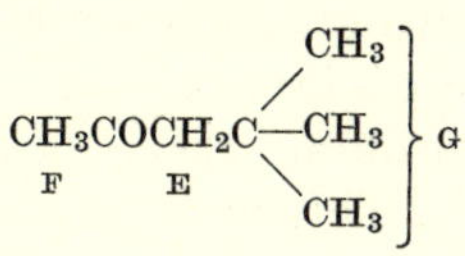

(XVI)

Spectrum No. 17 (ABX; see Chapter 6)

Method No. 1

The integral trace (reading from left to right) shows that the three groups of signals in the spectrum represent 15, 1 and 2 protons, respectively. The intense lowest field band, at $\sim 2{\cdot}7\ \tau$ is attributable to the three phenyl groups. The higher field portion of the spectrum comprises a quartet, centred at $6{\cdot}72\ \tau$, and two interleaving quartets, as appropriate for an ABX spin-coupled system.

It is concluded, therefore, that the reaction product probably has the structure below because in this there is an isolated ABX system, the methylene protons being non-equivalent as a result of the molecular geometry.

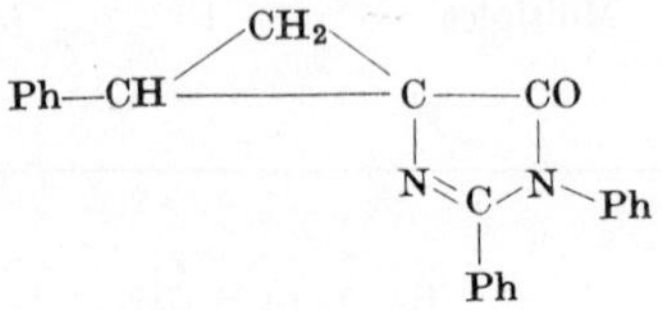

To analyse the ABX portion of the spectrum, the rules discussed in Chapter 6, pp. 91–94 are followed (cf also pp. 197–199).

Line assignment (i)

(a) Firstly, from insets A and B, the line positions in the twelve-line pattern are measured in c/s from T.M.S., with the following result.

X				A				B			
12	11	10	9	h	g	f	e	d	c	b	a
206·4	198·1	196·4	188·2	152·5	147·5	146	144·5	141	139·5	136	131

(b) The two "AB-type" quartets can then be picked out from the AB portion as lines a, b, c, e and d, f, g, h.

The spacing $|J_{AB}|$ occurs four times as $b-a=e-c=f-d=h-g=5$ c/s.

The mid-points of the two quartets are at $(b+c)/2=137{\cdot}75$ c/s and $(f+g)/2=146{\cdot}75$ c/s (a, b, c, e is then the high field quartet and d, f, g, h is the low field quartet). So we obtain $\frac{1}{2}|J_{AX}+J_{BX}|=146{\cdot}75-137{\cdot}75=9$ c/s giving $|J_{AX}+J_{BX}|=18$ c/s. This cross-checks with the X part of the spectrum, in which the separation between the outermost visible lines is 18·2 c/s. (Small discrepancies result from experimental error and/or from the fact that the spectrum is probably not strictly ABX but contains something of the ABC case).

The "1,3"-spacing in the "low-field quartet" (defined above from the relative positions of the quartet centres) $= 2D_+ = \text{g} - \text{d} = \text{h} - \text{f} = 6{\cdot}5$ c/s.

The "1,3"-spacing in the "high-field quartet" $= 2D_- = \text{c} - \text{a} = \text{e} - \text{b} = 8{\cdot}5$ c/s.

(c) The 18 c/s spacing in the X part of the spectrum, which is $|J_{AX} + J_{BX}|$ is necessarily between lines 9 and 12. The spacing between the other pair of lines, 10 and 11, is 1·7 c/s and this is $2|D_+ - D_-|$ which should equal 2 c/s on the basis of the measurements made on the AB spectrum. (Note once more the use of modulus bars to indicate that the enclosed quantity is to be treated as a numerical value without sign significance). Again there is a reasonable agreement between the X part of the spectrum and a spacing calculated from the AB portion.

Substitution of $|J_{AB}| = 5$, $2D_+ = 6{\cdot}5$ and $2D_- = 8{\cdot}5$ in the equations

$$(\nu_A - \nu_B) \pm \tfrac{1}{2}|J_{AX} - J_{BX}| = (4D_{\pm}^2 - J_{AB}{}^2)^{\frac{1}{2}}$$

where the choice of signs is + with D_+ and − with D_-, gives $(\nu_A - \nu_B) = 5{\cdot}5$ c/s and $J_{AX} - J_{BX} = -2{\cdot}7$ c/s. Combining the last value with $|J_{AX} + J_{BX}| = 18$, we obtain

$$J_{AX} = 7{\cdot}65 \text{ c/s}$$

and

$$J_{BX} = 10{\cdot}35 \text{ c/s}$$

From the AB part of the spectrum, $\nu_0 = \dfrac{137{\cdot}75 + 146{\cdot}75}{2} = 142{\cdot}25$ c/s, hence the origin positions of the protons A and B are obtained as

$$\nu_A = 142{\cdot}25 + \tfrac{1}{2}(\nu_A - \nu_B) = 145 \text{ c/s}$$

and

$$\nu_B = 142{\cdot}25 - \tfrac{1}{2}(\nu_A - \nu_B) = 139{\cdot}5 \text{ c/s.}$$

The remaining feature of the spectrum which may be settled is the relative sign of J_{AX} and J_{BX}. Because the ratio $\left(\dfrac{|J_{AB}|}{\nu_A - \nu_B}\right) = \dfrac{5}{5{\cdot}5} = 0{\cdot}9$, it is possible in the present case to make this determination unequivocally.

First we substitute in the equations

$$2D_{+} \cos 2\varphi_{+} = (\nu_A - \nu_B) + \tfrac{1}{2}(|J_{AX}| - |J_{BX}|)$$

and
$$2D_{-} \cos 2\varphi_{-} = (\nu_A - \nu_B) - \tfrac{1}{2}(|J_{AX}| - |J_{BX}|)$$

which have the built-in implication that J_{AX} and J_{BX} are of the *same* sign. We obtain

$$\cos 2\varphi_{+} = \frac{5{\cdot}5 - 1{\cdot}35}{6{\cdot}5} = 0{\cdot}638, \text{ and therefore } \varphi_{+} = 25° \; 10'$$

and
$$\cos 2\varphi_{-} = \frac{5{\cdot}5 + 1{\cdot}35}{8{\cdot}5} = 0{\cdot}806, \text{ and therefore } \varphi_{-} = 18° \; 9'$$

From the Table of line intensities (Table 6.1, p. 91) we may then calculate

Line	Intensity (relative)	Line	Intensity (relative)
9	1	12	1
10	0·98	14	0·01 (not observable)
11	0·98	15	0·01 (not observable)

Within experimental error this is in agreement with the observed spectrum.

It is concluded, therefore, that J_{AX} and J_{BX} are of the same sign.

As a final check, we may substitute in the second pair of equations. This operation has the built-in implication that J_{AX} and J_{BX} have *opposite* sign.

From
$$2D_{+} \cos 2\varphi_{+} = (\nu_A - \nu_B) + \tfrac{1}{2}(|J_{AX}| + |J_{BX}|)$$

and
$$2D_{-} \cos 2\varphi_{-} = (\nu_A - \nu_B) - \tfrac{1}{2}(|J_{AX}| + |J_{BX}|)$$

we obtain

$$\cos 2\varphi_{+} = \frac{5{\cdot}5 + 9}{6{\cdot}5} = \frac{14{\cdot}5}{6{\cdot}5}$$

which is impossible. (It is then unnecessary to proceed and find that $\cos 2\varphi_{-} = 3{\cdot}5/8{\cdot}5$).

Here, we conclude that J_{AX} and J_{BX} cannot have opposite sign, in agreement with the conclusion already reached.

Method No. 2

The various parameters are abstracted from the spectrum as in method No. 1 above. Equations (6–62) of Pople, Schneider and Bernstein, "High Resolution Nuclear Magnetic Resonance", p. 133, are then used.

Thus

$$D_+ \sin 2\varphi_+ = \tfrac{1}{2}J_{AB}$$

$$\sin 2\varphi_+ = \frac{2\cdot5}{3\cdot25} = 0\cdot769$$

$$\therefore \cos 2\varphi_+ = 0\cdot638$$

$$D_- \sin 2\varphi_- = \tfrac{1}{2}J_{AB}$$

$$\sin 2\varphi_- = \frac{2\cdot5}{4\cdot25} = 0\cdot588$$

$\therefore$

$$\cos 2\varphi_- = 0\cdot806$$

Thus

$$\phi_+ = 25^\circ\ 10'; \quad \varphi_- = 18^\circ\ 9'$$

$$\tfrac{1}{2}\Delta\nu + \frac{\Delta J}{4} = D_+ \cos 2\varphi_+ = 3\cdot25 \times 0\cdot638^*$$

$$= 2\cdot073$$

$$\tfrac{1}{2}\Delta\nu - \frac{\Delta J}{4} = D_- \cos 2\varphi_- = 4\cdot25 \times 0\cdot806$$

$$= 3\cdot426$$

$$\Delta\nu = 5\cdot499$$

$$\Delta J = -2\cdot706$$

i.e.

$$J_{AX} - J_{BX} = -2\cdot706$$

and since

$$J_{AX} + J_{BX} = 18\cdot0$$

$\therefore$

$$J_{AX} = 7\cdot65 \text{ and } J_{BX} = 10\cdot35$$

From the AB spectrum $\nu_0 = \dfrac{137\cdot75 + 146\cdot75}{2} = 142\cdot25$ c/s

$\therefore$

$$\nu_A = \nu_0 + \tfrac{1}{2}(\Delta\nu_B) = 145\cdot25 \text{ c/s}$$

and

$$\nu_B = \nu_0 - \tfrac{1}{2}(\Delta\nu_B) = 139\cdot75 \text{ c/s}$$

To summarize therefore

$$J_{AB} = 5 \text{ c/s}, \quad J_{AX} = 7\cdot7 \text{ c/s}, \quad J_{BX} = 10\cdot4 \text{ c/s}$$

$$\nu_A = 145\cdot25 \text{ c/s}; \quad \nu_B = 139\cdot75 \text{ c/s}; \quad \nu_X = 197\cdot3 \text{ c/s}$$

As a check, these figures may be used to calculate the transition energies and relative intensities (Table 6.1, p. 91).

*$\Delta\nu = \nu_A - \nu_B$; $\Delta J = J_{AX} - J_{BX}$

Line	Transition energy (frequency c/s)	Relative intensity	Line	Transition energy (frequency c/s)	Relative intensity
1 B	131·2	0·41	8 A	152·8	0·23
2 B	141·2	0·23	9 X	188·3	1·0
3 B	136·2	1·59	10 X	196·3	0·98
4 B	146·2	1·77	11 X	198·3	0·98
5 A	138·8	1·59	12 X	206·4	1·0
6 A	147·3	1·77	14 X	189·6	0·01
7 A	144·8	0·41	15 X	204·9	0·01

Inspection shows that we have an excellent fit between the observed and calculated spectra.

In terms of frequency, the two quartets and their assignments are as follows.

1	3	5	7
131·2	136·2	138·8	144·8
2	4	6	8
141·2	146·2	147·5	152·5

Line assignment (ii)

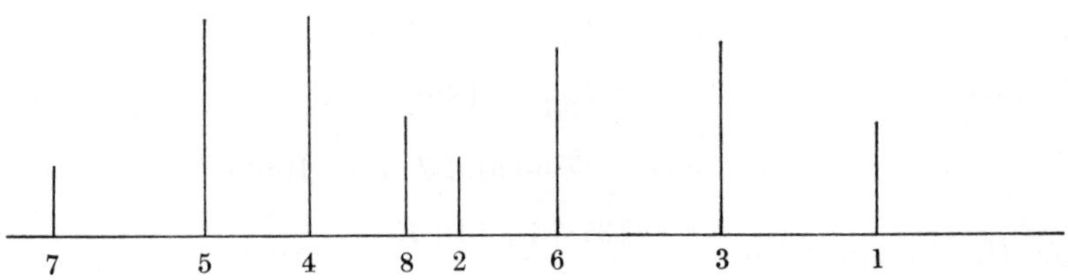

1	3	5	7
131·2	136·2	147·5	152·5
2	4	6	8
141·2	146·2	138·8	144·8

from which $J_{AB} = 5$ c/s, i.e.

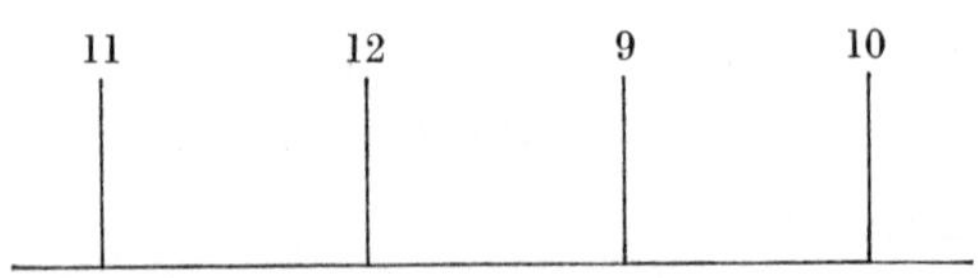

It should be noted, however, that lines 2, 4, 6, 8 do not have the normal appearance (weak, strong, strong, weak) of the AB quartet.

$$2D_- = 16{\cdot}3 \quad \therefore D_- = 8{\cdot}2$$

$$2D_+ = -2{\cdot}4 \;\therefore D_+ = -1{\cdot}2$$

$$|J_{AX}+J_{BX}| = 1{\cdot}2$$

which is impossible by definition.
Since both D^+ and D^- are *defined* as positive quantities, this assignment of lines is not correct.

Line assignment (iii)

Another alternative might be considered from a first inspection, namely

1	3	5	7
131·2	136·2	138·8	144·8
2	4	6	8
147·5	152·5	141·2	146·2

and from this assignment we would obtain

$$D_- = 3{\cdot}8 \text{ and } D_+ = -3{\cdot}15$$

from which $2|D_+ - D_-| = 13{\cdot}9$, a value inconsistent with the X spectrum. Moreover D^+ must by definition be a *positive* quantity.

Method No. 3

(i) Line assignments are made as before. The table of transition energies and relative intensities (6–15) on p. 134 of Pople, Bernstein and Schneider is used

$$J_{AB} = 3\text{–}1,\ 4\text{–}2,\ 7\text{–}5,\ 8\text{–}6 = 5 \text{ c/s}$$

$$\tfrac{1}{2}(J_{AX}+J_{BX})+(D_- - D_+) = 2\text{–}1,\ 4\text{–}3,\ 11\text{–}9 = 10$$

$$\tfrac{1}{2}(J_{AX}+J_{BX})-(D_- - D_+) = 6\text{–}5,\ 8\text{–}7,\ 10\text{–}9 = 8{\cdot}2$$

$$\therefore \quad (J_{AX}+J_{BX}) = 18{\cdot}2$$

$$2D_+ = 6\text{–}2,\ 8\text{–}4 = 6{\cdot}5$$

$$2D_- = 5\text{–}1,\ 7\text{–}3 = 8{\cdot}4$$

$$(\nu_A - \nu_B)+\tfrac{1}{2}(J_{AX}-J_{BX}) = \sqrt{(4D_+^2 - J_{AB}^2)} = 4{\cdot}2$$

$$(\nu_A - \nu_B)-\tfrac{1}{2}(J_{AX}-J_{BX}) = \sqrt{(4D_-^2 - J_{AB}^2)} = 6{\cdot}75$$

$$\therefore \quad (J_{AX}-J_{BX}) = -2{\cdot}55$$

$$\nu_A-\nu_B = 5{\cdot}5$$

$$\therefore \quad J_{AX} = 7{\cdot}8$$

$$J_{BX} = 10{\cdot}4$$

(ii) If the line assignments (ii) Method No. 2 are used, then we find the following:

$$J_{AB} = 5 \text{ c/s}$$

$$\tfrac{1}{2}(J_{AX}+J_{BX})+(D_- - D_+) = 10$$

$$\tfrac{1}{2}(J_{AX}+J_{BX})-(D_- - D_+) = -8{\cdot}2$$

$$\therefore \quad J_{AX}+J_{BX} = 1{\cdot}8$$

$$2D_+ = -1{\cdot}6$$

This is impossible by definition and this assignment of lines must be rejected.

Spectrum No. 18

The product from the addition of bromine to methyl vinyl ketone is obviously 1,2-dibromo-3-oxobutane $\overset{1}{C}H_2Br.\overset{2}{C}HBr.CO.CH_3$, the most likely conformation of which on general grounds is as follows:

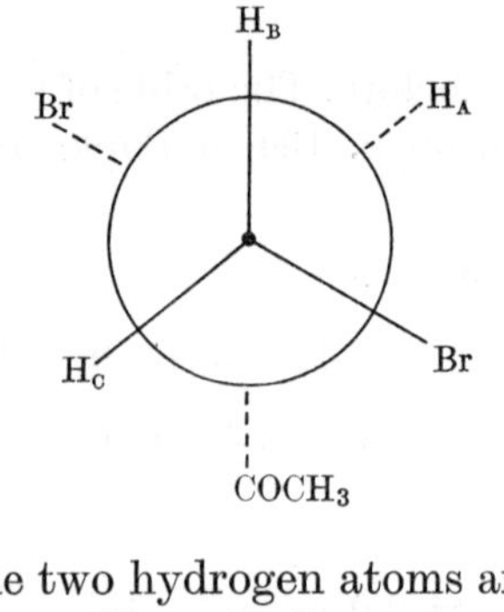

$CH_2Br.CHBr.CO.CH_3$

For C-1 the two hydrogen atoms are not in equivalent environments and geminal coupling will thus take place between protons (B) and (C): in addition, each will be split to a different extent by the vicinal proton (A). For C-2, proton (A) will likewise be split to different extents by vicinal coupling with both (B) and (C). Consideration of the data of Conroy p. 188, moreover, would suggest that for a dihedral angle $\phi = 180°$, the vicinal coupling constant (J_{AC}) would be about 10 c/s; for $\phi = 60°$ (J_{AB}) would be approximately 2 c/s. With geminal coupling (J_{BC}) a figure between -10 and -18 c/s might be expected (Table 8, p. 186).

H_A is adjacent to both the carbonyl group and a bromine atom, and should appear furthest downfield. Therefore, the low-field quartet is assigned to this proton. The chart positions for the peaks of this quartet are as follows (Inset A).

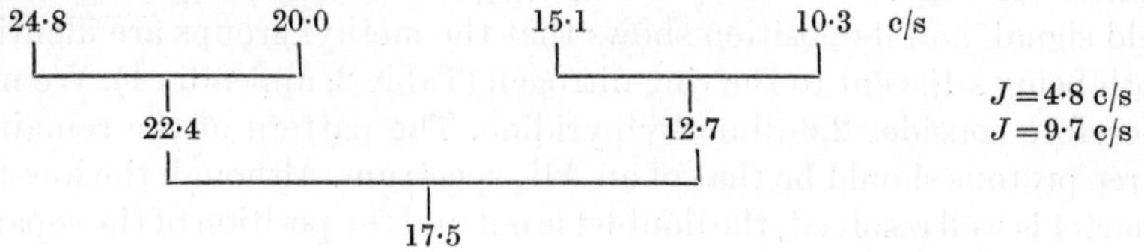

These figures for J suggest $J_{AC} = 9{\cdot}7$ c/s and $J_{AB} = 4{\cdot}8$ c/s. Since this multiplet has been recorded with an offset of 251 c/s, $\nu_A = 251 + 17{\cdot}5 = 268{\cdot}5$ c/s i.e. 5·53 τ. The other two signals, a triplet and an adjacent quartet are dealt with in that order. They have chart positions as follows (Inset B).

Triplet

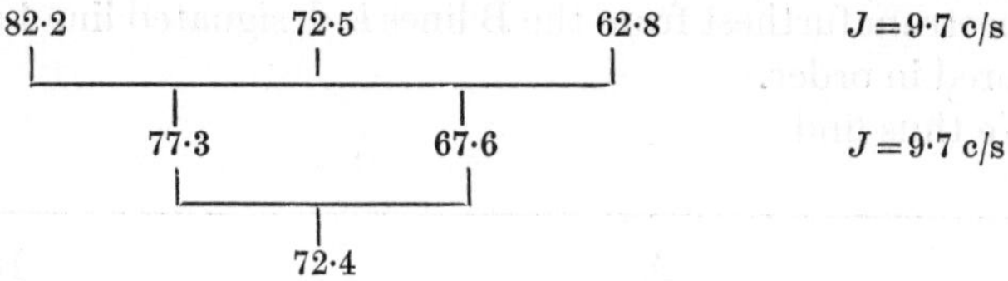

Quartet

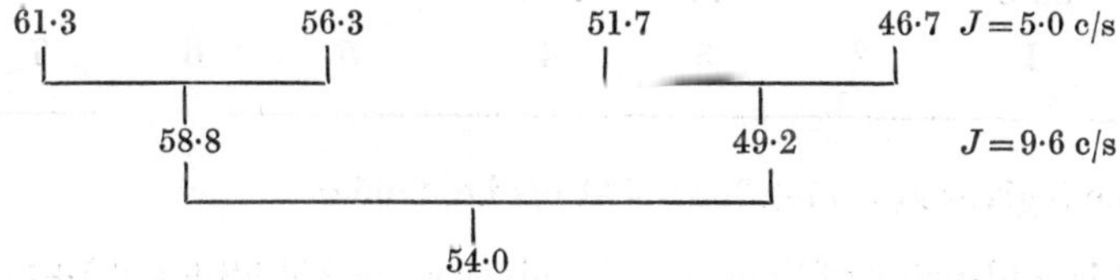

Since of these two signals only the quartet contains a $J = 5{\cdot}0$ c/s this must be J_{AB} which cross checks with the value of 4·8 c/s from the H_A quartet. This second quartet, therefore, arises from H_B and, since the instrumental offset is 161 c/s, $\nu_B = 54{\cdot}0 + 161 = 215$ c/s i.e. 6·42 τ. The remaining triplet then arises from H_C and the splitting $J_{AC} = J_{BC} = 9{\cdot}7$ c/s. $\nu_C = 72{\cdot}4 + 161 \equiv 233$ c/s i.e. 6·12 τ.

This interpretation is consistent with the conformation initially proposed.

Spectrum No. 19

Since this compound is a substituted pyridine (C_5) the other two C atoms must be present either as an ethyl group or as two methyl groups. The absence of the characteristic A B_3 pattern for the former suggests that it is indeed a dimethyl pyridine.

The integral trace shows that the lowest field quartet arises from one proton, the doublet from two, and the single sharp resonance at 7·56 τ from six protons. Two methyl groups are thus confirmed by this last high field signal, and its position shows that the methyl groups are identical, both being adjacent to the ring nitrogen (Table 2, appendix I). We must therefore consider 2,6-dimethylpyridine. The pattern of the remaining three protons should be that of an AB_2 spectrum. Although the low-field quartet is well resolved, the doublet is not and the position of the separate lines of this is subject to some uncertainty. Note also that because the doublet lines are not resolved, their width at half height is greater than in the low-field multiplet (compare also the methyl resonance). The weak combination line cannot be seen (cf. Chapter 5, p. 80). Starting from the low field end of the spectrum, the integral trace shows that the first four lines represent one proton (defined as the A proton) and the "doublet" represents two protons (defined as the B protons). That line of the A spectrum furthest from the B lines is designated line 1 and the rest numbered in order.
We thus find

A				B_2			
85·5	78·9	77	70·4 c/s	49·8	48·9	42·2	41·8 c/s
1	2	3	4	5	6	7	8

line 3 gives $\nu_A = 77 + 364 = 441$ c/s i.e. 2·65 τ

This mid-point of lines 5 and 7 gives $\nu_B = \frac{1}{2}(49{\cdot}8 + 42{\cdot}2) + 364$

$$= 410 \text{ c/s i.e. } 3{\cdot}17\ \tau$$

$$J_{AB} = \tfrac{1}{3}(\nu_4 + \nu_8 - \nu_1 - \nu_6) = \tfrac{1}{3}(70{\cdot}4 + 41{\cdot}8 - 85{\cdot}5 - 48{\cdot}9)$$

$$= 7{\cdot}4 \text{ c/s}$$

$$J/\Delta\gamma = 0{\cdot}239$$

The values for $\nu_B = 3{\cdot}17\ \tau$ and $\nu_A = 2{\cdot}65\ \tau$ should be compared with the values for the β proton (3·01 τ) and the γ proton in pyridine itself (2·64 τ)

(Table 4, p. 184); they are in reasonable agreement. $J_{AB} = 7{\cdot}4$ c/s is also the appropriate value given for J_{34} (7·5 c/s) (compare also, J. A. Elvidge and L. M. Jackmann, *J. Chem. Soc.* 860 (1961)).

Compare this with the discussion of the spectrum by H. J. Bernstein, J. A. Pople and W. G. Schneider (*Canad. J. Chem.* **35**, 65 (1957)).

Spectrum No. 20

The low field group of protons only is shown on the chart and the integral trace shows that the first four lines represent two protons, the second four, one proton only. It should be noted not only that the lines are better resolved than in No. 19, but also that the spectrum runs in the opposite direction. Following the usual convention, we may then set out the line positions as follows.

B_2				A			
56·8	53·9	48·1	47·5 c/s	40·2	34·4	31	25 c/s
8	7	6	5	4	3	2	1

whence there follows

$$\nu_B = \tfrac{1}{2}(\nu_5+\nu_7)$$
$$= \tfrac{1}{2}(47{\cdot}5+53{\cdot}9)+450$$
$$= \tfrac{1}{2}(101{\cdot}4)+450$$
$$= 500{\cdot}7 \text{ c/s}$$
$$= 1{\cdot}65\ \tau$$

$$\nu_A = \nu_3$$
$$= 484{\cdot}4 \text{ c/s}$$
$$= 1{\cdot}93\ \tau$$

$$J_{AB} = \tfrac{1}{3}(\nu_4+\nu_8-\nu_1-\nu_6)$$
$$= \tfrac{1}{3}(40{\cdot}2+56{\cdot}8-25-48{\cdot}1)$$
$$= \tfrac{1}{3}(97{\cdot}0-73{\cdot}1) = \tfrac{1}{3}(23{\cdot}9)$$
$$= 8 \text{ c/s}$$

$$J/\Delta\nu = \frac{8}{16{\cdot}8} = 0{\cdot}475$$

Thus for Spectra 19 and 20 we have

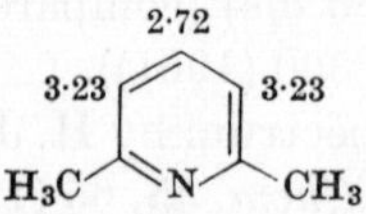

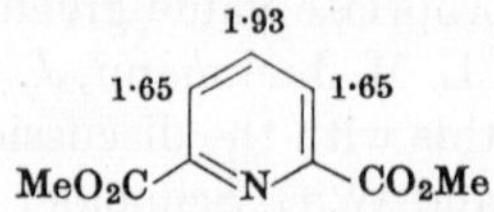

These τ values for the β and γ protons immediately clarify the reason for the reversal in position of the A and B protons as we pass from the pyridine substituted with two methyl groups to that substituted with two methoxycarbonyl groups.

Spectrum No. 21

Two line assignments are possible: these are as follows.

AB	Relative position:	32·2	30·8	24·5	23·1	12·2	7·5	4·8	0
	Assignment No. 1:	7	8	5	6	3	4	1	2
	Assignment No. 2.	8	7	6	5	3	4	1	2

Relative position:	6·0	4·6	1·4	0
Assignment No. 1:	12	11	10	9
Assignment No. 2:	11	12	9	10

For both assignments, $|J_{AB}| = 1-3,\ 5-7,\ 2-4,\ 6-8 = 7{\cdot}58$ c/s
ν_0 = mean of all eight AB lines = 16·89 c/s
From assignment No. 1,

$$\left.\begin{aligned} 2D_+ &= 2-6,\ 4-8\ .\ 23{\cdot}20 & \therefore D_+ &= 11{\cdot}60 \\ 2D_- &= 1-5,\ 3-7 = 19{\cdot}58 & \therefore D_- &= 9{\cdot}93 \end{aligned}\right\} \therefore 2(D_+ - D_-) = 3{\cdot}34$$

$$\text{As a cross-check, } 2(D_+ - D_-) = 10 - 11 = 3{\cdot}2$$

AB_+ quartet is centred at the mean of lines 2,4,6,8 = 15·35
AB_- quartet is centred at the mean of lines 1,3,5,7 = 18·43

$$\therefore \tfrac{1}{2}|J_{AX} + J_{BX}| = 3{\cdot}08 = \tfrac{1}{2}(9\text{–}12)$$

$$|J_{AX}| + J_{BX} = 6{\cdot}16$$

$$\Delta\nu + \tfrac{1}{2}\Delta J = \sqrt{4D_+^2 - J_{AB}^2} = 21{\cdot}93$$

$$\Delta\nu - \tfrac{1}{2}\Delta J = \sqrt{4D_-^2 - J_{AB}^2} = 18{\cdot}35$$

$$\therefore \Delta\nu = 20{\cdot}14;\ \Delta J = 2{\cdot}58$$

$$\text{Since } \left.\begin{aligned} J_{AX} - J_{BX} &= 2{\cdot}58 \\ J_{AX} + J_{BX} &= 6{\cdot}16 \end{aligned}\right\} \therefore J_{AX} = 4{\cdot}37 \text{ and } J_{BX} = 1{\cdot}79$$

$$\nu_A = \nu_0 + \tfrac{1}{2}\Delta\nu = 16{\cdot}89 + 10{\cdot}07 = 26{\cdot}96$$

$$\nu_B = \nu_0 - \tfrac{1}{2}\Delta\nu = 16{\cdot}89 - 10{\cdot}07 = 6{\cdot}82$$

Summary

$$|J_{AB}| = 7{\cdot}6, J_{AX} = 4{\cdot}4, J_{BX} = 1{\cdot}8 \text{ c/s}$$

$$\nu_A = 2{\cdot}73\ \tau;\ \nu_B = 3{\cdot}07\ \tau;\ \nu_X = 1{\cdot}77\ \tau$$

J_{AX} and J_{BX} are the *same* sign

Calculations on line assignment No. 2 give the following

$$|J_{AB}| = 7{\cdot}6;\ J_{AX} = 5{\cdot}0;\ J_{BX} = -1{\cdot}6 \text{ c/s}$$

$$\nu_A = 2{\cdot}73\ \tau;\ \nu_B = 3{\cdot}07\ \tau;\ \nu_X = 1{\cdot}77\ \tau$$

J_{AX} and J_{BX} are of *opposite* sign

Although it is possible in principle to distinguish between these two possibilities by calculating line intensities, the observed spectrum (lines 5,6 and 7,8) is not sufficiently well resolved to allow of the accurate intensity measurement from the chart.

Spectrum No. 22

This spectrum clearly consists of four well-defined quartets and the oxidation of methyl to carboxyl groups, followed by conversion to amides, would appear to have changed the ABX spectrum of problem 21 to the AMX condition. The appearance of the two high-field quartets is curious, but since we know that amidic protons give rise to broad signals (Table 5, p. 184), it may be deduced that the broad "envelope" underlying the high-field quartets arises from this feature of the molecule. The lines of the quartets have the following positions (c/s from TMS).

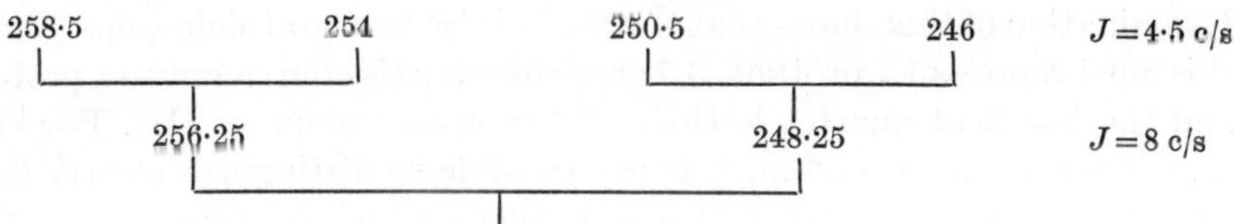

252·25
Offset 200
452·25 c/s = 2·47 τ

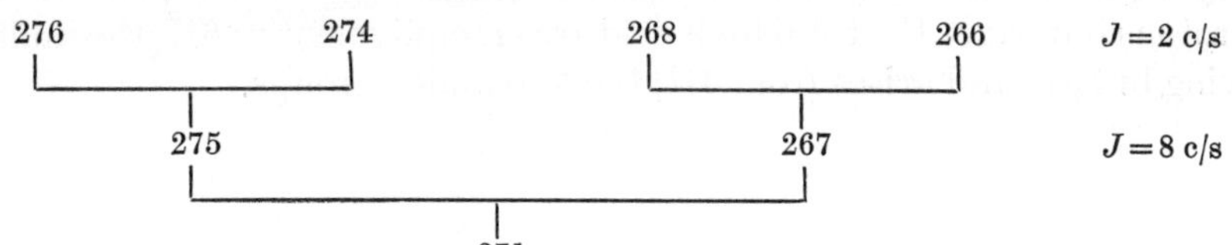

271
offset 200
471 c/s = 2·15 τ

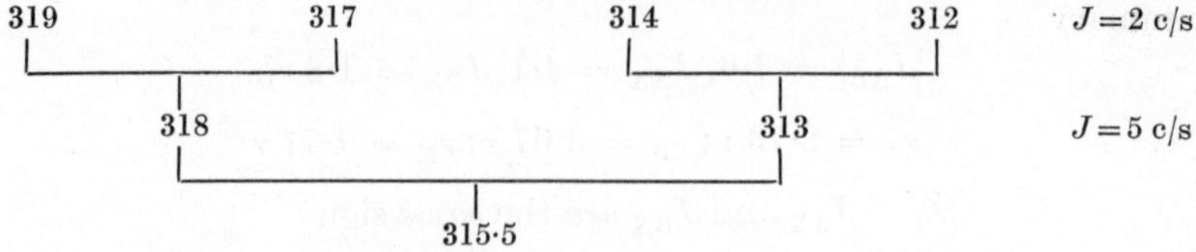

offset 200

515·5 c/s = 1·41 τ

It is to be noted that each J value occurs twice. Table 8, (p. 188) shows that of the three values observed, that of 8 c/s represents $J_{3,4}$ and the value 5 c/s, $J_{2,3}$. Since both appear in the high-field quartet this arises from the proton on C-3 (X proton below). The 2 c/s splitting (middle and low-field quartets) is likewise $J_{2,4}$ of the pyridine ring and by a simple extension of the above argument, the middle quartet may be assigned to the proton on C-4 (M proton below) while the low-field quartet arises from H_A below. The relative τ values for these protons agrees with the values expected on the basis of Table 4. If we follow the convention of labelling as A, the lowest field proton in the spectrum, and proceed through the alphabet as we pass to higher field, the formula below summarizes the analysis.

Broad signal centre ~5·75τ

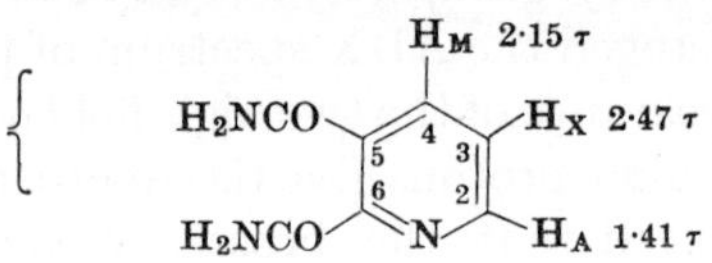

$J_{MX}=8$ c/s $=J_{3,4}$
$J_{AX}=5$ c/s $=J_{2,3}$
$J_{AM}=2$ c/s $=J_{2,4}$

As in Spectrum 13, a splitting diagram may be drawn.

Analysis is thus possible without recourse to the integral trace. Examination of this shows that the overall rise is 26 graticule steps. Since this must represent 7 protons, 3·7 graticule steps is equivalent to 1 proton and the low-field quartet is thus 0·76 protons (taken as 1H). For the remainder of the spectrum, it is not possible to distinguish clearly any separate steps and the integral is best taken as a whole equivalent to 6·1 protons.

It is useful to recall that an AMX spectrum consists of three quartets. The high field group of lines (2 quartets) must therefore represent two protons but, since the total high field area is equivalent to 6H, the underlying broad band arises from 4H, the two amide groups.

AUTHOR INDEX

Numbers in parentheses are reference numbers, and are given to assist locating in the text references where the authors' names are not given. Numbers in italics are the pages on which the references are listed.

R

S

T

SUBJECT INDEX